5th International Seminar on Science and Technology (ISSTEC 2025)

BIO Web of Conferences Volume 226 (2026)

Yogyakarta, Indonesia
23 September 2025

Editors:

Is Fatimah
Normah Binti Awang
Hiroyuki Nakamura
Imam Sahroni
Ganjar Fadillah
Halil Durak

ISBN: 979-8-3313-3288-4

Printed from e-media with permission by:

Curran Associates, Inc.
57 Morehouse Lane
Red Hook, NY 12571

CURRAN ASSOCIATES INC.
proceedings.com

Some format issues inherent in the e-media version may also appear in this print version.

For additional information, please contact EDP Sciences – Web of Conferences at the address below.

EDP Sciences – Web of Conferences
17, Avenue du Hoggar
Parc d'Activité de Courtabœuf
BP 112
F-91944 Les Ulis Cedex A
France

Phone: +33 (0) 1 69 18 75 75
Fax: +33 (0) 1 69 28 84 91

contact-edps@webofconferences.org

Additional copies of this publication are available from:

Curran Associates, Inc.
57 Morehouse Lane
Red Hook, NY 12571 USA
Phone: 845-758-0400
Fax: 845-758-2633
Email: curran@proceedings.com
Web: www.proceedings.com

TABLE OF CONTENTS

edp sciences

BIO Web of Conferences

Statement of Peer review

In submitting conference proceedings to BIO Web of Conferences, I certify to the Publisher that I adhere to the **Policy on Publishing Integrity** of the journal in order to safeguard good scientific practice in publishing.

1. All articles have been subjected to peer review administered by the proceedings editors.
2. Reviews have been conducted by expert referees, who have been requested to provide unbiased and constructive comments aimed, whenever possible, at improving the work.
3. Proceedings editors have taken all reasonable steps to ensure the quality of the materials they publish and their decision to accept or reject a paper for publication has been based only on the merits of the work and the relevance to the journal.

Title, date and place of the conference

The 5th International Seminar on Science and Technology (ISSTEC) 2025
September 23, 2025, Universitas Islam Indonesia, Yogyakarta, Indonesia

Proceedings editor(s):

Prof. Dr. Is Fatimah
Assoc. Prof. ChM. Dr Normah Binti Awang
Prof. Hiroyuki Nakamura
Imam Sahroni, Ph.D.
Ganjar Fadillah, Ph.D.
Prof. Halil Durak

Date and editor's signature

February 19, 2026

Imam Sahroni

Analysis of thermal radiation shielding of vacuum furnace for brazing and bake-out of accelerator

Suprapto[1,2,*], *Saefurrohman*[2], *Kurnia* Wibowo[3], *Angga Dwi* Saputra[2], *Taufik*[2], and *Alexander* Agung[1]

[1]Department of Nuclear Engineering and Engineering Physics, Faculty of Engineering, Gadjah Mada University, Indonesia
[2]Research Center for Accelerator Technology, Research Organization for Nuclear Energy, National Research and Innovation Agency (BRIN)
[3]Directorate of Laboratory Management, National Research and Innovation Agency (BRIN)

Abstract. An accelerator is a device that accelerates charged particles from low to high energy. It consists of several main components, which require specialized manufacturing processes, including brazing and bake-out carried out in a vacuum furnace at a pressure of 10^{-5} mbar. However, such a vacuum furnace is not yet available domestically and therefore needs to be designed and constructed locally. The design and construction began with an initial analysis of the vacuum furnace's thermal radiation shield. The vacuum furnace is planned to have a diameter of 105 cm and a length of 150 cm. Analysis was conducted using analytical calculations and simulations using ANSYS software for steady-state and transient conditions. The analysis results show that the use of three thermal radiation shields can significantly reduce electrical power and wall temperature at steady-state conditions, namely from 40,657 kW and 306.9°C without shields to 9,869 kW and 141.3°C for analytical calculations and 10,149 kW and 140.45°C for simulations using ANSYS software. Transient analysis shows that the heating power for a heating rate of about 40°C/min is 25 kW. This analysis can support the construction of vacuum furnaces for brazing and bake-out of accelerator components in Indonesia.

1 Introduction

An accelerator is a device for accelerating and guiding positively charged particles (ions) or negatively charged particles (electrons) from low speed (low energy) to high speed (high energy). There are two types of acceleration principles in accelerators, namely, (1) electrostatic accelerators, and (2) electrodynamic or electromagnetic accelerators. In an electrostatic accelerator, a static electric field or high-voltage DC, which is commonly supplied by a Cockcroft-Walton generator, a Dynamitron generator, or a Van de Graaff generator, is used for acceleration [1]. The energy of the particle beam is determined by the

* Corresponding author: suprapto1959@mail.ugm.ac.id

strength of the accelerating field, which depends on the high voltage level. On the other hand, an electrodynamic accelerator uses an oscillating electromagnetic field to accelerate the particles. The particle's acceleration is achieved by repeatedly passing the charged particles into the oscillating electric field, so that the right timing of the phase will accelerate the particles. Some examples of electrodynamic accelerators are: Linac, Cyclotrons, Microtrons, and Rhodotrons [1].

Research Centre for Accelerator Technology (PRTA-ORTN), National Research and Innovation Agency (BRIN), is developing a medium-energy accelerator, a 13 MeV ion accelerator (Cyclotron) named DECY 13 (Development of Experimental Cyclotron in Yogyakarta with 13 MeV proton energy). The next plans for accelerator research and development in the future are (1) a 30 MeV Cyclotron and (2) a 6 MeV Linear accelerator (Linac). The main parts of the accelerator are: the electron source/ion source, and accelerator system (e.g., accelerating tube, accelerating cavity, coupling cavity, RF Dee, and electromagnetic system) [2,3]. The research and development stages are design, fabrication (making accelerator components), construction, conditioning, and commissioning. The design is carried out by making concepts, analytical calculations, and simulations using software. Accelerator components fabrication is carried out by machining, brazing, and bake-out processes. The machining process can be performed in the workshop, but the brazing and bake-out processes cannot be done domestically, because, at the moment, there is no vacuum furnace specialized for accelerator components in Indonesia. The bake-out process is carried out in a vacuum chamber with a vacuum level of 10^{-5} mbar at a temperature of up to 450°C and a duration of up to 24 hours [4]. The temperature and the duration of this process depend on the type and properties of the material. The brazing process of accelerator components is performed at a temperature of 830°C to 860°C and even higher, depending on the type and properties of the material of both the material being brazed and the filler. The brazing process must be in a vacuum level of around 5×10^{-5} mbar [5].

Based on those needs, it is necessary to design a vacuum furnace for brazing and the bake-out process of the accelerator components. The vacuum furnace for brazing must be able to operate at a temperature of 860°C with a vacuum level of 5×10^{-5} mbar, and a temperature rise rate of more than 20°C/min with the uniformity (homogeneity) of 5°C to 10°C at a temperature of more than 600°C [5]. Fedchak, J.A., et al., discussed a 4-inch-thick layer of Fiberfrax insulation surrounding a wire-woven furnace for safety, operating only up to 450°C [6]. Furthermore, they did not discuss the details of the thermal analysis and focused more on the control system and operation. Badshah S et al. discussed a 15 mm thick graphite felt insulation layer, with a 0.5 mm thick molybdenum inside and a 2 mm thick stainless-steel outside [7]. A detailed thermal analysis was conducted to examine heat transfer by conduction at a temperature of 1000°C in a furnace. However, it did not discuss the relationship with the vacuum seal used. Wang Y. and Liu Z. discuss the heat insulation layer of graphite so that heat transfer is by conduction and is equipped with water cooling, as well as the control system [8]. The discussion is quite detailed and emphasizes temperature distribution (Ansys simulation), temperature measurement as a function of time, and the control system. Rao V. V. and Ravindra V., discuss the basic equations of radiation heat transfer, the magnitude of furnace wall cooling, heating with heating elements, and its instrumentation, and do not discuss in more detail the heat transfer that occurs [9].

This study analyzes the utilization of thermal radiation shielding in detail through analytical calculations and simulations using ANSYS software. Calculations are performed for both without shielding and with shielding. Simulations are only performed on the use of shielding, both in steady-state and transient conditions. Calculations and simulations are based on furnace operation at 1000°C at a vacuum of 5×10^{-5} mbar. The final analysis can be used as a reference to construct the vacuum furnace that is specifically designed for accelerator components.

2 Methodology

Thermal radiation shielding analysis is crucial for minimizing heat transfer to the environment. The study is performed through analytical calculations and simulations using ANSYS software. These calculations and simulations are based on the specific concept and shielding material used. The shielding material is selected based on its high temperature resistance and low emissivity, which are essential properties for effectively confining heat and inhibiting heat flow to the environment. The results obtained from the calculations and simulations are then rigorously compared to ensure their validity. Three stages are carried out in the analysis, including: (1) studying the brazing and bake-out process of accelerator components, (2) determining the basis for calculations, and (3) simulation using ANSYS software. Thus, the results obtained can be implemented in actual vacuum furnace construction.

2.1 Study of Brazing and Bake-out Processes and Vacuum Furnaces

The complete heating and cooling cycles are crucial for both the brazing and bake-out processes performed within a vacuum furnace, as these represent the final stages in the fabrication of accelerator components. The heating phase commences at room temperature and continues until the furnace reaches the melting point of the brazing filler, with a heating rate of 20-30°F/min, taking into account any increase in pressure or decrease in vacuum. This is influenced by both the heating rate and the quantity and quality of the workpiece. These variables must be recognized to ensure that the minimum vacuum is in the region of 10^{-5} mbar [5]. Subsequently, the temperature is maintained at this level for several minutes to allow for initial heating, stabilization, and homogenization of the workpiece temperature. Heating then resumes until the brazing temperature is achieved; this specific value depends on the filler material used, typically ranging approximately 50°C above its melting point [5]. This temperature is then held for an additional few minute to ensure an even distribution and wetting of the molten filler liquid across the surface of the brazed object. The subsequent cooling process is executed slowly to prevent defects such as deformation or damage. These defects can arise from shrinkage or differences in the thermal expansion coefficients between the workpiece (main material) and the filler material. The brazing process for the Ti-38Zr-24Ni filler material involved an initial heating rate of 20°C/min, starting from room temperature and reaching a temperature of 550°C. At this temperature, the material was held for 35 minutes to achieve homogenization. Subsequently, the temperature was increased to the brazing temperature of 930°C at a rate of 10°C/min [10].

In this context, precise temperature control and accurate measurement are paramount to ensure the quality of the process. Furthermore, the vacuum level is equally critical; during furnace heating for brazing, pressure can increase, or the vacuum can degrade, which would adversely affect the process performance. For instance, brazing processes using Ag-Cu alloy-based filler materials typically require temperatures between 830°C and 860°C within a vacuum level of approximately 5×10^{-5} mbar [11]. During preheating, an increase in pressure or a decrease in vacuum typically occurs. The vacuum level must be maintained at least at approximately 10^{-5} mbar to avoid affecting the subsequent brazing process [11]. K, R, S, and N types of thermocouples were employed for temperature measurement due to their suitable operational ranges. To ensure accurate temperature uniformity measurement, these sensors are strategically placed at points identified from prior simulations. The sensors continuously collect data throughout both the preheating and brazing processes. For vacuum level measurement, the vacuum sensors operate during the pre-vacuum phase, preheating, and brazing.

During the pre-vacuum process, a rotary pump is employed to achieve a vacuum level ranging from 10^{-2} to 10^{-3} mbar. A Pirani vacuum sensor is utilized for measuring this vacuum level. For the subsequent preheating and brazing phases, a diffusion or turbomolecular pump is used to attain a higher vacuum level of 10^{-5} to 10^{-6} mbar. In these stages, a Penning vacuum sensor measures the vacuum level. The selection of these specific vacuum sensors and pumps is based on their high applicability to vacuum furnace systems.

2.2 The basis for analytical heat transfer calculations

The concept of the vacuum furnace is designed with thermal radiation shielding to confine heat within the furnace and inhibit heat transfer from the furnace to the atmospheric environment, thereby reducing the required heating power. At atmospheric pressures around 10 mbar, heat transfer is predominantly governed by convection. Conversely, at pressures ranging from 1 mbar to 10^{-3} mbar, heat transfer is dominated by conduction resulting from the movement of gas atoms. This exhibits a linear relationship with pressure. Furthermore, under high vacuum conditions at pressures below 10^{-4} mbar, heat transfer is primarily driven by radiation [9]. Heat transfer also occurs by conduction through the supporting structure. This designed vacuum furnace is operated at a vacuum of 5×10^{-5} mbar or higher so that heat transfer only occurs by radiation [9]. Thus, the equation used for analytical calculations inside the furnace is based on radiation. Outside the furnace, however, it is based on radiation and convection at atmospheric pressure. Figure 1 illustrates this heat transfer mechanism. The heat transfer can be calculated using the following equation (1) [12].

$$\dot{Q} = \varepsilon\sigma A_1\left(T_{s,1}^4 - T_1^4\right) = kA_{2,cond}\,\frac{T_1 + T_2}{L} = h_{c,D}A_{2,conv}\left(T_2 - T_{\infty,2}\right) + \varepsilon\sigma A_{2,rad}\left(T_2^4 - T_{\infty,2}^4\right) \quad (1)$$

where ($\dot{Q}$) is the heat transfer that occurs (W), σ is the Stefan-Boltzmann constant (5.67×10^{-8} W/m^2.K^4), ε is the emissivity of the heat emitter surface, and T_1, T_2 are the surface temperatures (K) where $T_1 > T_2$, and $T_{s,1}$, $T_{\infty,2}$ are the temperatures inside and outside the furnace, or the environment (K), k is the coefficient of conduction (W/m.K), $h_{c,D}$ is the coefficient of convection (W/m^2.K), L is the thickness of the furnace shell/cover wall (m), and A_1 is the surface area of heat emitter (m^2), $A_{2,cond}$, $A_{2,conv}$, $A_{2,rad}$ each the area of heat transfer: conduction, convection, and radiation of the furnace shell or cover wall (m^2).

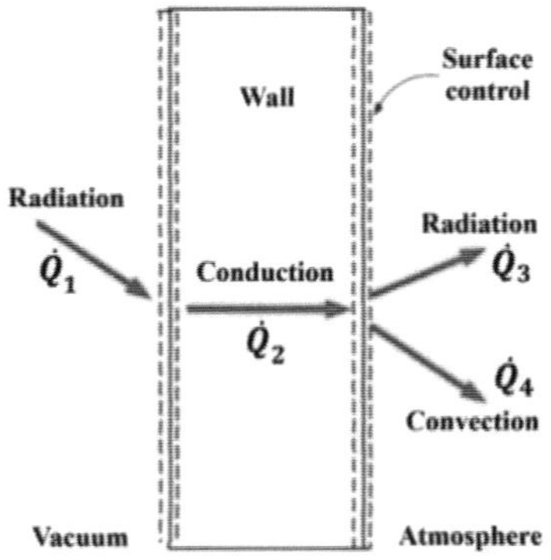

Fig. 1. Heat transfer mechanism on the vacuum furnace wall.

2.2. 1 Heat transfer by convection

The heat transferred by convection from the furnace wall (furnace shell and cover) to the environment is free convection, as there is no air blowing. The calculation equation is as follows [12]:

$$\dot{Q} = h_{c,D}\, A\, (T_s - T_\infty) \tag{2}$$

and

$$h_{c,D} = \frac{k}{D}\, N_{u,D} \tag{3}$$

For the furnace shell [12]

$$N_u = \left\{0{,}6 + \frac{0{,}387\, R_{a,D}^{1/6}}{\left[1 + (0{,}559/P_r)^{9/16}\right]^{8/27}}\right\}^2 \tag{4}$$

For the furnace cover [12]

$$N_u = \left\{0{,}825 + \frac{0{,}387\, R_{a,D}^{1/6}}{\left[1 + (0{,}492/P_r)^{9/16}\right]^{8/27}}\right\}^2 \tag{5}$$

and

$$R_{a,D} = \frac{g\, \beta\, (T_s - T_\infty)\, D^3}{\nu^2}\, P_r \tag{6}$$

where D is the diameter of the furnace shell or cover (m), N_u is the Nusselt number, R_a is the Rayleigh number, P_r is the Prandtl number, g is the gravity (m/s^2), β is the coefficient of volume expansion of the gas (air), and ν is the kinematic viscosity (m^2/s).

2.2.2 Heat transfer by radiation

The solution of heat transfer in the furnace begins with solving the equations related to the thermal radiation shield. Heat transfer for radiation without thermal radiation shielding is determined by Equation (7) [12].

$$\dot{Q}_{1,2} = \frac{E_{b,1} - E_{b,2} \text{ or } A_1 \sigma (T_1^4 - T_2^4)}{\left(\frac{1-\varepsilon_1}{\varepsilon_1 A_1} + \frac{1}{A_1 F_{1,2}} + \frac{1-\varepsilon_{2,1}}{\varepsilon_{2,1} A_2}\right)} \tag{7}$$

The furnace concept is designed to use three shields so that there are five walls in total, namely one inner furnace wall, three layers of shield walls, and one outer furnace wall (shell and cover). Figure 2 and Figure 3 illustrate this concept. Figure 2 shows the design concept of three thermal radiation shields in a vacuum furnace, and Figure 3 shows the schematic of the heat transfer circuit to complete the analysis analytically. Analytical calculations and simulations were carried out based on this concept. Equation (7) is developed to solve the heat transfer that occurs in the vacuum furnace due to the presence of thermal radiation shielding, so that the equation becomes Equation (8).

$$\dot{Q}_{1,2} = \frac{E_{b,1} - E_{b,2} \text{ or } A_1 \sigma (T_1^4 - T_2^4)}{\left(\frac{1-\varepsilon_1}{\varepsilon_1 A_1} + \frac{1}{A_1 F_{1,3}} + \frac{1-\varepsilon_{3,1}}{\varepsilon_{3,1} A_3}\right) + \left(\frac{1-\varepsilon_{3,4}}{\varepsilon_{3,4} A_3} + \frac{1}{A_3 F_{3,4}} + \frac{1-\varepsilon_{4,3}}{\varepsilon_{4,3} A_4}\right) + \left(\frac{1-\varepsilon_{4,5}}{\varepsilon_{4,5} A_4} + \frac{1}{A_4 F_{4,5}} + \frac{1-\varepsilon_{5,4}}{\varepsilon_{5,4} A_5}\right) + \left(\frac{1-\varepsilon_{5,2}}{\varepsilon_{5,2} A_5} + \frac{1}{A_5 F_{5,2}} + \frac{1-\varepsilon_{2,5}}{\varepsilon_{2,5} A_2}\right)} \tag{8}$$

Fig. 2. The three-layer design concept of thermal radiation shielding of a vacuum furnace.

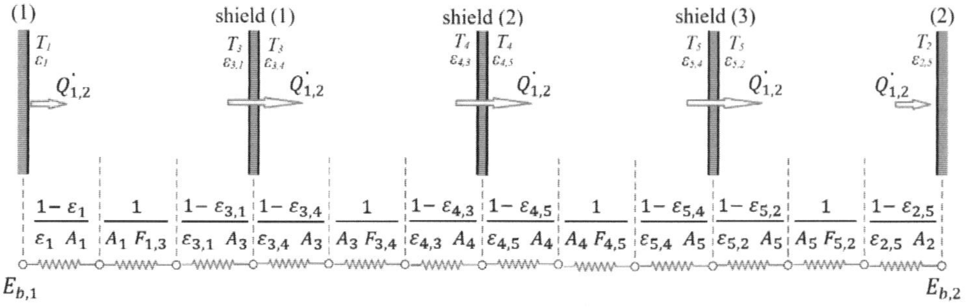

Fig. 3. Schematic of the radiation heat transfer network of the three-layer thermal radiation shielding of a vacuum furnace.

Because of infinitely long concentric cylinders, the thermal radiation from surface 1 can be assumed to be all received on surface 2, and so on, then $F_{1,3}$; $F_{3,4}$, $F_{4,5}$, $F_{5,2}$ = 1, and $A = \pi\, r\, L$, and $\frac{A_1}{A_3} = \frac{r_1}{r_3}, \frac{A_3}{A_4} = \frac{r_3}{r_4}, \frac{A_4}{A_5} = \frac{r_4}{r_5}$, and $\frac{A_5}{A_2} = \frac{r_5}{r_2}$, (cylinder shell), then Equation (8) can be simplified to Equation (9).

$$\dot{Q}_{1,2} = \frac{A\sigma(T_1^4 - T_2^4)}{\left\{\frac{1}{\varepsilon_1}+\frac{1-\varepsilon_{3,1}}{\varepsilon_{3,1}}\left(\frac{r_1}{r_3}\right)\right\}+\left\{\frac{1}{\varepsilon_{3,4}}+\frac{1-\varepsilon_{4,3}}{\varepsilon_{4,3}}\left(\frac{r_3}{r_4}\right)\right\}+\left\{\frac{1}{\varepsilon_{4,5}}+\frac{1-\varepsilon_{5,4}}{\varepsilon_{5,4}}\left(\frac{r_4}{r_5}\right)\right\}+\left\{\frac{1}{\varepsilon_{5,2}}+\frac{1-\varepsilon_{2,5}}{\varepsilon_{2,5}}\left(\frac{r_5}{r_2}\right)\right\}} \quad (9)$$

Equations (7) and (9) are used for analytical calculations. Equation (7) for calculations without shielding and Equation (9) for using three shields. In addition to analytical calculations, a special simulation was also carried out for three shields using ANSYS software to determine the wall temperature distribution, which is also used as a comparison for analytical calculations. The shielding material used is Monel with emissivity values as in Table 1 [13].

Table 1. Monel unoxidized emissivity values [13].

	Emissivity values (ε)					
Temperature (°C)	25	100	500	1000	1500	2000
$\varepsilon_{unoxidation}$	0.02	0.03	0.08	0.15	0.23	0.28

2.3 Simulation using ANSYS software

ANSYS-Mechanical Enterprise Software licensed by the National Research and Innovation Agency (BRIN) is used to simulate heat transfer from within the furnace to the surrounding environment, as well as to determine heat transfer or heat flow, temperature distribution, and transient conditions. Simulations are carried out for the utilization of three layers of thermal radiation shielding in the furnace, both at steady-state and transient conditions. Simulations for steady-state conditions assume a homogeneous temperature inside the furnace of 1000°C and are carried out with three mesh variations, namely 20, 30, and 40, while for transient conditions, two variations of heating power are based on simulation results at steady-state conditions and achievements up to a temperature of 1000°C.

The results of heat flow from analytical calculations and simulations using ANSYS software are the electrical power supply requirements of the vacuum furnace.

3 Results And Discussion

This discussion focuses primarily on the challenges of using thermal radiation shielding in vacuum furnaces to minimize heat transfer for efficient use of electrical power supplies. Thermal radiation shielding confines heat within the furnace and inhibits heat transfer from the furnace to the atmosphere, thereby reducing furnace heating power and increasing the efficiency of electrical power supplies.

3.1 Conceptual design of the vacuum furnace

The conceptual design of the vacuum furnace features a vessel with a diameter of 105 cm and a length of 150 cm. This size has been determined based on the brazing requirements and the treatment (bake-out) of the accelerator components, as illustrated in Figure 4. These requirements will be met through an initial basic analysis study, followed by a more detailed analysis that aligns with the design concept and its specific demands. Table 2 presents the calculations performed to determine the appropriate type and capacity of the vacuum pump pair. Based on the calculation results (Table 2), the Diffstak 160 pump cannot meet the requirements, while the Diffstak 250 pump and the Agilent VHS 10 pump can meet the requirements. Thus, the pumps that can be used are the Diffstak 250 and VHS 10 Agilent pumps or equivalent capacity, and the larger the pump capacity, the better.

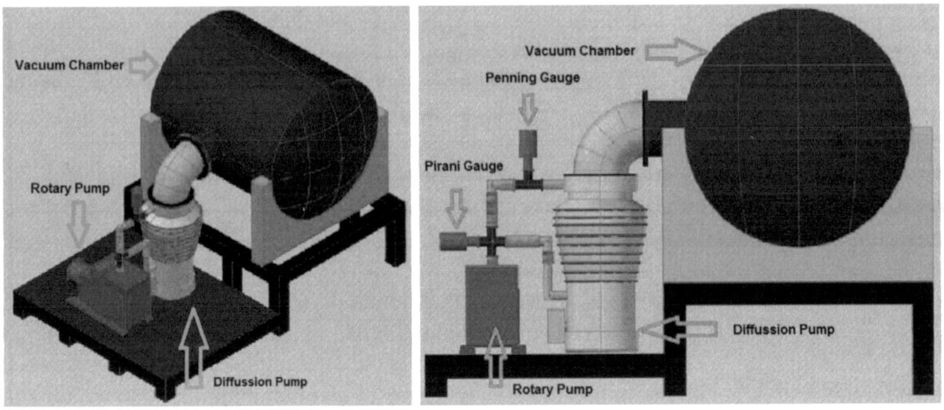

(a). perspective view (b) side view

Fig. 4. Vacuum furnace concept.

Table 2. Results of vacuum calculations for 3 types of pump capacities.

Pump type	S_p (l/s)	C_{inst} (l/s)	S_{ef} (l/s)	Q_g (mbar l/s)	P_u (mbar)
Diffstak160 (25 °C)	700	1763.786	501.119	1.3191×10^{-2}	2.63×10^{-5}
Diffstak250 (25 °C)	2130	5449.301	1531.409	1.3598×10^{-2}	8.88×10^{-6}
VHS10 Agilent (25 °C)	3650	9675.498	2650.225	1.4344×10^{-2}	5.41×10^{-6}
Diffstak160 (1000 °C)	700	1763.786	501.119	4.4840×10^{-2}	1.25×10^{-4}
Diffstak250 (1000 °C)	2130	5449.301	1531.409	4.5800×10^{-2}	4.19×10^{-5}
VHS10 Agilent (1000°C)	3650	9675.498	2650.225	4.8067×10^{-2}	2.54×10^{-5}

3.2 Analytical heat transfer calculations and simulation using ANSYS software at steady-state conditions

The calculations of analytical heat transfer are divided into two categories: inside and outside the furnace. This model assumes that the temperature inside and outside the furnace wall (shell) is the same, given that the wall thickness is only around 5 mm. The calculation is based on equation (1). The calculation of the free convection heat transfer constants to the environment uses equations (3) to (6), and the results are presented in Table 3. Next, the calculation of the free convection heat transfer to the environment uses equation (2). For radiation heat transfer to the environment, equation (7) is used.

Table 3. Results of the calculation of the convection heat transfer constants.

	P_r	$R_{a,D}$	$N_{u,D}$	$h_{c,D}$ (W/m^2 K)
Shell	0.697	2.78E+09	159.514	5.578
Cover	0.697	2.78E+09	168.305	5.886

For radiation heat transfer inside the furnace without a shield, equation (7) is used; with three shields, equation (9) is used. In addition to the analytical calculations, a special simulation was carried out using ANSYS software for three shields to determine the wall temperature distribution and heating power requirements under steady-state conditions. This simulation can be used for comparison with the analytical calculations. The shielding material used is Monel, with the emissivity values shown in Table 1 [13]. Monel emissivity values (Table 1) are entered directly into the simulator system (ANSYS software) for simulation purposes. However, for analytical calculations, emissivity data must be used according to the shield and walls temperature (shield 1, shield 2, shield 3, outer and inner walls), based on the temperature obtained from the ANSYS simulation results. The emissivity data in the reference do not match the needs for analytical calculations, so regression must be carried out to determine the emissivity. Furthermore, the emissivity value for analytical calculations is obtained using the regression equation obtained from Figure 7. The regression equation, as shown in Figure 5, is as follows:

$$y = 0{,}000138\,x + 0{,}016 \tag{10}$$

where y equals ε, emissivity, and x equals T, temperature (°C). Based on the regression equation (10), the emissivity (ε) can be determined using the shielding temperature simulated with ANSYS software (Table 4), as shown in Table 4.

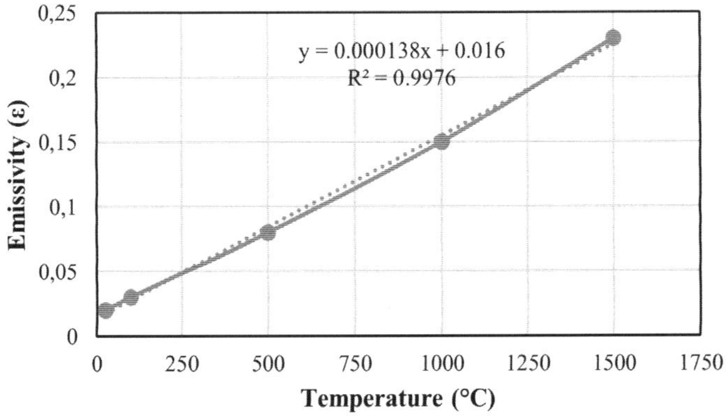

Fig. 5. Monel emissivity regression

Table 4. Emissivity regression results for analytical calculations

	outer walls	Shielding 1	Shielding 2	Shielding 3	inner walls
T (°C)	140	665	864	952	1000
ε	0.035	0.108	0.135	0.147	0.154

The analytical calculations and the ANSYS simulations were performed under the assumption of homogeneous furnace temperatures at 1000°C. The results are summarized in Table 5 and Table 6, while the results of the temperature and heat flux distribution simulations are presented in Fig. 6 and Fig. 7. Table 5 presents a summary of the results of analytical calculations and simulations using ANSYS software related to the heating power requirements and furnace wall (shell) temperatures, both without shields and with three thermal radiation shields. Simulations were performed only for the furnace configuration with three thermal radiation shields. Based on the result of analytical calculations, the furnace heating power requirement is 40.657 kW and the wall temperature is 306.9°C without shields, and 9.869 kW with three shields. Meanwhile, the result of the simulation shows that the heating power requirement with three shields is 10.149 kW. The wall and flange temperatures, determined through analytical calculations, were found to be 141.3°C. This value falls within the range observed in ANSYS simulations, which showed a minimum temperature of 133.8°C and a maximum of 147.08°C. This wall temperature meets the operating temperature limit of the seal made of Viton material, which is a maximum of 200°C [14]. Thus, the use of three layers of thermal radiation shielding is sufficient to contain heat up to an operating temperature of 1000°C in the vacuum furnace. Under steady-state conditions, the heat transferred to the environment directly corresponds to the heat that must be generated by the heater, or the heating power. The difference between analytical calculations and ANSYS simulation results for determining heating power is not significant, at 475.3 watts (4.8%).

Table 5. Summary of analytical calculation results and ANSYS simulations at steady-state conditions.

	P_{total} (kW) Analytic	P_{total} (kW) ANSYS	ΔP (kW)	%	T_{wall} (°C) Analytic	T_{wall} (°C) ANSYS min	ANSYS max	ANSYS (max+min)/2
Without shielding	40.657	-	-	-	306.9	-	-	-
With shielding	9.869	10.149	0.475	4.8	141.3	133.8	147.1	140.45

Table 6. Summary of analytical calculation results and ANSYS simulations with mesh variations.

	ANSYS Simulation				Analytical Calculation			
Mesh	P_{tot} (kW)	P_{konv} (kW)	P_{rad} (kW)	$(T_{min}+T_{max})/2$ (°C)	P_{tot} (kW)	P_{konv} (kW)	P_{rad} (kW)	T (°C)
20	10.246	4.558	5.688	142.1				
30	10.149	4.525	5.624	141.1	9.869	4.395	5.474	141.3
40	10.046	4.490	5.556	140.3				

Table 6 shows a summary of the simulation results obtained using ANSYS software, featuring three mesh variations for the heating power requirements and wall (shell) temperature of the furnace with three thermal radiation shields. This simulation ensures that the mesh size determination is appropriate. The difference in power requirements is 0.097 kW (0.96%) for mesh 20-20 and 30-30; 1.034 kW (1.03%) for mesh 30-30 and 40-40, and the temperature is 0.92°C (0.68%) for mesh 20-20 and 30-30; 0.87°C (0.74%) for mesh 30-30 and 40-40.

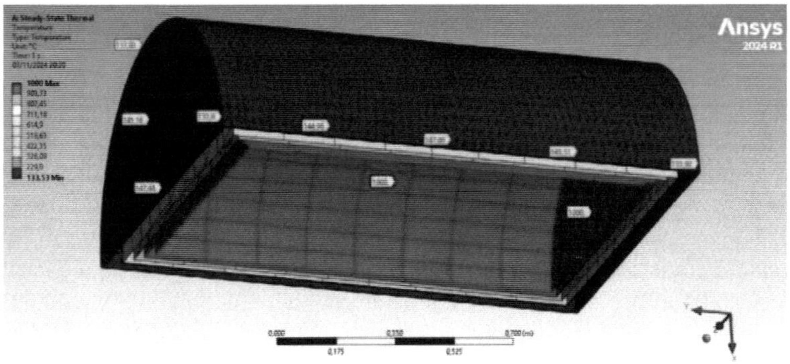

Fig. 6. Furnace wall temperature resulting from the simulation using ANSYS.

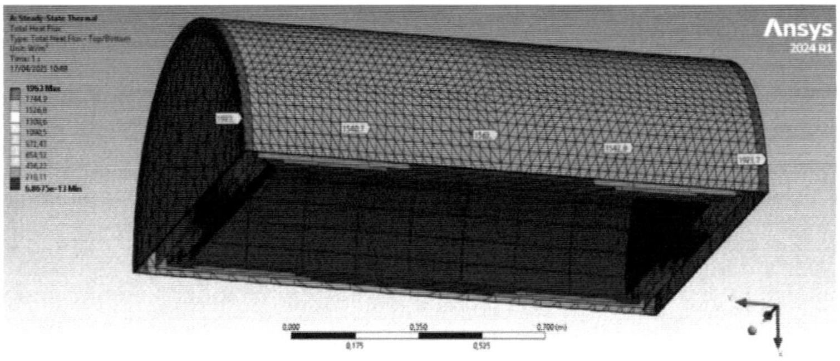

Fig. 7. Heat flux on the furnace wall resulting from the simulation using ANSYS.

Based on the simulation results, the mesh size variation does not show a significant difference. These findings demonstrate that utilizing three layers of thermal radiation shielding made from Monel plate material is highly effective in confining heat within the furnace. This effectiveness drastically reduces both the rate of heat transfer to the environment and the overall heating power requirements, specifically from 40.657 kW to 9.869 kW.

The furnace is designed to operate at a vacuum of approximately 10^{-5} mbar and must be easy to open and close for inserting and removing workpieces. Consequently, the furnace must be equipped with a strong, tight flange and seal. The flange is made of stainless steel to match the furnace wall material, while the vacuum seal is made of Viton. Viton is capable of operating at temperatures up to 200°C [14]. This operating temperature capability is far above the calculated wall or flange temperature of 141.3°C when the furnace operates at 1000°C. Based on these operating conditions, Viton can be used as a seal on the furnace flange. To prevent damage to the seal (Viton) due to heating, the flange can be cooled, for example, by using water from a chiller.

The radiation shielding arrangement consists of three layers, so the furnace has five walls in total. Based on the ANSYS simulation results (Figure 6), the temperature distribution on the outer wall surface is uneven, namely a minimum temperature of 133.08°C and a maximum of 147.08°C when the temperature inside the furnace is assumed to be homogeneous at 1000°C. Similarly, the distribution of heat flow per unit area (heat flux) on the outer wall surface of the vessel is also not homogeneous (Figure 7). The minimum heat flux of 1540.7 W/m^2 occurs in the center, while the maximum of 1923.7 W/m^2 occurs at the end. These results imply that if the furnace heat generation is homogeneous, the temperature will be lower at locations with higher heat flux compared to those with lower heat flux. This phenomenon occurs because the heat flux represents the heat flow from inside the furnace to the environment. Considering the specified temperature uniformity requirement of 5°C-10°C for temperatures above 600°C [5], proper control of heat generation and heater layout becomes very important. This control is necessary to achieve a balance between the heat flow from the furnace to the environment and the heat generated by the heater so that temperature uniformity within the furnace meets requirements.

Simulations using ANSYS software were also performed to determine the heating power requirements to reach the design temperature of 1000°C under transient conditions. The heating time must be by heating rates of 20-30°F/min [5], and 20°C/min [10]. The heating rate must meet these requirements, so it was determined to be 40°C/min. Thus, the heating time required to reach 1000°C is 25 minutes. The simulation results under transient conditions, namely during initial heating, are shown in Figure 8.

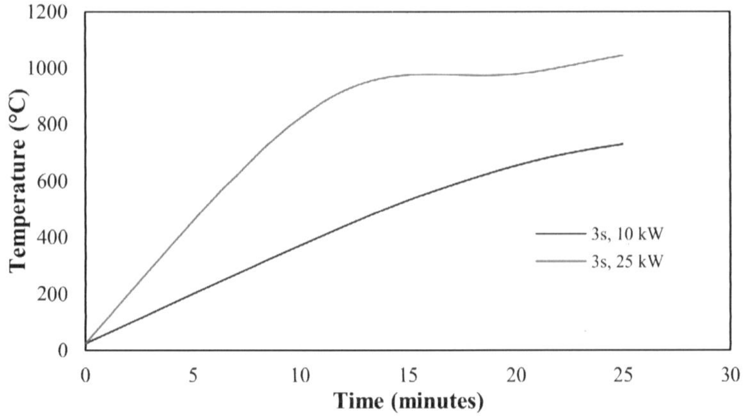

Fig. 8. Temperature transients in the furnace are simulated using ANSYS software.

The simulation began with a heating power requirement at steady-state conditions of 10 kW and continued with 25 kW. The simulation results show that during 25 minutes of heating, the achievable temperature is 730.43°C for a heating power of 10 kW and 1043.9°C for a heating power of 25 kW. Based on these results, the electrical power supply requirement for heating up to 1000°C and meeting a heating rate of 40°C/min is 25 kW.

4 Conclusion

This study successfully conducted a comprehensive analysis of the utilization of thermal radiation shielding for high-temperature vacuum furnaces. The main success of the analysis shows that the use of three layers of thermal radiation shielding can significantly minimize the need for an electrical power supply. This reduction is due to the decrease in heat transfer to the environment at steady-state conditions, thus reducing the electrical power supply requirement and the vessel wall temperature from 40,657 kW and 306.9°C without shielding to 9,869 kW and 141.3°C for analytical calculations and 10,149 kW and 140.45°C for simulations using ANSYS software with three-layer shielding. The analysis/simulation also shows that, while the inner wall temperature is assumed to be homogeneous, the outer wall surface exhibits a non-homogeneous temperature and heat flux distribution. This means that a heat generation control system and a specific heater layout are required to achieve uniform temperatures. Transient analysis confirms that the heating power supply required for the brazing process, which requires a heating rate of around 40°C/min, is 25 kW. This analysis provides valuable insights into the construction of a specialized vacuum furnace for brazing and bake-out processes for accelerator components, which are required for manufacturing purposes in Indonesia.

5 Acknowledgement

This research was supported by the RIIM LPDP Grant and BRIN, grant numbers B-4131/II.7.5/TK.01.03/02/2025 and B-1703/III.2/TK.01.03/1/2025. We also thank to the Research Center for Accelerator Technology, Research Organization for Nuclear Energy, as well as Mr. B.A. Tjipto Sujitno for his guidance throughout the research. We also express our gratitude to the Department of Nuclear Engineering and Engineering Physics for the collaboration.

References

1. S.Y. Lee, Accelerator Physics. 4th Editio. New Jersey: World Scientific Publishing Co. Pte. Ltd; 2019
2. S.V. Kutsaev, R. Agustsson, A. Araujo-Martinez, S. Boucher, R. Berry, O. Chimalpopoca, A. Diego, E. Ivanov, K. Kaneta, A.-L. Lamure, A. Pronikov, A.Yu. Smirnov, Radioisotope replacement with compact electron linear accelerators. Nucl Instruments Methods Phys Res Sect B Beam Interact with Mater Atoms. **540**, 12-18 (2023). https://doi.org/10.1016/j.nimb.2023.04.004
3. L. Faillace, D. Alesini, G. Bisogni, F. Bosco, M. Carillo, P. Cirrone, G. Cuttone, D. De Arcangelis, A. De Gregorio, F. Di Martino, V. Favaudon, L. Ficcadenti, D. Francescone, G. Franciosini, A. Gallo, S. Heinrich, M. Migliorati, A. Mostacci, L. Palumbo, V. Patera, A. Patriarca, J. Pensavalle, F. Perondi, R. Remetti, A. Sarti, B. Spataro, G. Torrisi, A. Vannozzi, L. Giuliano. Perspectives in linear accelerator for FLASH VHEE: Study of a compact C-band system. Phys Medica. **104**, 149-159 (2022) https://doi.org/10.1016/j.ejmp.2022.10.018
4. Z. Chai, N. Wei, J. Jiao, J. Meng, C. Luo, C. Li, X. Ma, X. Lin, X. Zhu, W. Xie, J. Liu, Y. Wan, Z. Niu, Z. Hu, W. Yang, X. Zhang, Study on baking deformation of the thin-walled arc chamber. Vacuum **210**, 111856 (2023). https://doi.org/10.1016/j.vacuum.2023.111856
5. F. Motschmann., *Vacuum Brazing*, Mini-CAS Course: Mechanical and Materials Engineering for Accelerators, CERN, 6 November 2020 – 9 April 2021. Available at: https://indico.cern.ch/event/958382/
6. J. A. Fedchak, J. Scherschligt, D. Barker, S. Eckel, A. P. Farrell, M. Sefa. Vacuum furnace for degassing stainless-steel vacuum components. J. Vac. Sci. Technol. A **36**, 023201 (2018) https://doi.org/10.1116/1.5016181
7. S. Badshah, M. Atif, I.U. Haq, S.A. Malik, M. Badshah, S. Jan, Thermal analysis of vacuum resistance furnace. Processes. **7,** 907 (2019) https://doi.org/10.3390/pr7120907
8. Y. Wang, Z. Liu, Development of numerical modeling and temperature controller optimization for internal heating vacuum furnace. IEEE Access **9**, 126765–126773 (2021)
9. V.V. Rao, V. Ravindra, Vacuum furnaces for metallurgical processing. J. Metall. Mater. Sci. **62**, 97–101 (2020)
10. M. Liang, Y. Qin, D. Zhang, F. Zhao, Effect of brazing clearance on the microstructure and mechanical properties of TC4/TC4 joints brazed in vacuum with Ti–Zr–Ni filler metal. J. Mater. Eng. Perform. **32**, 2973–2982 (2023)

11. R. Singh, K. K. Pant, S. Lal, D. P. Yadav, S. R. Garg, V. K. Raghuvanshi, G. Mundra, Vacuum brazing of accelerator components. J. Phys. Conf. Ser. **390**, 012025 (2012)
12. Y. A. Cengel. Heat Transfer A Practical Approah. Second Edition, editor. (1997)
13. Omega.co.uk. Emissitivity of Common Materials. :1–8. Available from: https://www.omega.co.uk/literature/transactions/volume1/emissivitya.html
14. ERIKS. Fluorelastomers, Viton sealing elements An overview. 2000; In. Available from: https://static.eriksgroup.com/en/brochures/eriks - viton fluoroelastomers an overview.pdf

Calculation and measurement of electron emission current on plasma generator tube for pulse electron irradiator

Agus Purwadi*[1,2], *Sihana*[1], *Taufik*[1], *Saefurrochman*[1], *Elin* Nuraini[1], *Suprapto*[1], *Ihwanul* Aziz[1], *Wijono*[1], *Darsono*[1], *Syarip*[1]

[1]Research Center for Accelerator Technology, Nuclear Energy Research Organization (ORTN), National Research and Innovation Agency (BRIN), Yogyakarta, Indonesia
[2]Department of Nuclear Engineering and Engineering Physics, Faculty of Engineering, Universitas Gadjah Mada, Yogyakarta, Indonesia

Abstract. Calculation and measurement of electron emission current in Plasma Generator Tube (PGT) specifically for electron pulse irradiation have been performed. At an electron density of $n_e = 10^{17}$ m^{-3}, a plasma temperature of $T_e = 7$ eV, and an arc discharge current $I_d = 80$ A (pulse width $\tau = 100$ μs), the measurement and calculation show that the electron emission current is 29.4 A and 29.8 A, respectively, in the form of pulses. Because the current is alternating and pulsed, it can directly enter the first cavity in a linear accelerator (Linac) experiment. As a result, the current has passed through the grid and may replace the electron collector. On the surface of the PGT coated with a square SS grid/mesh, the emission window area measures (600×15) mm^2 with a grid radius of 0.3 mm and a grid spacing of 0.25 mm. In addition to contributing to plasma characteristics, the grid hole size also affects the electron emission current I_e. The PGT grid holes are best square in shape and size with a side length of $p \approx 0.5$ mm. The electron emission current is influenced by the plasma properties as well as the grid hole size. The optimal grid hole size and shape for PGT is one with a grid radius of $r_e = 0.3$ mm.

1 Introduction

The advances in accelerator technology are enabled by advances in beam physics, radio frequency (RF) sources, materials science, and fabrication techniques [1]. Electron plasma is emitted into the accelerator area through a grid hole (extraction system) mounted on the Plasma Generator Tube (PGT) wall. In the accelerator area, the electron beam is accelerated to thermal velocities by a Coulomb field using an external high voltage, and then extracted after passing through a foil window in the vacuum chamber wall.

The PGT is expected to produce a plasma arc discharge in the order of tens to hundreds of Amperes (A), so only an arc discharge is suitable [2]. A system capable of efficient electron production, stable extraction, and optimal electron density distribution within the beam can only be achieved by using an extraction system, namely by installing a grid as the

*Corresponding author: agus004@brin.go.id

emitter anode on the PGT wall [3]. The dimensions of the electron pulse irradiator and the extraction voltage must be tailored to the size and shape of the PGT, as these parameters are closely related to whether electrons can be optimally extracted from the PGT.

The design activity for this electron extraction device begins with the design of the PGT for the plasma arc discharge current I_d = 80 A (pulse width τ = 100 μs) with a plasma/electron density in the range of $n_e = 10^{10}$ cm^{-3}, where the optimum density value is in the range of $n_e = (10^9\text{-}10^{12})$ cm^{-3}. The determination of the discharge current parameters I_d and electron density n_e was selected based on the results of literature studies, initial coordination agreements that have been carried out in the PRTA-BRIN research subgroup, Yogyakarta. The purpose of this study is to design and build a PGT tool system that begins with the design/calculation which is then built by the PGT tool. From the results of the manufacture of this PGT tool, the electron emission current data I_e (1-6) A can be presented at a plasma electron density ma 10^{16} m^{-3} for various plasma electron temperatures between (1-11) eV or (1-11) 11.600 °K. The amount of electron emission current that has passed through the grid (0.3 - 1.0) mm is in the form of a pulse and is an alternating current that can function as a buncher in an electron linear accelerator (linac) as its main component. Previously it was known that the electron linac device consists of an electron source, buncher and acceleration cavity up to the requested power. The PGT device can function as a buncher with an electron beam current that can directly enter the accelerator cavity which is only assisted by other auxiliary system systems as we previously knew that the linac usually consists of three main stages (1) electron emission from the cathode and its pre-acceleration with a DC field to an energy of tens to keV, (2) grouping the DC electron beam into a beam and synchronizing it with the correct phase of the high-frequency electromagnetic field and (3) accelerating the relativistic electron beam to the required energy [4]. In previous research, the pulse electron irradiator device can be used in the processing of toxic materials (natural rubber latex), surface modification in the semiconductor and polymer industries, and the food industry for pasteurization without damaging the texture and nutrition, as well as for waste neutralization [5, 6]. Meanwhile, pulsed electron irradiator devices in current research are very promising for use in various industries, including the health sector.

2 Theory

The electron current density of the plasma thermal j_{e0} can be written as :

$$j_{e0} = en_e\sqrt{\frac{kT_e}{2\pi m}} \quad (1)$$

(1)

with e = electron charge = 1.602 x 10^{-19} C, n_e = plasma density, k = Boltzmann constant = 1.381 × 10^{-23} J/K , T_e = plasma temperature and m = electron mass = $9.10^9 \times 10^{-31}$ kg.

For electron emission voltage Ua without extraction voltage (U_a = 0). When there is no extraction voltage (U_a = 0) or collector potential φ_k = 0, the electrons must be able to overcome the potential barrier, therefore the electron current density j_{ek} to the collector is given by the equation of Boltzmann relation:

$$j_{ek} = j_{e0} \exp\left(\frac{-e\varphi_p}{kT_e}\right) \quad (2) \qquad (2)$$

Where j_{e0} is the electron thermal current density, e is the electron charge, φ_p is the plasma potential, k is Boltzmann constant and Te is electron temperature. It is shown by the equation

(2) even though there is no extraction voltage ($U_a = 0$), there is still current flowing to the collector.

For the extraction voltage U_a is increased so that U_a less than plasma potential φ_p or $0 < U_a < \varphi_p$ then the equation will be obtained:

$$j_{ek}=j_{e0}\ exp\left(\frac{-e(\varphi_p-\varphi_k)}{kT_e}\right) \quad (3) \qquad (3)$$

Where j_{ek} the value of electron current density on the equation (3) is larger than the j_{ek} value on the equation (2) namely when $U_a = 0$ or $\varphi_k = 0$. It because this can be said the equation in general, the current density of electron emission j_{ek} through a potential is determined by the Boltzmann relation [7]. If the j_{ek} value increases then the φ_p value also increases which will increase the potential barrier on the surface of the PGT wall and pass the electron plasma to other electrodes (which are not to the collector). Electron emission with extraction voltage ($0 < U_a < \varphi_p$) will experience an increase in plasma potential φ_p .

For the extraction voltage Ua equal to the plasma voltage φ_p ($U_a = \varphi_p$) then the equation is shown as follow:

$$j_{ek}=j_{e0}\ .\ 1=\frac{1}{4}\ e\ n_e\ v_e \quad (4)$$

Where the electron plasma current density j_{e0} is equal with the electron current density j_{ek} ($j_{e0} = j_{ek}$) as equation (1) and $v_e=\sqrt{\frac{kT_e}{\pi m}}$

For electron emission with an extraction voltage Ua greater than potential plasma φ_p (Ua > φp), electron current density size je will still be obtained. The electron emission current density j_e reaches a saturation value (maximum) with a much higher value than the initial electron emission current ($U_a = 0$) or can be expressed in the equation:

$$j_e=exp\left(\frac{-e\varphi_p}{kT_e}\right) \quad (5)$$

It can be shown to the equation (6) i.e. the increasing factor of the electron emission current N is

$$N=exp\left(\frac{e\varphi_p}{kT_e}\right) \quad (6)$$

Where e is electron charge, φ_p is the plasma potential, k is Botzmann constant and Te is electron temperature. Thus, at the extraction voltage $U_a > 0$, for $U_a < \varphi_p$, $U_a = \varphi_p$ and $U_a > \varphi_p$ the value of j_e will be obtained which is the same as the value of j_e in equation (5), what is different is the magnitude of the beam energy which is proportional to the difference in the increase in the extraction stress.

To plan the value of the electron emission current in the IEP device, in addition to being determined by the magnitude of the extraction voltage U_a to increase the electron beam power, it is also necessary to know the use of the magnitude of the plasma density n_e in the BGP to determine the magnitude of the thermal electron current. The magnitude of the plasma density n_e depends on the dimensions (volume) and the magnitude of the plasma arc discharge current I_d that occurs in the BGP, while the value of the arc discharge current I_d depends on the operation of the trigger discharge voltage system and the arc discharge voltage system using the Ignitor Discharge Power Supply (IDPS) and Arc Discharge Power Supply (ADPS) tool. The plasma temperature parameter T_e related to electron power is influenced by the magnitude of the extraction voltage U_a. Electron beam emission in the emission window or the formation of electron beams in the accelerator area will be optimal if the plasma density in the BGP is in the range of $n_e = 10^9$ cm^{-3} to $n_e = 10^{12}$ cm^{-3} [8]. In a rectangular BGP with volume V (length p, width l and height t), if the plasma arc discharge current is I_d with a pulse width of τ, then the plasma density n_e, which is the amount of charge Q divided by the volume of the vessel V, is formulated as

$$n_e = \frac{Q}{V} = \frac{I_d\tau}{plt}(C\,cm^{-3}) = \frac{I_d\tau}{plte}(particle\;cm^{-3}) \tag{7}$$

Where n_e = electron density, e = electron charge = 1.602 x 10-19 C. If the desired arc current pulse width τ = 100 μs and the base area of the BGP = $(p \times l)$ = 80 cm × 20 cm = 1600 cm^2, then by substituting the quantities τ, p, and l in equation (7), the cylinder length t is obtained in the equation.

$$t = \frac{I_d\tau}{pln_e e}(cm) = 39.014\left(\frac{I_d}{n_e}\right)(cm) \tag{8}$$

The plasma density n_e in a cylindrical BGP with radius r and length l is

$$n_e = \frac{Q}{V} = \frac{I_d\tau}{\pi r^2 l}(Ccm^{-3}) = \frac{I_d\tau}{\pi e r^2 l}(\text{particle } cm^{-3}) \tag{9}$$

Or the cylinder radius r can be written in the equation.

$$r = \sqrt{\frac{I_d\tau}{\pi e n_e l}}\ (\text{cm}) \tag{10}$$

If the desired arc discharge current in a cylindrical BGP is I_d = 80 A with a pulse width τ = 100 μs, equation (10) can be written as

$$\text{r} = 10^1\sqrt{\frac{1{,}5889}{\text{n}_\text{e}\text{l}}}\ (\text{cm}) \tag{11}$$

Variations in the cylinder length r in equation (11) can be obtained i.e. with large variations in plasma density n_e (in the range of 10^9 cm^{-3} to 10^{12} cm^{-3}) and the cylinder length l can be varied to obtain the value of the cylinder radius r at a certain plasma density n_e.

To determine the magnitude of the electron emission current through the grid with a certain emission current magnitude, this can be done by varying the size of the grid radius r_e and the value of the plasma parameters in PGT (electron density n_e and electron temperature T_e). For plasma cathode electron source, using a total grid/emitter area of (15 × 600) mm^2, it is hoped that an optimum electron beam current I_e can be obtained, although of course it will also depend on the value of the electron emission efficiency α. Or in full for determination the emission electron current I_e according to the formula:

$$I_e = \alpha \times I(t) \tag{12}$$

Where $I(t)$ is the total emission current passing through the grid and α is the electron emission efficiency which is usually formulated as:

$$\propto = \frac{S_e}{S_e + S_a} \quad \text{for } r_{e \ll l_s} \tag{13}$$

In Figure 1 is showed the schematic of electron emission/extraction sheath stabilization namely how the electron beam is emitted from its surface through the grid radius r_e (as an anode) with a sheath thickness l_e, which is collected in the collector with an accelerating voltage of V_a :

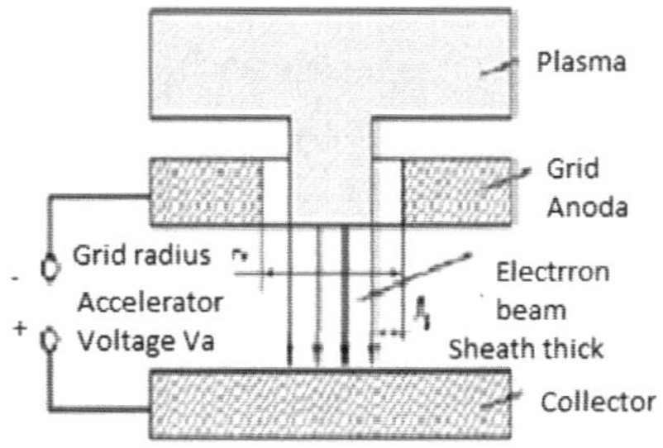

Fig. 1. Schematic of electron extraction sheath stabilization

where S_e = total area of grid holes, S_a luas total lubang grid,
S_a = total area of anode 1
r_e = grid radius,
l_s = sheath thickness .
The optimum alpha size is $\alpha_{opt.} = 0{,}50$ for $r_e \approx l_s$
If G = number of grid holes, then the electron extraction current is:

$$I_e = G\,A\,j_{e0} = G(\pi r_e^2)\ e\ n_e \sqrt{\frac{k\,T}{2\,\pi\,m}} \tag{14}$$

The electron emission current I_e will be determined as a function of r_e, n_e and T_e.

3 Methodology

In a rectangular BGP with volume V (length p, width l and height t), if the plasma arc discharge current is I_d with a pulse width of τ, then the plasma density n_e, which is the amount of charge Q divided by the volume of the vessel V, is formulated as in equation (7). And if the desired arc current pulse width $\tau = 100$ µs and the base area of the BGP = $(p\times l)$ = 80 cm × 20 cm = 1600 cm^2 then by substituting the quantities τ, p and l in equation (7) the cylinder length t will be obtained as in equation (8) while the plasma density n_e in a cylindrical BGP with radius r and length l is as shown in equation (9) or the length of the cylinder radius r can be written in equation (10) and if in a cylindrical BGP the desired plasma arc discharge current is I_d = 80 A with a pulse width $\tau = 100$ µs then equation (10) is written as equation (11) and the plasma density n_e (in the range of 10^9 cm^{-3} to 10^{12} cm^{-3}) and the cylinder length l can be varied to obtain the value cylinder radius r at a certain plasma density n_e. To determine the electron emission current density through a grid area of A = (15 x 600) mm^2 is by using equation (12) to (14) in the theory chapter while the realization for the Pulse Elecron Irradiator (PEI) to be created/realized, the shape and output of the beam current can be adjusted according to the equations we use or can use the methodology that has been used in the results and analysis.

4 Results and Discussions

The Pulse Electron Irradiator (PEI) device which is a high vacuum chamber consists of a PGT and a high voltage electron extractor. The emitted electron beam is accelerated from the emission window (on the PGT wall) towards the foil window on the wall of the irradiator vessel The PGT where the plasma occurs of the trigger electrode, isolator, cathode, grid, trigger source, plasma generator consists of the components source, accelerating voltage, spot plasma, plasma and a foil window which is an pulse electron beam output [9].

The plasma electron source is inside the PGT so that the PGT is the most important vessel in the pulse electron irradiator device. The characteristics of the electron beam produced (output) of the PGT device are greatly influenced by the extraction voltage U_a and plasma parameters inside the PGT. The plasma inside the PGT as a whole is considered homogeneous and the electron energy distribution is Maxwellian. In the most specific situation, the plasma has a positive potential towards the discharge electrode and this means that ions are extracted from the plasma surface, while electrons must overcome a potential barrier to escape from the plasma surface to the collector electrode [10]. In the process of forming a plasma beam, a large increase in the extraction voltage must cause a large increase in the speed and energy of the ions and electrons. The application of the extraction ion voltage will not change the conditions of ions emitted from the plasma into the acceleration area. Therefore, the phenomenon of emission or injection of plasma electrons is not as simple as the injection of electrons into the surface of the plasma.

Table 1 shows the calculation results for the variation in height (t) of a rectangular BGP (base area = p × l = 80 cm × 20 cm) as a function of plasma density (n_e) and arc discharge current (I_d). The plasma density ne was varied within the optimum plasma density range in the electron source vessel, namely between $n_e = 10^9$ cm^{-3} to $n_e = 10^{12}$ cm^{-3}, while the plasma arc discharge current I_d was varied from $I_d = 30$ A to $I_d = 60$ A.

Table 1. Height t of a rectangular BGP (base area = $p \times l$ = 80 cm × 20 cm) as a function of plasma density (n_e) and arc discharge current (I_d) for a pulse width of $\tau = 100$ s. (n_e varied from 10^9 cm^{-3} to 10^{12} cm^{-3} and I_d varied from 30 A to 60 A).

No.	Density n_e $\times 10^{10}$ cm^{-3}	High t (cm) for $I_d = 30$ A	High t (cm) for $I_d = 40$ A	High t (cm) for $I_d = 50$ A	High t (cm) for $I_d = 60$ A
1	0,1	11704,120	15605,490	19506,870	23408,240
2	0,2	5852,060	7802,747	9753,433	11704,120
3	0,3	3901,373	5201,831	6502,289	7802,747
4	0,4	2926,030	3901,373	4876,717	5852,060
5	0,5	2340,824	3121,099	3901,373	4681,648
6	0,6	1950,687	2600,916	3251,144	3901,373
7	0,7	1672,017	2229,356	2786,695	3344,034
8	0,8	1463,015	1950,687	2438,358	2926,030
9	0,9	1300,458	1733,944	2167,430	2600,916
10	1	1170,412	1560,549	1950,687	2340,824
11	2	585,206	780,275	975,343	1170,412
12	3	390,137	520,183	650,229	780,275
13	4	292,603	390,137	487,672	585,206
14	5	234,082	312,111	390,137	468,165
15	6	195,069	260,092	325,114	390,137
16	7	167,202	222,936	278,669	334,403
17	8	146,301	195,069	243,836	292,603
18	9	130,046	173,394	216,743	260,092
19	10	117,041	156,055	195,069	234,082
20	20	58,521	78,0275	97,5343	117,041
21	30	39,014	52,018	65,023	78,027
22	40	29,2603	39,014	48,767	58,521
23	50	23,408	31,211	39,014	46,816

24	60	19,507	26,009	32,511	39,014
25	70	16,720	22,293	27,867	33,440
26	**75**	**15,605**	**20,807**	**26,009**	**31,211**
27	76	15,400	20,533	25,667	30,800
28	77	15,200	20,267	25,333	30,400
29	78	15,005	20,007	25,009	30,010
30	79	14,815	19,754	24,692	29,631
31	80	14,630	19,507	24,383	29,260
32	90	13,004	17,339	21,674	26,009
33	100	11,704	15,605	19,507	23,408

For a cylindrical BGP with radius r and length l, the plasma density n_e is determined by equation (9) and the determination of the cylinder radius r uses equation (10). Because the desired plasma arc discharge current is $I_d = 80$ A with a pulse width $\tau = 100$ µs, if the plasma density n_e is the same as the n_e that occurs in a rectangular BGP and using equation (11), the radius r of the BGP (cylindrical shape) that must be used for a certain cylinder length l can be determined. Table 2 shows the results of calculating the value of the cylinder radius r for a cylindrical BGP (plasma arc discharge current $\boldsymbol{I_d} = 80$ A, pulse width $t = 100$ µs) as a function of the plasma density n_e and the cylinder length l (the plasma density n_e is varied from $n_e = 10^9$ cm^{-3} to $n_e = 10^{12}$ cm^{-3} and the cylinder length l is varied from 60 cm to 90 cm).

Table 2. Radius r of cylindrical BGP (for plasma arc discharge current $I_d = 80$ A, $t = 100$ µs) as a function of plasma density n_e and cylinder length l (n_e varied from 10^9 cm^{-3} to 10^{12} cm^{-3} and l varied from 60 cm to 90 cm).

No.	Density n_e × 10^{10} cm^{-3}	Radius r (cm) for $l = 60$ cm	Radius r (cm) for $l = 70$ cm	Radius r (cm) for $l = 80$ cm	Radius r (cm) for $l = 90$ cm
1	0,1	514,6034	476,4302	445,6596	420,1719
2	0,2	363,8796	336,887	315,1289	297,1064
3	0,3	297,1064	275,0671	257,3017	242,5864
4	0,4	257,3017	238,2151	222,8298	210,086
5	0,5	230,1376	213,0661	199,305	187,9066
6	0,6	210,086	194,5018	181,9398	171,5345
7	0,7	194,5018	180,0737	168,4435	158,8101
8	0,8	181,9398	168,4435	157,5645	148,5532
9	0,9	171,5345	158,8101	148,5532	140,0573
10	1	162,7319	150,6605	140,9299	132,87
11	2	115,0688	106,533	99,65252	93,9533
12	3	93,9533	86,98385	81,36594	76,71255
13	4	81,36594	75,33023	70,46497	66,43501
14	5	72,77591	67,3774	63,02579	59,42128
15	6	66,43501	61,50687	57,53441	54,24396
16	7	61,50687	56,9443	53,26651	50,22015
17	8	57,53441	53,26651	49,82626	46,97665
18	9	54,24396	50,22015	46,97665	44,29001
19	10	51,46034	47,64302	44,56596	42,01719

20	20	36,38796	33,6887	31,51289	29,71064
21	30	29,71064	27,50671	25,73017	24,25864
22	40	25,73017	23,82151	22,28298	21,0086
23	50	23,01376	21,30661	19,9305	18,79066
24	60	21,0086	19,45018	18,193978	17,15345
25	70	19,45018	18,00737	16,84435	15,88101
30	**75**	**18,79066**	**17,39677**	**16,273189**	**15,34251**
31	76	18,66663	17,28194	16,165774	15,24124
32	77	18,54502	17,16935	16,060458	15,14195
33	78	18,42576	17,05894	15,957174	15,04457
34	79	18,30877	16,95063	15,855858	14,94905
35	80	18,19398	16,84435	15,756447	14,85532
36	90	17,15345	15,88101	14,855321	14,00573
37	100	16,27319	15,06605	14,092995	13,287

Table 3 shows the results of calculating the number of grid holes in the emission anode area A = (15 × 600) mm^2 against variations in grid hole radius r_e with a distance between grid holes h = 0.25 mm.

Table 4 shows the calculation results of the electron emission current (I_e) as a function of grid radius (0.28 mm to 0.90 mm) at a plasma electron density of $n_e = 10^{17}$ m^{-3} for various electron temperature values T_e = (1 eV to 5 eV), and continued in Table 5 for T_e = (6 eV to 10 eV).

Table 3. Calculation results of the number of grid holes in the emission anode area A= 15 × 600 mm^2 with respect to variations in grid hole radius r_e with a distance between grid holes h = 0.25 mm.

Number of grid holes (at emission anode A = 15 × 600 mm2 against variation of grid hole radius re with distance between grid holes h = 0.25 mm				
No	Grid radius re (mm)2	Square side s = (π re2)1/2	The media area of each grid L = (s + h)2	Number of grid holes n = A/L
1	0.28	0.49616	0.55675	16165
2	0.30	0.53160	0.61090	14732
3	0.40	0.70880	0.91930	9790
4	0.50	0.88600	1.29050	6974
5	0.60	1.06320	1.72450	5219
6	0.70	1.24040	2.22130	4052
7	0.80	1.41760	2.78090	3236
8	0.90	1.59480	3.40330	2644

Table 4. Value of electron emission current (I_e) as a function of hole radius r_e (n_e = 10^{17}m^{-3}) for electron temperature T_e = (1, 2, 3, 4, 5) eV.

Grid radius (m)	Number of gride holes	Plasma Electron Temperature Te (oK) / Electron Emission Current				
		1×11600/ Ie (A)	2×1600 / Ie (A)	3×11600 / Ie (A)	4×11600/ Ie (A)	5×11600 / Ie (A)
2.80E-04	16165	1.06E+01	1.50E+01	1.84E+01	2.12E+01	2.37E+01
3.00E-04	14732	1.11E+01	1.57E+01	1.92E+01	2.22E+01	2.48E+01
4.00E-04	9790	1.31E+01	1.85E+01	2.27E+01	2.62E+01	2.93E+01
5.00E-04	6974	1.46E+01	2.06E+01	2.53E+01	2.92E+01	3.26E+01
6.00E-04	5219	1.57E+01	2.22E+01	2.72E+01	3.15E+01	3.52E+01
7.00E-04	4052	1.66E+01	2.35E+01	2.88E+01	3.32E+01	3.72E+01
8.00E-04	3236	1.73E+01	2.45E+01	3.00E+01	3.47E+01	3.88E+01
9.00E-04	2644	1.79E+01	2.54E+01	3.11E+01	3.59E+01	4.01E+01

Table 5. Value of electron emission current (I_e) as a function of hole radius r_e (n_e=10^{17}m^{-3}) for electron temperature T_e = (6, 7, 8, 9, 10) eV.

Grid radius (m)	Number of gride holes	Plasma Electron Temperature Te (oK) / Electron Emission Current				
		6×11600/ Ie (A)	7×1600 / Ie (A)	8×11600 / Ie (A)	9×11600/ Ie (A)	10×11600 / Ie (A)
2.80E-04	1.615	2.60E+01	2.81E+01	3.00E+01	3.18E+01	3.36E+01
3.00E-04	14732	2.72E+01	2.94E+01	3.14E+01	3.33E+01	3.51E+01
4.00E-04	9790	3.21E+01	3.47E+01	3.71E+01	3.93E+01	4.15E+01
5.00E-04	6974	3.58E+01	3.86E+01	4.13E+01	4.38E+01	4.62E+01
6.00E-04	5219	3.85E+01	4.16E+01	4.45E+01	4.72E+01	4.97E+01
7.00E-04	4052	4.07E+01	4.40E+01	4.70E+01	4.99E+01	5.26E+01
8.00E-04	3236	4.25E+01	4.59E+01	4.91E+01	5.20E+01	5.48E+01
9.00E-04	2644	4.39E+01	4.75E+01	5.07E+01	5.38E+01	5.67E+01

From the experimental results it can be shown that there are two discharges that occur sequentially in the PGT, namely the plasma spot discharge and the arc discharge. The plasma spot in the PGT which functions as an arc discharge trigger occurs at a location 1 mm in front of the ignitor electrode surface which is diluted by a 12 kV IDPS voltage power source with an argon working gas pressure of the order of 10^{-4} mBar. Mean while, arc discharge plasma appears due to the potential difference between the surface of the plasma spot volume and the potential of the wall of the PGT which is installed with a 1 kV ADPS voltage power source. Photo images of the arc discharge plasma and the electron beam emission current coming out of the beam window on the wall of the PGT are shown in Figure 2.

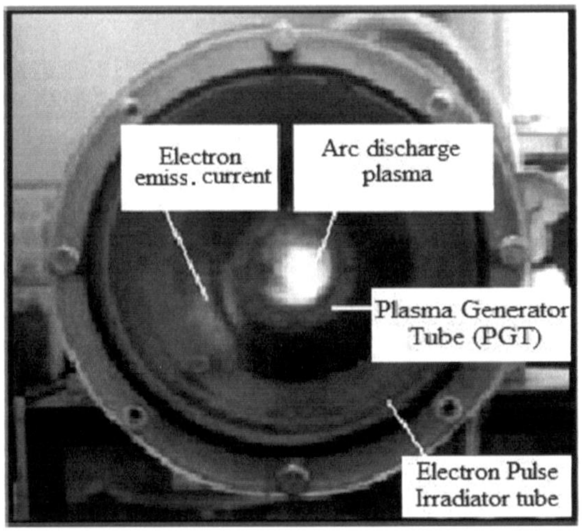

Fig. 2. Photograph of arc discharge plasma in the PGT.

In Figure 3 is shown a photo image of a partial cross-section of the electron emission current beam coming out of a beam window of (15 × 600) mm^2 (on the wall of the PGT) which is observed through the beam window on the wall of the MBE vessel. The magnitude of the plasma arc current which is the electron current discharge I_e in the PGT is measured using a Rogowski coil and an oscilloscope.

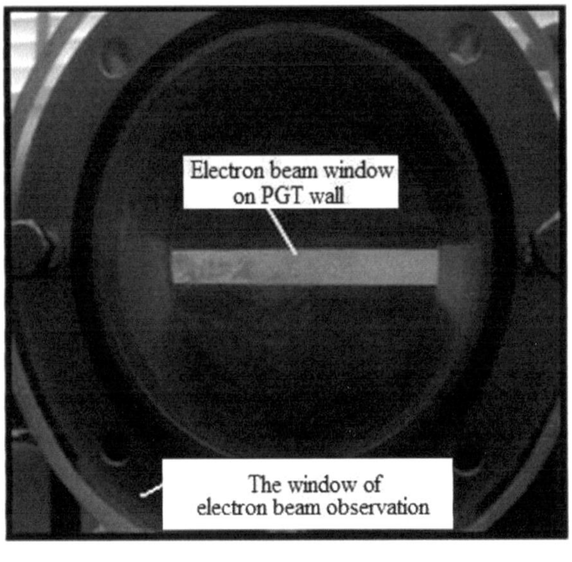

Fig. 3. Cross-sectional photo of the electron emission current beam coming out of PGT.

The plasma electron emission current I_e in the PGT is formed by the potential difference between the surface potential of the plasma volume φ_p (plasma is formed due to the high voltage of Ignitor discharge power suppy in the trigger electrode system) and the positive electrode potential on the wall of the plasma generator φ_c (which functions as a collector). In the experiment, the plasma potential value φp is greater than the potential of the collector/wall of the of the PGT (Ignitor discharge power supply voltage ~ 12 kV, Arc discharge power supply voltage ~ 1 kV).

The grid is planned to be installed on the PGT wall which functions as a plasma emitter that extracts electrons currents from the PGT in the form of pulse. The PGT is covered with a grid made of SS material in a square shape with an optimum hole size of (15×600) mm^2. If the distance between one grid hole and another is 0.25 mm, then for a PGT emitter media/area of (0.75×0.75) mm^2 there will be 1 grid hole, as shown in Figure 4.

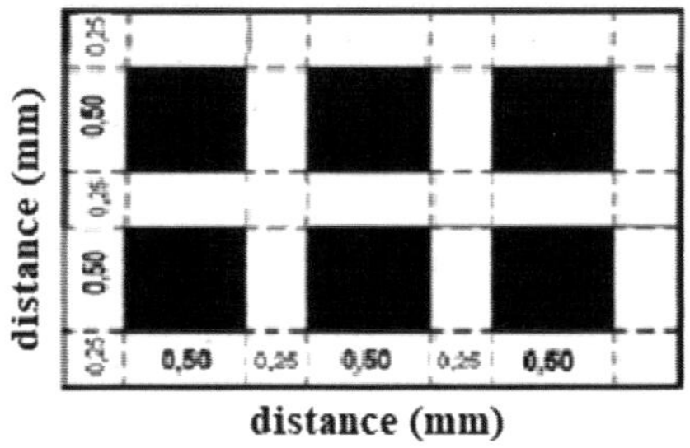

Fig. 4. Grid holes (0.50×0.50) mm^2 with distance between grid holes one to the other is 0.25 mm.

The maximum electron emission current I_e through the grid hole is the result of multiplying the maximum electron emission current density in equation (4) with the area of the grid hole, so the size of I_e will also depend on plasma parameters such as electron density n_e, electron temperature T_e also depends on size. grid grid r_e. Meanwhile, the other quantities in this equation are just constants whose values are certain, namely for e = electron charge = 1.602×10^{-19} C, k = Boltzmann's constant = 1.37×10^{-23} J/K and m_e = electron mass = 9,11 $\times 10^{-31}$ kg. Because the number of grid holes is known and the magnitude of the electron emission current I_e through each grid hole is also calculated, the total electron emission current exiting through the electron emitter on the PGT can be determined.

A square grid measuring (15×600) mm^2 is identical to a circular grid with a radius of 0.28 mm or 2.8×10^{-4} m because the area of both geometric shapes has the same value. Furthermore, this can be used as a standard that each grid hole is circular with a radius of 0.28 mm $\times$ 0.30 mm is the optimum grid hole radius size.

Because the optimum grid hole radius (circular shape) for plasma cathode electron source is 0.30 mm and the distance between adjacent grids is 0.25 mm, for an emitter surface area of (0.85×0.85) mm^2 will be occupied by 1 (one) hole grid.

According to theory, the electron extraction efficiency will apply to a grid hole radius (r_e) that is much smaller than the plasma sheath thickness (l_e) and the optimum value is αopt. = 0.50. This may be interpreted that the efficiency value α is in the range (0 – 50) %. The extraction electron efficiency (α) at the extraction voltage (maximum) that will be obtained is $\alpha = I_e/I_d$ = 29.80 A / 80 A = 37.25 %. Furthermore, it can be seen from the table (calculation results), the magnitude of the electron beam emission current I_e, whose value is close to 29.80 A, is I_e = 29.40 A, so that the size of the grid hole and the corresponding plasma parameters and or the best one to use in the electron irradiator device is a square shape with a side length of p = 0.53 mm or p close to 0.50 ($p \approx 0.50$ is a common size on the market) and the plasma parameters in PGT use an electron density of n_e=10^{16} m^{-3} and the plasma electron temperatur T_e=7 eV.

Figure 5 and Figure 6 show the graph of the electron emission current I_e as a function of grid radius r_e with a density (n_e = 10^{16} m^{-3}) and temperature T_e = (1 to 6) eV and T_e = (6 to 11) eV.

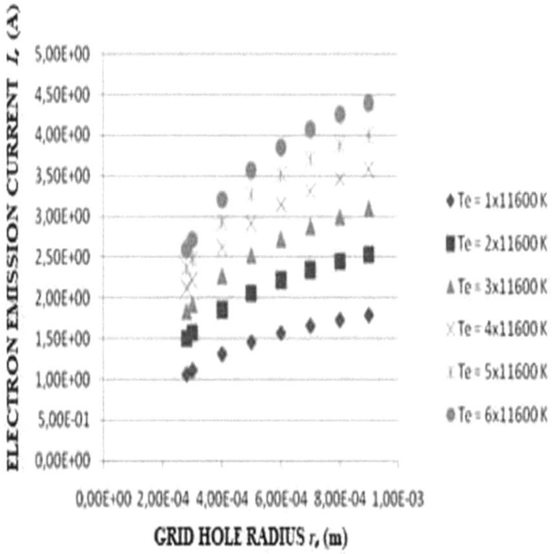

Fig. 5. Electron emission current graph I_e function of grid radius r_e at electron density $n_e = 10^{16}$ m^{-3} for temperature T_e = (1-6) eV.

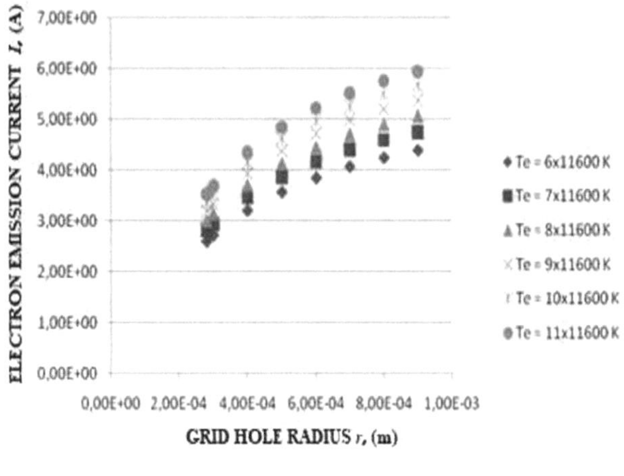

Fig. 6. Graph of electron emission current I_e as a function of grid radius r_e on density electron n_e = 10^{16} m^{-3} for temperature T_e .= (6-11) eV.

If compared with what other people have done by R. Purwadi, et al [11], the design of this pulse electron source equipment is very different, namely without using a pressure vessel, which will be cheaper and simpler. Therefore, to plan the value of the electron extraction current on the PGT device, besides being determined by the extraction voltage U_a, to increase the power of the electron beam.

The system that is able to produce efficientelectrons, stable extraction, and optimum distribution of electron density in the beam is only by using an extraction system by installing a grid as an emitter anode on the PGT wall. The emission window on the PGT is covered with a square grid sheet (cassette) made of Stainless Steel (SS) with an optimum hole size. The maximum electron emission current I_e through the grid hole is the result of multiplying the maximum electron emission current density in equation (4) with the area of the grid hole. The magnitude of I_e besides depending on plasma parameters such as the electron density n_e, the temperature of the electrons T_e also depends on the size of the r_e grid radius. Since the number of grid holes is known and the magnitude of the electron emission current I_e through each grid hole can be calculated, the total electron emission current flowing out through the electron emitter in the PGT can be determined.

5 Conclusion

Plasma density n_e depends on the dimensions (volume) of the PGT or linear accelerator (linac) electron buncher and the magnitude of the plasma arc discharge current I_d and the pulse width τ that occurs in the PGT, while the values of I_d and τ depend on the operation/setting of the trigger power source system and the plasma arc discharge power source. Based on the calculation and measurement results, it can be concluded that for the use of I_d = 80 A with τ = 100 μs and the area of the electron beam emission hole (600 × 15) cm^2, the dimensions of the PGT are rectangular with dimensions (80×20×40) cm^3. From the calculation and measurement results, the electron emission currents were obtained respectively of 29.4 A and 29.8 A in the form of pulses that occurred at an electron density of n_e=10^{17} m^{-3}, plasma temperature T_e = 7 eV, arc discharge current I_d = 80 A (pulse width τ = 100 μs), where the current obtained has passed through the grid and has been able to replace the function of the electron buncher because the current has been alternating and in the form of pulses so that it can directly enter the initial cavity in the linear accelerator (Linac) experiment. The size of the rectangular PGT with a volume of 0.064 m^3, an emission window area of (600×15)×10^{-6} m^2, a grid radius of 3 ×10^{-4} m (on the PGT surface coated with a square SS grid/gauze with a grid distance of 0.25 × 10^{-3} m). The magnitude of the electron emission current I_e depends not only on the plasma parameters but also on the size of the grid holes. The optimum shape and size of the PGT grid holes are square with side length $p \approx 5\times10^{-4}$ m. The value of electron emission efficiency (α) at injector voltage V = 3 kV is obtained as α = 37.25%. The magnitude of the electron emission current I_e depends not only on plasma parameters but also on the size of the grid holes, where the optimum shape and size of the PGT are grid holes with a grid radius of $r_e = 3\times10^{-4}$ m.

Acknowledgement

The authors would like to thank the Head of the PRTA-ORTN-BRIN Office and all colleagues who have helped design and construct as well as calculate and measure the Plasma Generating Tube (PGT) so that this paper can be completed and realized in written form with the title; Calculation and Measurement of The Electron Emission Current on The Plasma Generator Tube for The Pulse Electron Irradiator.

References

1. S. V. Kutsaev, Advanced technologies for applied particle accelerators and examples of their use (review). J. Tech. Phys. **66**, 161 (2021). https://doi.org/10.1134/S1063784221020158
2. Physical_background : Glow_Discharges: Discharge_Regimes [Internet]. 2014. Available from: http://www.glow-discharge.com
3. A. Krokhmal, J. Z. Gleizer, Ya. E. Krasik, V. Ts. Gurovich & J. Felsteiner, Grid-controlled electron emission from a hollow-anode electron source. J. Appl. Phys. **95**, 3304–3310 (2004). https://doi.org/10.1063/1.1651340

4. S. Kutsaev, Electron bunchers for industrial RF linear accelerators: theory and design guide. Eur. Phys. J. Plus **136**, 216 (2021). https://doi.org/10.1140/epjp/s13360-021-01312-3
5. J. Z. Gleizer, V. Vekselman, S. Yatom, J. Feildstein, Y. E. Krasik, Radiation Effects & Defects in Solids. first, pp. 1-10, 2011
6. J. Z. Gleizer, Y. Hadas, V. T. Gurovich, J, Felsteiner, and Y.E. Krasik, High-current electron beam generation in a diode with a multicapillary dielectric cathode. J. Appl. Phys. **103**, 043302 (2008)
7. F. G. Baksht and V. G. Ivanov. Dissociation rate for molecular hydrogen near the cathode in a low-temperature plasma. Soviet Phy. Tech. Phys. 31, 926 (1968)
8. E. Oks . Lecture 4 Electron emission from plasma. Yogyakarta, Indonesia: Presented in BATAN Accelerator School; 2011.
9. A. Purwadi, Suprapto, I. Aziz, Saefurrohman & A. A. Harumningtyas, Requirements for obtaining the maximum electron current in the pulse electron irradiator. AIP Conf. Proc. **2296**, 020109 (2020). https://doi.org/10.1063/5.0030543
10. J. A. Bittencourt, Fundamentals of Plasma Physics (Pergamon Press (Aust.), Potts Point, N.S.W., Australia, 2011)
11. P. Raharjo, K.Uemura, N.N.Koval, V. Shugurov, V. Denisov, V. Jakovlev, W. Setiawan, and M. Utama, Application of Large Area Plasma Cathode Elektron Beam for Natural Rubber Vulcanitation. ITAC Ltd, 8-2 Kamisuwa Tsubame City Niigata 959-0181 Niigata, Japan. 2010;497–501.

Kinetic study of essential oil extraction from nutmeg seeds (*Myristica fragrans* Houtt) using microwave air-hydrodistillation (MAHD)

Mustafiah Mustafiah[1], *Firman* Kurniawansyah[1], and *Mahfud* Mahfud[1*]

[1]Department of Chemical Engineering, Institut Teknologi Sepuluh Nopember, Jalan Arief Rahman Hakim, Surabaya 60111, Indonesia

Abstract. The extraction of essential oil from *Myristica fragrans* Houtt (nutmeg) seeds has gained increasing interest due to its pharmacological and industrial applications. This study investigates the extraction kinetics of nutmeg seed essential oil using Microwave Air-Hydrodistillation (MAHD) and compares the performance of five mathematical models: first-order, second-order, power law, weibull and elovich. Experimental extractions were performed at two microwave power levels (300 W and 450 W) with solid to liquid ratios of 0.20, 0.25, and 0.30 g/mL. Kinetic parameters were determined by fitting each model's equation to the experimental yield data, and model performance was evaluated using the coefficient of determination (R^2) and root mean square error (RMSE). Among all models, the Weibull model provided the best fit, with the highest R^2=0.9990 and lowest RMSE=0.0014, accurately capturing both the initial rapid extraction phase and the later diffusion-controlled phase. These findings confirm that the Weibull model offers a robust predictive framework for MAHD, supporting process optimization and scale-up of nutmeg essential oil extraction systems while preserving oil quality.

1 Introduction

Nutmeg (*Myristica fragrans* Houtt), a tropical evergreen tree native to the Banda Islands of Indonesia and now cultivated extensively across Southeast Asia and the Caribbean, is highly valued for its dual role as a culinary spice and a source of bioactive phytochemicals. Among its most commercially significant derivatives is nutmeg seed essential oil, which has attracted substantial attention due to its complex composition and broad spectrum of applications. This essential oil contains a wide array of volatile compounds, including sabinene, myristicin, safrole, and eugenol, all of which contribute to its well-documented antimicrobial, antioxidant, anti-inflammatory, and neuroactive properties [1]. These biological attributes have positioned nutmeg oil as an important raw material in various industries, including pharmaceuticals, cosmetics, food preservation, and perfumery.

* Corresponding author: mahfud@chem-eng.its.ac.id

Driven by the increasing global demand for plant-based natural products, recent research has emphasized the optimization of extraction techniques to enhance yield, selectivity, and essential oil quality. A critical challenge in this context is the preservation of thermolabile constituents, which are highly susceptible to thermal degradation during prolonged exposure to elevated temperatures. Conventional techniques such as hydrodistillation (HD) and steam distillation remain widely employed due to their operational simplicity and well-established protocols. However, these methods are inherently limited by extended extraction times, high energy consumption, and the potential degradation or volatilization loss of heat-sensitive compounds. In contrast, microwave-assisted hydrodistillation has been reported to significantly improve extraction efficiency through rapid volumetric heating and enhanced mass transfer, thereby reducing thermal degradation and shortening processing time [2]. In response, alternative "green" extraction methods are being explored for their potential to offer greater efficiency, reduced processing times, and lower environmental impact. One such technique is Microwave-Assisted Extraction (MAE), and in particular its variant, Microwave Air-Hydrodistillation (MAHD), which has demonstrated promising results in extracting essential oils from various botanical sources [3, 4].

MAHD operates by delivering microwave energy directly to the plant-water matrix, inducing rapid internal heating, causing cellular rupture, and facilitating the release of volatile compounds. This technique has been successfully applied to the extraction of essential oils from aromatic plants such as *Ocimum basilicum* (basil), *Cymbopogon citratus* (lemongrass), and *Syzygium aromaticum* (clove), resulting in enhanced extraction rates and improved oil quality. Despite these advantages, the application of MAHD to nutmeg seeds, whose dense matrix and particular oil distribution present unique challenges, remains inadequately explored in the scientific literature.

A major barrier to the broader adoption and scale-up of MAHD for nutmeg oil production is the limited understanding of the underlying mass transfer mechanisms and extraction kinetics specific to this seed matrix. Most existing research on nutmeg focuses on oil yield quantification and chemical profiling, with little attention to the kinetic behavior of oil release under microwave-assisted conditions. This represents a significant knowledge gap, considering the complex interplay between microwave energy absorption, moisture migration, structural disintegration, and internal diffusion that occurs during MAHD. Without a reliable kinetic framework tailored to nutmeg seeds, it is difficult to predict extraction performance, optimize the process, or design scalable industrial systems. Moreover, inherent variability in nutmeg seed properties such as size, oil content, and tissue porosity, adds further complexity to standardizing extraction processes [5, 6].

To address these limitations, recent studies have begun employing kinetic modeling as a tool to describe the dynamic process of essential oil extraction from plant matrices. Mathematical models enable the quantification of key process parameters, including extraction rates, equilibrium yields, and the transition between distinct extraction phases (e.g., washing vs. diffusion). These insights are invaluable for simulation, process control, and the design of industrial-scale equipment [4]. Models such as first-order, second-order, power law, Elovich, and Weibull equations have been successfully applied in studies involving lemongrass, basil, clove, and other medicinal plants under microwave- and ultrasound-assisted extraction systems [7, 8].

Each kinetic model offers unique advantages and limitations. First-order and second-order models, derived from reaction kinetics principles, are often insufficient to capture the multiphase nature of plant-based extractions. In contrast, power-law and elovich models are better suited for nonlinear kinetics and can be particularly useful when surface desorption or heterogeneous diffusion dominates the process. The Weibull model, with its flexible sigmoidal curve form, has emerged as a highly effective approach for modeling both the initial rapid release and the subsequent slower diffusion phases commonly observed in microwave-assisted extractions [4, 6].

Despite the growing application of these models in botanical extractions, a systematic comparison of their performance in the context of nutmeg seed oil extraction via MAHD has not yet been reported. While similar kinetic studies exist for other plant species, the unique physicochemical properties of nutmeg seeds necessitate a dedicated analysis to determine which model is most appropriate for this particular matrix under varying operational parameters such as microwave power and solid-to-liquid ratio. This gap in the literature underscores the need for empirical validation and model benchmarking to support the optimization and industrial implementation of MAHD for nutmeg oil production [3, 5].

Previous studies on microwave air-hydrodistillation (MAHD) of *Myristica fragrans* seeds have demonstrated that microwave power plays a significant role in enhancing essential oil yield through dielectric heating and structural disruption of plant matrices. In particular, parametric optimization studies evaluating microwave power (300–600 W) and feed-to-solvent ratio using response surface methodology (RSM) have reported improved extraction efficiency, chemical composition quality, and antibacterial activity by [9]. Nevertheless, these investigations primarily emphasized final yield optimization and compositional characteristics, without systematically elucidating the time-dependent extraction behavior or the underlying mass-transfer kinetics governing the MAHD process. To address this limitation, the present study advances beyond yield-based parametric analysis by explicitly investigating the dynamic extraction kinetics of nutmeg essential oil under different microwave power levels. The extraction process is quantitatively modeled using five kinetic models (first-order, second-order, power-law, Weibull, and Elovich), enabling comparative statistical evaluation and mechanistic interpretation of microwave-assisted mass transfer. By distinguishing between rapid surface washing and diffusion-controlled stages, this work provides deeper kinetic insight into the dominant extraction regimes and contributes predictive modeling tools that are relevant for process optimization and potential scale-up of MAHD systems.

In this study, we address the above knowledge gap by conducting a comprehensive kinetic analysis of essential oil extraction from nutmeg seeds using MAHD. The objective is to compare the performance of five widely used kinetic models: first-order, second-order, power law, Weibull, and Elovich in fitting experimental extraction data obtained under controlled conditions. Key statistical parameters, including R^2 and RMSE, are employed to assess the accuracy and reliability of each model. The evaluation is performed across multiple solid-to-solvent ratios and microwave power levels, providing a robust comparison of model performance under different operational regimes. The novelty of this research lies in its integrative, model-based approach to understanding nutmeg oil extraction kinetics under MAHD. By applying and comparing multiple models within a single experimental framework, this study offers new insights into the extraction dynamics of *Myristica fragrans* Houtt seeds and provides quantitative tools for process optimization. The results are expected to support the broader goal of advancing sustainable, efficient, and scalable technologies for recovering high-value plant-based essential oils.

2 Materials and Methods

2.1 Materials

Nutmeg seeds (*Myristica fragrans* Houtt) used in this study were obtained from Morowali Regency, Central Sulawesi, Indonesia. The seeds were first air-dried at ambient temperature (25–30 °C) for seven days to reduce moisture content. The dried seeds were then ground and sieved to obtain a uniform particle size of 20 mesh, ensuring consistent exposure of the material to the extraction process. The prepared ground nutmeg samples were divided into three batch sizes: 100 g, 125 g, and 150 g. Each batch was mixed with 500 mL of distilled water (as the extraction medium) in the distillation flask during the MAHD process. These batch masses correspond to solid-to-liquid (feed-to-solvent) ratios of approximately 0.20, 0.25, and 0.30 g/mL, respectively.

2.2 Schematic of the Experimental System

The Microwave Air-Hydrodistillation (MAHD) system was constructed using a modified domestic microwave oven (Electrolux model EMM20K22BA) as the primary heating unit. The distillation apparatus consisted of a 1000 mL double-neck round-bottom Pyrex® flask, which served as the main extraction vessel. One neck of the flask was connected to a vertical reflux condenser, while the other neck was fitted to a Clevenger-type apparatus for collecting the condensed essential oil. An air diffuser (sparger) was inserted through the flask's stopper to ensure uniform dispersion of fine air bubbles throughout the aqueous phase. Pressurized air was supplied by an air compressor (Lakoni IMOLA-75 model) and introduced at a regulated flow rate of 1 L/min using an in-line flow meter. This continuous stream of air, bubbling through the heated mixture, facilitated the co-distillation by carrying the volatile oil vapors upward. The vapor mixture of water and essential oil traveled to the condenser, where it cooled and liquefied. The condensate then flowed into the Clevenger separator, allowing the essential oil to separate from water due to immiscibility and density differences, and the oil was collected in the graduated arm of the apparatus Figure 1.

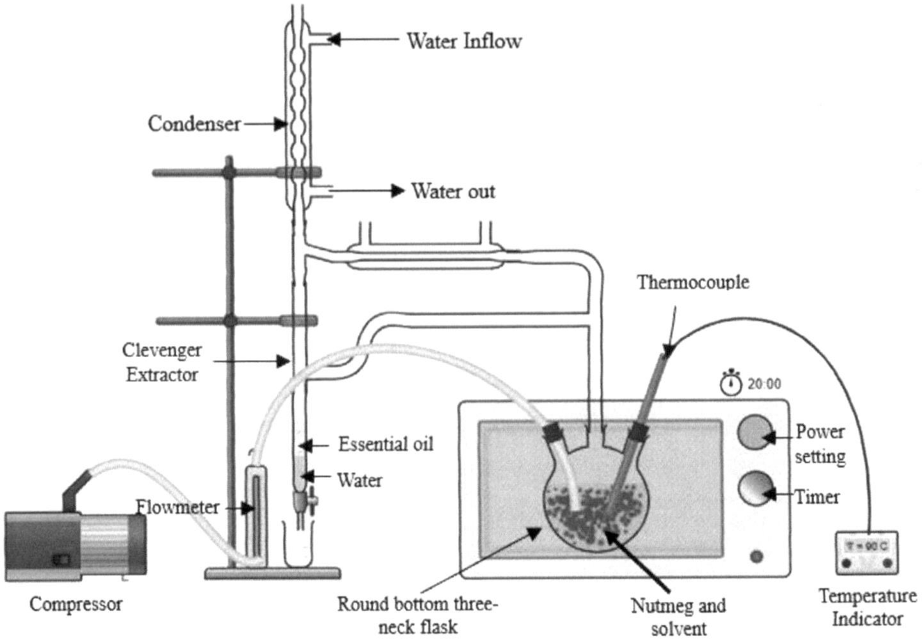

Fig. 1. Schematic diagram of the apparatus used for essential oil extraction from nutmeg seeds via the MAHD method.

2.3 Experimental Procedure

Each extraction experiment was conducted using the MAHD method under controlled conditions. Microwave irradiation was applied at two power levels: 300 W (medium power) and 450 W (higher power), to examine the effect of microwave intensity on extraction kinetics. Ground nutmeg seed samples of 100 g, 125 g, and 150 g were used, corresponding to feed-to-solvent ratios (F/S) of approximately 0.20, 0.25, and 0.30 g/mL, respectively, when combined with 500 mL of distilled water in the extraction flask. For each run, the loaded flask (containing the nutmeg sample and water) was placed inside the microwave oven, which acted as both the heating source and a containment for the process. The microwave unit was equipped with a temperature probe or indicator to monitor the internal vapor temperature during extraction, ensuring that operating conditions remained within desired ranges.

The extraction was carried out for a total duration of 180 minutes (3 hours) for each batch. Throughout the extraction period, distillate (a mixture of water and oil) was continuously condensed and separated. To obtain time-resolved kinetic data, condensate was collected at regular intervals of 20 minutes (i.e., at 20, 40, 60 minutes, etc., up to 180 minutes). At each interval, the distillate was removed from the Clevenger receiver and transferred to a separatory funnel. The essential oil layer was then carefully isolated from the water layer in the separatory funnel. The volume (and corresponding mass) of oil recovered at each interval was measured and recorded. All recovered essential oil samples were stored in sealed glass vials at 0 °C (in a freezer) to prevent any compositional changes before subsequent analysis. This time-interval sampling allowed for the construction of extraction yield curves (oil yield as a function of time) for each experimental condition.

2.4 Kinetic Modeling of the Extraction Process

A kinetic modeling approach was employed to analyze the extraction behavior of nutmeg seed essential oil under the MAHD process. Five kinetic models were selected for evaluation: the Weibull model, first-order model, second-order model, power law model, and Elovich model. Each of these models provides a mathematical expression that relates extraction yield to time, and each is often associated with different underlying mechanisms of extraction. A comparative summary of the mathematical formulations of these models is presented in Table 1.

Table 1. Presents a comparative mathematicals equations of five models.

Model's name	Basic equation	Model comparison	References
1st-Order	$\frac{dC_t}{dt} = k_1(C_s - C_t)$	$q_a = (1 - e^{-k1.t})$	[10]
2nd-Order	$\frac{dC_t}{dt} = k_2(C_s - C_t)^2$	$q_a = \frac{k_2 t}{1 + C_s k_2 t}$	[11]
Power Law	$q_t = B.t^n$	$q_a = B.t^n$	[12]
Weibull	$f(x;\lambda,\beta,\gamma) = \beta\gamma(\beta x)^{\gamma-1}\exp(-\lambda(\beta x)^{\gamma}$	$q_a = 1 - \exp\left(-\frac{t^m}{D}\right)$	[7]
Elovich's	$dy/_{dt} = \beta exp^{-\alpha\gamma}$	$q_a = E_0 + E_1.\ln t$	[6]

By fitting these models to the experimental extraction curves, we aimed to identify which model best represents the extraction kinetics of nutmeg oil under MAHD. A best-fit model would not only closely reproduce the experimental data but also reflect the underlying physical mechanisms (e.g., external mass transfer versus internal diffusion control) of the extraction process. Identifying an accurate kinetic model is important for optimizing the extraction process (e.g., predicting yields at given times, scaling up to larger systems) while ensuring maximum efficiency and preservation of oil quality.

2.5 Statistical Analysis

The performance of each kinetic model was evaluated using two statistical metrics: the coefficient of determination (R^2) and the Root Mean Square Error (RMSE). These metrics were calculated by comparing the model-predicted yields to the experimental yields at each time point. The equations used for these calculations are standard in model fitting:

2.5.1 *Coefficient of determination (R^2):*

$$R^2 = 1 - \frac{\Sigma\,(q_{exp} - q_{model})^2}{\Sigma\,(q_{exp} - \bar{q}_{exp})^2} \quad (1)$$

This equation represents the fraction of the total variance in the experimental data explained by the model. Here, $q_{exp,i}$ is the experimental yield at time point i, $q_{model,i}$ is the model-predicted yield at the same time, $\bar{q}_{exp}$ is the mean of all experimental yields, and N is the number of data points. R^2 values range from 0 to 1, with values closer to 1 indicating a better fit (i.e., the model explains most of the variability in the data).

2.5.2 *Root Mean Square Error (RMSE):*

$$RSME = \sqrt{\frac{1}{N}\sum_{i=1}^{n}\left(q_{exp,i} - q_{model,i}\right)^2} \quad (2)$$

RMSE provides an absolute measure of the average deviation between model predictions and experimental values (in the same units as q). Lower RMSE values indicate more accurate model predictions overall.

In these formulas, we define:
q_{exp} = Experimental yield at the i-th time interval (g/mL)
q_{model} = Yield predicted by the model for the i-th time interval (g/mL)
$\bar{q}_{exp}$ = Mean of all experimental yield values (g/mL)
N = Total number of experimental data points (time intervals)

The model with the highest R^2 and lowest RMSE was considered the most accurate in describing the extraction kinetics. Selecting an appropriate model based on these criteria forms the foundation for optimizing operating conditions and enhancing process efficiency and cost-effectiveness at scale [4]. All model fitting and statistical calculations were performed using standard nonlinear regression techniques. Model comparison was then conducted to determine which kinetic model provides the best representation of the extraction process for each experimental condition.

3 Results and Discussion

The kinetic parameters of essential oil extraction from nutmeg seeds (*Myristica fragrans* Houtt) using MAHD were determined by fitting the experimental data to the five selected models: first-order, second-order, power law, Weibull, and Elovich. This section presents a comparative analysis of these models under different operating conditions. The experiments were performed on nutmeg 20 mesh particle size at two microwave power settings; results at

300 W are discussed first, followed by those at 450 W. Key findings are illustrated in Figure 2 (300 W) and Figure 3 (450 W).

3.1 Extraction Kinetics at 300 W Microwave Power

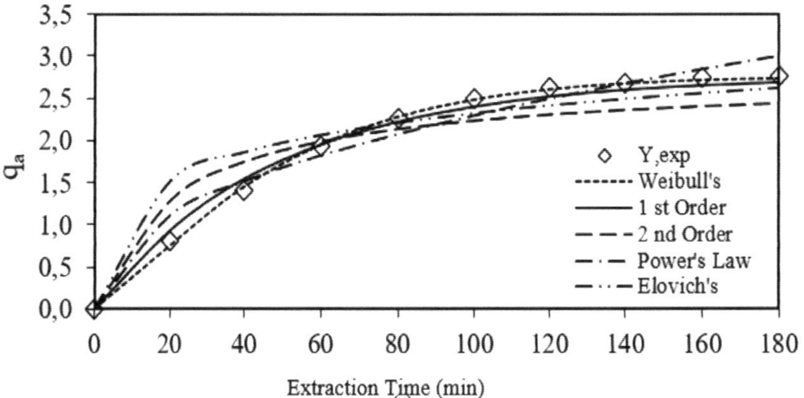

(a)

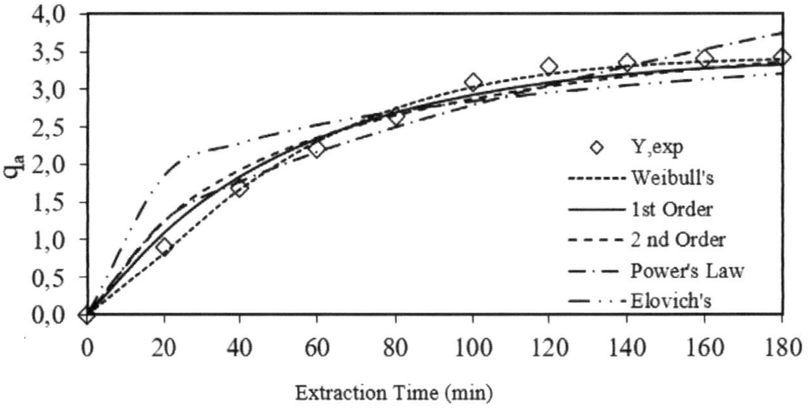

(b)

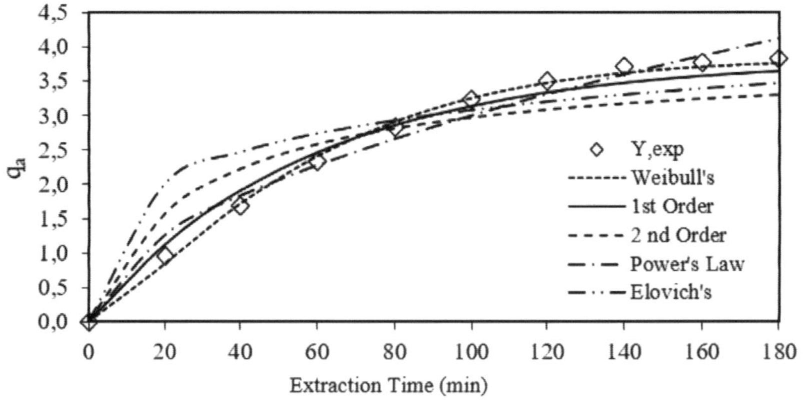

(c)

Fig. 2. Comparison between experimental data (symbols) and model predictions (lines) for the extraction of nutmeg seed essential oil under varying solid-to-solvent ratios at 300 W microwave power: (a) 0.20 g/mL, (b) 0.25 g/mL, and (c) 0.30 g/mL.

At 300 W, the extraction profiles for all three solid-to-solvent ratios (0.20, 0.25, and 0.30 g/mL) exhibited a characteristic biphasic behavior. An initial rapid extraction phase was observed, driven largely by external mass transfer (washing), whereby readily accessible oil on or near the surface of particles is quickly released. This was followed by a slower diffusion-controlled phase as the extraction approached equilibrium. Such two-step extraction patterns are typical in microwave-assisted extraction systems: the early stage is governed by the desorption of oil from the particle surface and convection into the bulk, while the later stage is dominated by the internal diffusion of oil from within the solid matrix [6], [13]. The rapid heating and cell rupture caused by microwaves accelerate the first phase, whereas the second phase reflects the time required for the remaining oil to migrate out of the solid fragments.

For the lowest solid-to-solvent ratio, 0.20 g/mL, Figure 2a, among the five models tested, the Weibull model provided the best agreement with the experimental data. The Weibull fit closely tracked the entire extraction curve, capturing both the steep initial rise and the gradual plateau. Quantitatively, the Weibull model achieved the highest coefficient of determination (R^2 = 0.9990) and the lowest root mean square error (RMSE = 0.0014) at this ratio Table 3. This excellent fit underscores the model's flexibility in describing complex mass transfer mechanisms and corroborates earlier studies that highlighted the Weibull model's superiority for microwave-assisted extractions that involve a rapid-to-slow transition [4, 7].

In comparison, the first-order and second-order models showed lower predictive accuracy at 0.20 g/mL. The first-order model, while reasonably capturing the general trend, had a slightly lower (R^2=0.9947) and a higher (RMSE=0.0100) value than the Weibull model. It tended to underestimate the yield during the very early minutes (due to its fixed exponential form), but followed the later part of the curve moderately well. The second-order model performed worse: it exhibited significant deviations, particularly as time progressed, yielding a substantially lower (R^2=0.9319) and a larger (RMSE=0.1053). The underperformance of these reaction-kinetic models suggests that the MAHD extraction of nutmeg oil cannot be adequately described by simple kinetic assumptions alone. The extraction involves more complex diffusion behavior within the plant matrix that first- and second-order models, which assume a uniform kinetic rate throughout, fail to capture [8].

The power law model showed intermediate performance at 0.20 g/mL. It performed better than the second-order model and could mimic the initial rapid increase in yield (since the power law can accommodate a fast-rising curve through its exponent n). However, it tended to overestimate the yield during the mid-to-late stages of extraction, indicating a poorer representation of the slowing kinetics as the system neared equilibrium. The Elovich model displayed the highest deviation among all models in this case, reflected in the lowest R^2 of 0.9073 and the highest RMSE of 0.1737 at 0.20 g/mL. The Elovich model is based on a chemical adsorption analogy; its poor performance implies that such an adsorption-driven kinetic assumption does not align with the actual mechanism of essential oil release under MAHD conditions [2]. In practice, nutmeg oil extraction is not limited by surface chemisorption, but rather by internal diffusion and solubilization processes, which Elovich's formulation does not adequately capture.

When the solid-to-solvent ratio was increased to 0.25 g/mL Figure 2b, similar trends were observed. The Weibull model remained the most accurate across the entire extraction curve. It closely matched the experimental data points at all times, yielding (R^2= 0.9966) and (RMSE = 0.0062), Table 3. The model effectively captured the biphasic extraction behavior: a rapid initial phase followed by a slower approach to equilibrium. This indicates that even as the amount of nutmeg increased relative to the solvent volume, the Weibull model's flexibility allowed it to adapt to the slightly prolonged diffusion tail and still accurately represent the process. This behavior is consistent with the notion that the Weibull model can account for a shift in the dominant mechanism from external mass transfer to internal diffusion by appropriately tuning its parameters [4, 7].

The first-order model at 0.25 g/mL provided a reasonable fit in the early and middle stages of extraction but began to deviate in the later stages. leading to a slight underestimation of the final yield. R^2 0.9905 and RMSE 0.0250 are the same as Weibull's but still indicate a decent overall fit for much of the process. This limitation highlights that as more material is present (a higher F/S ratio), the extraction's slowing phase becomes more pronounced (due to a greater amount of oil needing to diffuse out), and the simple exponential approach of the first-order model struggles to represent that deceleration fully.

The second-order model accurately predicts the extraction trend at 0.25 g/mL. Its reliance on a quadratic driving force and assumption of parabolic kinetics did not match the empirical curve, especially during the late phase. The second-order model's fit resulted in a relatively low RMSE of 0.0509 and R^2 of 0.9806, which improved from the 0.20 ratio scenario (perhaps because with more material, the second-order model's initial slope can be tuned), but still indicated noticeable discrepancies.

The power law model at 0.25 g/mL again tended to overshoot the extraction yield in the initial phase if parameters were tuned to fit later portions, or vice versa. It captured the high initial extraction rate (due to the nature of the power-law rising quickly for t^n with $n > 0$), but did not accurately emulate the curvature as the extraction slowed. The Elovich model remained the least accurate, showing large deviations throughout the extraction period. The underlying assumption of the Elovich model, that the rate decreases exponentially with the amount extracted (typical of chemisorption simply does not hold for this physical extraction scenario, which is more governed by diffusion and equilibrium partitioning than by surface site exhaustion [8].

For the highest solid-to-solvent ratio of 0.30 g/mL Figure 2c, the extraction profile at 300 W showed a slightly more prolonged second phase, as expected when more solid and oil are present. The Weibull model once again provided the best overall fit to the data. It achieved $R^2 = 0.9973$ and an RMSE of about 0.0069 Table 2, demonstrating strong agreement. The Weibull curve closely followed the empirical points from the rapid initial extraction through the gradual deceleration over 3 hours. The success of the Weibull model across all ratios underscores its adaptability. Its two-parameter sigmoidal form effectively accommodates both the fast and slow stages, which is critical for complex systems like MAHD that do not follow simple one-rate kinetics [4, 6].

In contrast, the first-order and second-order models at 0.30 g/mL showed only moderate agreement during the early to mid extraction periods and deviated more significantly in the later stage (particularly after 100 minutes). Both models struggled as the extraction continued and the easily accessible oil fraction was depleted. The first-order model's predicted curve started to level off too early or too late, depending on parameter fit, and the second-order model's predictions diverged even more once the diffusion limitation became dominant. These deviations are reflected in their statistics. The first-order model at 0.30 g/mL had R^2 0.9913 with a relatively low RMSE 0.0302. Note that this RMSE seems low, likely due to scale or perhaps a unit issue, but consistent comparison shows it is larger than Weibull's, and the second-order model had R^2 0.9244 with RMSE 0.2237, indicating a poor fit Table 3. Clearly, the internal diffusion and delayed release of volatile compounds in the latter half of the process are phenomena that these simpler kinetic models do not capture well.

The power law model, at the 0.30 g/mL ratio, again tended to overestimate yields in the early phase if fit overall, which reduced its accuracy during the deceleration phase. Its R^2 0.9785 and RMSE 0.0467 were slightly improved over the second-order model but still inferior to the Weibull's and even the first-order model in this case. The Elovich model continued to perform the worst, with the lowest R^2 value of 0.8717 and the highest RMSE of 0.4045 at 0.30 g/mL. This confirms that the Elovich equation's premise is not suitable for describing the MAHD of nutmeg oil, regardless of the operating conditions [7, 8].

Table 2. Estimated kinetic parameters of nutmeg seed essential oil extraction using MAHD at different feed-to-solvent ratios (g/mL) at a microwave power of 300 W.

Ratio (g/mL)	1st-order k_1 (min^-)	2nd-order k_2 ($g \cdot mg^{-1} \cdot min^-$)	Power Law B	Power Law n	Weibull β (min)	Weibull α (–)	Elovich E_0	Elovich E_1
0.20	0.0205	0.0157	0.2827	0.4542	1.2306	50.6976	0.0000	0.5059
0.25	0.0190	0.0074	0.2797	0.4996	1.2798	55.1637	0.0000	0.6188
0.30	0.0172	0.0090	0.2485	0.5410	1.2659	60.5084	0.0000	0.6675

Table 3. Comparison of model fit quality (R^2 and RMSE) for each kinetic model at different solid-to-solvent ratios using MAHD at 300 W.

Ratio (g/mL)	1st-order		2nd-order		Power Law		Weibull		Elovich	
	R^2	RSME	R^2	RSME	R^2	RSME	R^2	RSME	R^2	RSME
0.2	0.9947	0.0100	0.9319	0.1053	0.9657	0.0383	0.9990	0.0014	0.9073	0.1737
0.25	0.9905	0.0250	0.9806	0.0509	0.9681	0.0608	0.9966	0.0062	0.8854	0.3153
0.3	0.9913	0.0302	0.9244	0.2237	0.9785	0.0467	0.9973	0.0069	0.8717	0.4045

From the above results at 300 W, it is evident that the Weibull model provides the most accurate representation of nutmeg oil MAHD kinetics. It consistently yielded the highest R^2 and lowest RMSE, indicating an excellent fit. These findings are in line with previous research that has found the Weibull model effective in describing complex mass transfer dynamics in microwave-assisted essential oil extraction processes [4, 6]. The first-order model, while somewhat successful especially at lower ratios, showed limitations in the deceleration phase, suggesting it may be used as a simpler approximation, but with caution. The second-order, power law, and Elovich models were less suitable, with the Elovich model being the least accurate for this system. Therefore, for design and optimization purposes (such as predicting the extraction time needed to reach a certain yield or scaling up the process), the Weibull model would be recommended as the primary model, and perhaps the first-order model could serve as a secondary option for rough initial estimates in certain cases.

3.2 Extraction Kinetics at 450 W Microwave Power

Experiments were also carried out at a higher microwave power of 450 W to assess how increased energy input affects extraction kinetics and model performance. The results at 450 W for the solid-to-solvent ratios of 0.20, 0.25, and 0.30 g/mL are illustrated in Figure 3, and the model comparison metrics are summarized in Table 4.

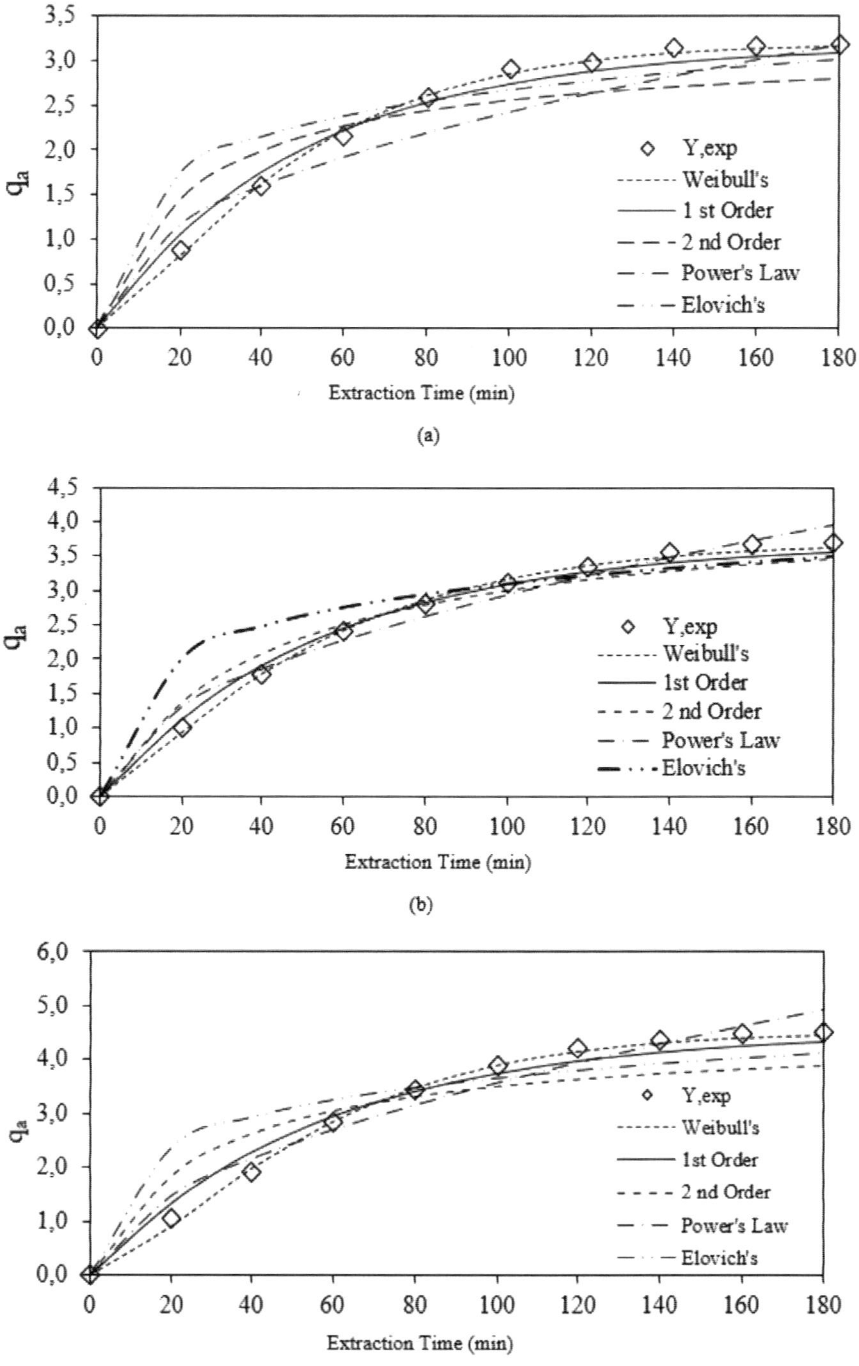

(c)

Fig. 3. Comparison of experimental data (symbols) with simulated kinetic model curves (lines) for nutmeg seed essential oil extraction at 450 W microwave power for solid-to-solvent ratios: (a) 0.20 g/mL, (b) 0.25 g/mL, (c) 0.30 g/mL.

At 450 W, the overall extraction process was noticeably faster, as expected due to the higher energy input. The biphasic profile was still observed at all ratios, but the initial phase was more rapid, and the transition to the slower phase occurred earlier in the extraction period. All three ratios (0.20, 0.25, 0.30) g/mL showed the majority of oil being extracted in the first 60–80 minutes, followed by a tapering off toward equilibrium. This behavior aligns with known effects of microwave power: higher power accelerates heating and cell rupture, thus speeding up the "washing" stage, but once the easy-to-extract oil is depleted, the remaining diffusion-controlled stage still takes time [4], [13].

Across all conditions at 450 W, the Weibull model again provided the best fit. Specifically, at 0.20 g/mL and 450 W, the Weibull model achieved R^2 = 0.9988 with RMSE 0.0020 Table 5, indicating an almost perfect representation of the data. For 0.25 g/mL, the Weibull model maintained a high accuracy R^2=0.9980, RMSE=0.0048, and at 0.30 g/mL, R^2 = 0.9986 with RMSE=0.0055. These consistently high R^2 values at 450 W reaffirm that the Weibull model's flexibility is effective even under enhanced extraction rates. Recent studies similarly highlight the Weibull model's robustness in modeling complex extraction kinetics under various microwave-assisted systems, crediting its ability to mimic both fast initial release and slower diffusion tailing [6, 14].

In contrast, the first-order model at 450 W showed good agreement with experimental data only up to the mid-stage of extraction, beyond which it deviated. At 0.20 g/mL (450 W), the first-order model had R^2=0.9929 and RMSE=0.0179. It fit the initial sharp increase reasonably well and even the middle portion, but in the later part of the curve, it predicted a slightly higher yield than observed, resulting in some mismatch. At 0.25 g/mL, the first-order model's R^2=0.9955 with RMSE=0.0165, likely because the larger amount of oil made the early stage slightly less dominant, giving the model more data in the exponential region to fit. However, at 0.30 g/mL, the first-order model's R^2=0.9889 and RMSE=0.0613, reflecting a significant deviation as extraction progressed. These results emphasize that while raising microwave power accelerates extraction, it does not fundamentally change the two-phase nature of the process. The first-order model, lacking a mechanism to handle the slowing phase adequately, struggles more as the system has more oil to release (higher ratios), even under high power.

The second-order model at 450 W demonstrated even greater deviation overall. Its R^2 values ranged roughly from 0.9265 at 0.20 g/mL to 0.9782 at 0.25 g/mL, but with a relatively high RMSE of 0.1561 to 0.0667. These statistics in Table 5 underscore that the second-order kinetic assumption (often interpreted as involving a quadratic term of unextracted fraction) is not suitable here. The physical process likely involves simultaneous heat and mass transfer complexities that are not captured by the simple assumptions of the second-order model. The particularly low R^2 at 0.20 and 0.30 g/mL indicates poor fits, supporting the conclusion that the second-order model is not appropriate for representing the diffusion-dominated behavior in later stages of MAHD.

The power law model at 450 W again tended to overestimate extraction yield in the initial phase. At high microwave power, extraction is so rapid initially that the power law model, if tuned to get the early points right, will overshoot later, or if tuned for later, will undershoot early. The R^2 values for the power law were between 0.9649 at 0.20 g/mL and 0.9811 at 0.25 g/mL, but RMSE values were relatively large, 0.0405–0.1074. This indicates that while the model can capture some portion of the curve (especially the initial steepness for a certain n exponent), it lacks accuracy for the latter part of the extraction. Essentially, the power law can mimic a rapidly decelerating process only to a limited extent and fails once the curve flattens significantly.

The Elovich model remained the poorest performer at 450 W. It yielded the lowest R^2 (around 0.8611–0.8954 across the ratios) and the highest RMSE (0.2638–0.6679). The large errors confirm that the Elovich model's underlying assumption of a chemical adsorption mechanism is incompatible with the actual extraction mechanism under microwave heating. The microwave-driven extraction of essential oils involves physical disruption and diffusion, rather than adsorption-desorption kinetics, and thus Elovich's exponential decay form does not match the data [7, 8].

Table 4. The estimated kinetic parameters for MAHD-assisted extraction of nutmeg seed essential oil at various feed-to-solvent ratios (g/mL) under a microwave power of 450 W.

Ratio (g/mL)	1st-order k_1 (min^{-1})	2nd-order k_2 ($g \cdot mg^{-1} \cdot min^{-1}$)	Power Law B	Power Law n	Weibull β (min)	Weibull α (–)	Elovich E_0	Elovich E_1
0.20	0.0199	0.0131	0.2912	0.4600	1.2750	52.6055	0.0000	0.5806
0.25	0.0181	0.0074	0.2916	0.5022	1.1777	57.0008	0.0000	0.6753
0.30	0.0175	0.0077	0.2853	0.5486	1.3507	60.1212	0.0000	0.7937

Table 5. Coefficient of determination (R^2) and RMSE for each kinetic model in nutmeg seed oil MAHD at 450 W for different solid-to-solvent ratios.

Ratio (g/mL)	1st- Order		2nd Order		Power's Law		Weibull's		Elovich's	
	R^2	RSME	R^2	RSME	R^2	RSME	R^2	RSME	R^2	RSME
0.20	0.9929	0.0179	0.9265	0.1561	0.9649	0.1074	0.9988	0.0020	0.8954	0.2638
0.25	0.9955	0.0165	0.9782	0.0667	0.9811	0.0405	0.9980	0.0048	0.8948	0.3723
0.30	0.9889	0.0613	0.9173	0.3642	0.9697	0.0946	0.9986	0.0055	0.8611	0.6679

From Table 5, it is evident that increasing the microwave power to 450 W does not change the qualitative ranking of model performance; it only accentuates differences in some cases. The Weibull model remains superior, with near-perfect R^2 across all ratios. The first-order model still performs second-best, particularly at an intermediate ratio (0.25 g/mL), but its fit degrades at the highest ratio due to pronounced diffusion effects. The second-order model's accuracy is comparatively low, reinforcing that its kinetic basis is not aligned with the actual mechanism here. The power law model has respectable R^2 values at higher ratios (where the initial phase is a bit less dominant relative to total time), but its RMSE indicates non-negligible prediction errors. The Elovich model consistently fails to provide an acceptable fit.

These results strongly indicate that the Weibull model is the most appropriate kinetic model for describing the MAHD extraction of essential oil from nutmeg seeds, under both moderate and high microwave power conditions and across a range of solid loadings. The Weibull model's success can be attributed to its flexible sigmoidal shape, which effectively mirrors the two-stage extraction kinetics: its parameters allow it to adjust the curvature to fit a sharp rise and a long tail. In practical terms, this means the Weibull model can be reliably used for process design and optimization.

The first-order model, while not as precise, could still be used for approximate descriptions or when simplicity is desired (since it has fewer parameters and a straightforward exponential form). However, one must keep in mind that it will likely be the time required to reach near-complete extraction (because it doesn't flatten out as slowly as reality) and overestimate extraction in later stages at high solid loadings [13]. If we were to use it, it may be most applicable for the initial design of shorter extraction durations where the process is largely in the exponential phase.

The other models (second-order, power law, Elovich) are not recommended for this system. The second-order model, which worked in some contexts of solid-liquid extraction, did not capture the plateau behavior well here. The power law model, lacking an asymptotic maximum, is inherently limited for processes that have a clear equilibrium yield, as all extractions do. It can fit an initial segment, but will either diverge or flatten incorrectly. The Elovich model's assumptions simply do not hold, as evidenced by its poor fit; hence, it is unsuitable for the MAHD of nutmeg oil.

Raising the microwave power from 300 W to 450 W increased the rate of extraction but did not change the fundamental mechanism; a rapid release followed by diffusion-limited release is seen in both cases. The Weibull model adeptly captures this mechanism under all tested conditions. These findings are consistent with prior research on microwave-assisted extraction kinetics, which often report the Weibull as being most effective for complex extractions [6, 14, 15]. The results here extend that understanding to nutmeg seed oil, reinforcing the value of the Weibull model for modeling and potentially controlling or optimizing such extraction processes.

3. Conclusion

This study demonstrated that the extraction of essential oil from *Myristica fragrans* Houtt (nutmeg) Seeds using the Microwave Air-Hydrodistillation (MAHD) method follows a distinct two-phase kinetic profile: an initial rapid extraction stage dominated by external mass transfer (washing of readily available oil), followed by a slower internal diffusion stage as the extraction nears completion. Five kinetic models, first-order, second-order, power law, Weibull, and Elovich, were applied to describe the extraction behavior under varying conditions of microwave power (300 W and 450 W) and solid-to-liquid ratios (0.20, 0.25, and 0.30 g/mL). Among these models, the Weibull model consistently provided the best fit to the experimental data across all tested conditions, achieving the highest coefficients of determination with R^2=0.9990 and the lowest RMSE values as low as 0.0014. This indicates its strong ability to accurately predict both the fast initial extraction regime and the subsequent slow regime.

The findings confirm that the Weibull model offers a flexible and reliable mathematical framework for modeling essential oil extraction from nutmeg seeds using MAHD. Its ability to accommodate complex, multi-stage mass transfer phenomena makes it highly suitable for process design, simulation, and optimization in this context. We recommend the adoption of the Weibull model in future scale-up studies and in developing industrial process control strategies for nutmeg oil extraction, especially in scenarios where high extraction efficiency and quality preservation of the oil are paramount. The first-order model may serve as a useful secondary model for approximate kinetics in early stages or low-loading scenarios, but it is less accurate during diffusion-controlled periods.

References

1. A. Muchtaridi, A. Subarnas, A. Apriyantono, and R. Mustarichie, Identification of compounds in the essential oil of nutmeg seeds (Myristica fragrans Houtt.) that inhibit locomotor activity in mice. Int. J. Mol. Sci. **11**, 4771–4781 (2010). https://doi.org/10.3390/ijms11114771
2. A. Farhat, A. Fabiano-Tixier, M. El, and J. Maingonnat, Microwave steam diffusion for extraction of essential oil from orange peel : Kinetic data, extract 's global yield and mechanism, Food Chem. **125**, 255–261 (2011). https://doi: 10.1016/j.foodchem.2010.07.110

3. F. Chemat, N. Rombaut, A.-G. Sicaire, A. Meullemiestre, A.-S. Fabiano-Tixier, M. Abert-Vian, Ultrasound assisted extraction of food and natural products. Mechanisms, techniques, combinations, protocols and applications. Ultrason. Sonochem. **34**, 540–560 (2017). https://doi.org/10.1016/j.ultsonch.2016.06.035
4. A. Radivojac, O. Bera, D. Micić, S. Đurović, Z. Zeković, S. Blagojević, and B. Pavlić, Conventional versus microwave-assisted hydrodistillation of sage herbal dust: Kinetics modeling and physico-chemical properties of essential oil. Food Bioprod. Process. **123**, 90–101 (2020). https://doi.org/10.1016/j.fbp.2020.06.015
5. T. P. Dao, N. Q. Tran, T. T. Tran, and V. T. Lam, Assessing the kinetic model on extraction of essential oil and chemical composition from lemon peels (Citrus aurantifolia) by hydro-distillation process. Mater. Today Proc. **51**, 172–177 (2022). https://doi.org/10.1016/j.matpr.2021.05.069
6. D. Darnengsih, H. Ni'mah, M. Mahfud, and A. C. K. Fitri, Kinetics model of Cananga (Cananga odorata) fresh flower extraction using microwave hydrodistillation method. Makara J. Sci. **28**, 287–295 (2024). https://doi.org/10.7454/mss.v28i4.2146
7. H. Haqqyana, V. F. W. Tania, A. M. Suyadi, H. S. Kusuma, A. Altway, and M. Mahfud, Kinetic study in the extraction of essential oil from clove (Syzygium aromaticum) stem using microwave hydrodistillation. Moroccan J. Chem. **8**, 64–71 (2020) https://doi.org/10.48317/IMIST.PRSM/morjchem-v8i1.19126
8. H. Benmoussa, I. Béchohra, S. He, W. Elfalleh, and R. Chawech, Optimization of sonohydrodistillation and microwave-assisted hydrodistillation by response surface methodology for the extraction of essential oils from Cinnamomum cassia barks. Ind. Crops Prod. **192**, 115995 (2023). https://doi.org/10.1016/j.indcrop.2022.115995
9. M. Mustafiah, F. Kurniawansyah, and M. Mahfud, Enhanced extraction of Myristica fragrans Houtt seed essential oil via microwave air-hydrodistillation: Parametric study, optimization, and antibacterial evaluation. Results Eng. **29**, 109074 (2026). https://doi.org/10.1016/j.rineng.2026.109074
10. H. S. Kusuma and M. Mahfud, Kinetic studies on the extraction of essential oil from sandalwood (Santalum album) by microwave air-hydrodistillation method. Alexandria Eng. J. **57**, 1163–1172 (2018). https://doi.org/10.1016/j.aej.2017.02.007
11. Y.-S. Ho, H. A. Harouna-Oumarou, H. Fauduet, and C. Porte, Kinetics and model building of leaching of water-soluble compounds of Tilia sapwood. Sep. Purif. Technol. **45**, 169–173 (2005). https://doi.org/10.1016/j.seppur.2005.03.007
12. S. Kitanović, D. Milenović, and V. B. Veljković, Empirical kinetic models for the resinoid extraction from aerial parts of St. John's wort (Hypericum perforatum L.). Biochem. Eng. J. **41**, 1–11 (2008). https://doi.org/10.1016/j.bej.2008.02.010
13. P. T. Dao, N. Y. T. Tran, Q. N. Tran, G. L. Bach, and T. V. Lam, Kinetics of pilot-scale essential oil extraction from pomelo (Citrus maxima) peels: Comparison between linear and nonlinear models. Alexandria Eng. J. **61,** 2564–2572 (2022). https://doi.org/10.1016/j.aej.2021.07.002
14. X. Peng, N. Liu, M. Wang, B. Liang, C. Feng, R. Zhang, X. Wang, X. Hu, H. Gu, and D. Xing, Recent advances of kinetic model in the separation of essential oils by microwave-assisted hydrodistillation. Ind. Crops Prod. **187**, 115418 (2022). https://doi.org/10.1016/j.indcrop.2022.115418
15. A. Radivojac, O. Bera, Z. Zeković, N. Teslić, Ž. Mrkonjić, D. Bursać Kovačević, P. Putnik, and B. Pavlić, Extraction of peppermint essential oils and lipophilic compounds: Assessment of process kinetics and environmental impacts with multiple techniques. Molecules **26**, 2879 (2021). https://doi.org/10.3390/molecules26102879

Unlocking green efficiency: optimized solvent-free microwave extraction of essential oils from *zingiber officinale*

Citra Wahyu Rizkita[1], *Lailatul* Qadariyah[1], *Mahfud* Mahfud[1*], *Sukardi* Sukardi[2]

[1]Department of Chemical Engineering, Institut Teknologi Sepuluh Nopember (ITS)
[2]Department of Agricultural Technology, Brawijaya University

Abstract. SFME has emerged as a green and efficient technique for essential oil recovery; however, its application to ginger rhizomes and the identification of kinetically operating conditions remain limited. This study investigates the time-dependent extraction behavior and extraction kinetics of ginger essential oil using SFME under varying operating conditions. Microwave power and sample mass were systematically varied to generate yield–time profiles, which were subsequently analyzed using several kinetic models, including first-order, second-order, power law, hyperbolic, Elovich, and Weibull models. The extraction process exhibited a characteristic two-stage behavior, consisting of a rapid initial extraction followed by a slower diffusion-controlled stage. Kinetic analysis revealed that empirical models, particularly the power law and Weibull models, effectively described the non-linear extraction behavior, indicating mixed mass transfer control mechanisms. At a constant microwave power of 450 W, sample mass significantly influenced extraction kinetics, with intermediate solid loading providing a balanced extraction rate and favorable characteristic extraction time. Based on kinetic efficiency and extraction stability rather than final yield alone, the optimal operating condition was identified at 450 W with a sample mass of 200 g. This kinetic-based evaluation provides mechanistic insight into SFME of ginger and offers a rational efficient operating conditions for green extraction.

1 Introduction

Essential oils (EOs) are volatile, aromatic substances extracted from various parts of plants, including leaves, seeds, rhizomes, and flowers. They are widely valued for their multifaceted biological activities such as antimicrobial, anti-inflammatory, antioxidant, and anticancer effects, making them important ingredients in the pharmaceutical, cosmetic, and food industries [1]. In recent years, global consumer interest in plant-based natural products has driven significant research into developing more efficient, sustainable, and scalable extraction technologies for essential oils.

Among the many plants exploited for their essential oils, ginger (*Zingiber officinale*) is particularly important due to its rich chemical composition and established health benefits.

*Corresponding author: mahfud@its.ac.id

Ginger rhizomes contain a wide variety of bioactive compounds, notably terpenes and phenolics, which contribute to their pharmacological and sensory properties [2]. The essential oil of ginger is reported to exhibit strong antioxidant and antimicrobial potential, which supports its broad application in traditional medicine, aromatherapy, and nutraceutical formulations. Within this species, the *Zingiber officinale* stands out due to its larger rhizome size, pale-yellow color, and high essential oil content [3]. These features make it an attractive varietal for industrial processing; however, there is a noticeable gap in research regarding the systematic optimization of its oil extraction process.

Traditionally, essential oils are extracted using hydrodistillation, steam distillation, or solvent extraction methods. While these techniques are widely accepted, they are associated with various drawbacks such as high energy consumption, long extraction times, and risk of thermal degradation of sensitive compounds [4]. In addition, the use of organic solvents poses environmental and health concerns due to solvent residue in the final product. These limitations have stimulated interest in modern extraction technologies that offer greener, faster, and more selective processes.

Solvent-Free Microwave Extraction (SFME) is one such emerging green technique that has gained attention for its efficiency and sustainability. SFME employs microwave dielectric heating to rapidly and selectively heat water within plant cells, generating internal pressure that causes cell rupture and the subsequent release of essential oils [5].The major advantages of SFME include solvent elimination, reduced extraction time, lower energy use, and better retention of thermolabile compounds compared to conventional techniques [6]. Furthermore, SFME can be performed at atmospheric pressure, and its scalability makes it suitable for both laboratory and industrial settings.

While SFME has been successfully applied to various aromatic plants such as lemongrass, peppermint, and cananga, its application to *Zingiber officinale* remains limited. Previous studies have demonstrated the feasibility of microwave-assisted extraction on ginger rhizomes [10]; however, most investigations emphasize extraction performance under fixed operating conditions or focus primarily on kinetic modeling without explicitly identifying optimum extraction conditions. As a result, the time-dependent extraction behavior and kinetically favorable operating conditions for SFME of ginger remain insufficiently explored.

The novelty of this study lies in the kinetic evaluation of solvent-free microwave extraction of ginger rhizomes to identify optimum operating conditions based on yield maximization and extraction rate behavior. Rather than employing multivariate numerical optimization, the optimum condition in this work is defined as the operating condition that achieves the highest extraction yield with efficient kinetic performance, thereby avoiding unnecessary energy input and excessive extraction time.

Microwave power and extraction time were systematically varied to generate yield–time profiles, which were subsequently analyzed using kinetic models to determine extraction rate constants and characteristic extraction stages. Microwave power governs the internal heating rate and cell rupture intensity, directly influencing the initial extraction rate, while extraction time controls the approach toward equilibrium yield [8]. Through kinetic interpretation, the optimum SFME condition was identified as the condition that provides the maximum oil yield while maintaining favorable extraction kinetics.

This kinetic-based determination of optimum operating conditions provides a rational framework for evaluating SFME performance for ginger extraction and offers fundamental insight for future process optimization and scale-up studies [9].

Accordingly, this study aims to investigate the effects of microwave power, extraction time, and sample mass on the essential oil yield from *Zingiber officinale* using the solvent-free microwave extraction (SFME) method. In addition, the extraction kinetics of SFME are analyzed through mathematical modeling to describe the time-dependent extraction behavior and to determine kinetically favorable operating conditions based on yield maximization.

Furthermore, this study seeks to identify and experimentally validate the optimum SFME operating condition that produces the highest essential oil yield while maintaining efficient extraction kinetics. The study aims to address the current knowledge gap in the application of SFME to ginger and provide a mechanistic foundation for selecting efficient operating conditions. The outcomes are expected to support future process optimization and scale-up studies, as well as contribute to the development of sustainable and high-efficiency green extraction technologies for essential oil–rich medicinal plants.

2 Research Methods

2.1 Material and Equipment

Fresh rhizomes of *Zingiber officinale* were sourced from a local agricultural market in Surabaya, Indonesia. The rhizomes were selected based on visual uniformity, maturity (7–9 months post-planting), and the absence of physical damage or microbial spoilage. Upon arrival, the samples were washed thoroughly with distilled water to remove dirt and debris, and then sliced into thin uniform sections approximately 2–3 mm thick. No drying or pre-treatment was applied to maintain the native moisture level for microwave-assisted extraction.

The extraction was performed using a domestic microwave oven (Electrolux EMM20K22BA, 2450 MHz, maximum power 800 W), modified and fitted with a 1000 mL double-necked round-bottom flask connected to a Clevenger-type apparatus for essential oil collection (Figure 1). The condenser was maintained at low temperature using a refrigerated water circulator to ensure complete condensation of volatiles. The sample mass was weighed using a digital analytical balance (±0.01 g precision), and a stopwatch was used to monitor extraction time. The essential oils were collected in glass vials, sealed tightly, and stored in the dark at 4 °C until further quantification.

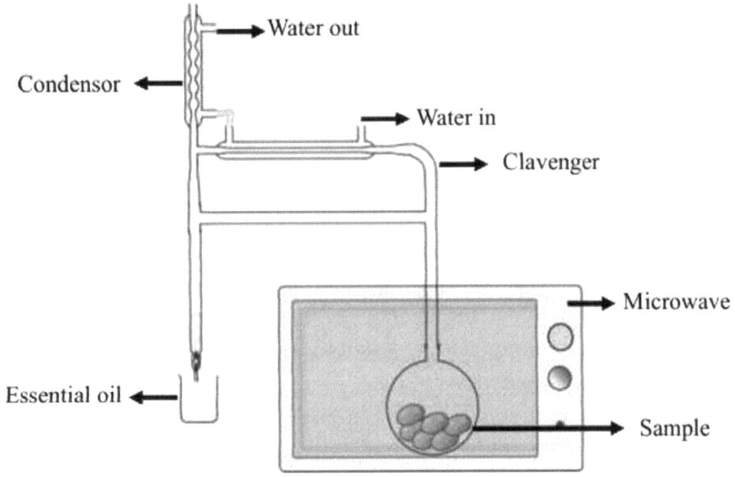

Fig. 1. SFME experimental set-up

2.2 Solvent – Free Microwave Extraction (SFME) Procedure

The SFME method was conducted without the addition of any solvent or water. A predetermined amount of fresh ginger sample (100–300 g) was placed in the round-bottom flask. The system was irradiated with microwave energy at designated power levels (300–600 W) for varying durations (60–180 minutes), according to the experimental design matrix. During extraction, the water content naturally present in the plant matrix was vaporized by microwave heating, generating internal pressure that disrupted plant cell walls and released volatile compounds (Figure 2).

The released vapors were passed through the condenser, where they condensed into a two-phase liquid consisting of water and essential oil. The oil layer was manually separated using a pipette or separatory funnel, measured gravimetrically, and expressed as yield (% w/w) relative to the initial mass of fresh rhizome. Each experimental condition was performed in duplicate to ensure reproducibility.

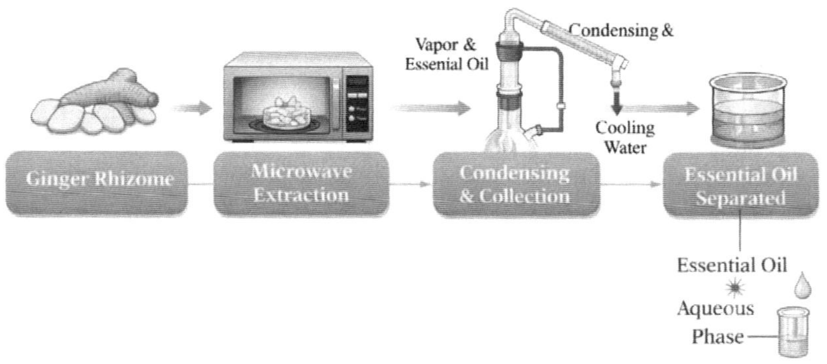

Fig. 2. Solvent Free Microwave Extraction Scheme

2.3 Kinetic Modelling

The extraction kinetics were evaluated using various mathematical models, including pseudo-first-order, pseudo-second-order, power law, hyperbolic, and Elovich models. These models were applied to the experimental data to estimate rate constants and identify the underlying extraction mechanisms. In addition, Peleg's model—a type of hyperbolic model—was utilized to characterize the mass transfer behavior during the extraction process. The kinetic models assessed in this study are presented in Table 1. Nonlinear regression analysis was employed to estimate the parameters for each model. The most appropriate model was selected based on goodness-of-fit indicators such as the coefficient of determination (R^2) and root mean square error (RMSE) [13].

Table 1. Kinetics model of extraction

Kinetic Model	Equation model	Refs
1. 1st-Order	$\frac{C_t}{C_s} = (1 - e^{-kt})$	1
2. 2nd-Order	$\frac{C_t}{C_s} = \frac{C_s k_2 t}{1 + C_s k_2 t}$	2

3.	Power Law	$\frac{C_t}{C_S} = B.t^n$	3
4.	Hyperbolic	$\frac{C_t}{C_S} = \frac{k_1 t}{1 + k_2 t}$	4
5.	Elovich's	$\frac{C_t}{C_S} = E_0 + E_1 x \ln t$	5

3 Result and Discussion

3.1 Time – Dependent Extraction Behavior of SFME

The time-dependent extraction behavior of essential oil from *Zingiber officinale* using solvent-free microwave extraction (SFME) was evaluated based on yield–time data obtained under various operating conditions. The experimental results reveal a non-linear extraction profile, indicating that the extraction rate changes progressively with time and operating parameters. Variations in microwave power and sample mass significantly influence the shape of the extraction curves and the time required to approach equilibrium, suggesting the involvement of multiple mass transfer phenomena during the SFME process [8].

Overall, the SFME process exhibited a characteristic two-stage extraction behavior, consisting of an initial fast extraction stage followed by a slower diffusion-controlled stage. Under all investigated conditions, a rapid increase in yield was observed during the early extraction period, indicating efficient microwave-induced cell rupture and fast release of readily accessible essential oil [9]. This stage was particularly pronounced under moderate microwave power, where internal heating was sufficient to enhance oil liberation without causing excessive thermal degradation.

Among the investigated conditions, the highest essential oil yield was obtained at a microwave power of 450 W with a sample mass of 200 g, reaching approximately 1.5–1.6% after 150–180 minutes of extraction. Under this condition, the extraction curve exhibited a steep initial slope followed by a gradual approach toward equilibrium, reflecting a favorable balance between extraction rate and yield. Lower microwave power (300 W) resulted in slower yield development and lower final yield, whereas higher power (600 W) accelerated the initial extraction but tended to shorten the fast extraction stage, likely due to rapid depletion of easily accessible oil and potential thermal effects.

Sample mass also influenced the extraction behavior by affecting microwave energy distribution and mass transfer resistance. A moderate sample mass (200 g) provided more uniform volumetric heating and efficient vapor diffusion, whereas excessive biomass loading delayed the approach to equilibrium. These observations indicate that both microwave power and sample mass govern the relative contributions of the fast and slow extraction stages.

The observed yield–time behavior confirms that SFME of ginger is governed by a combination of microwave-induced cell rupture and subsequent diffusion-driven mass transfer. The condition yielding the highest extraction efficiency was therefore selected as the kinetically favorable operating condition, which was subsequently analyzed using kinetic models to quantitatively describe the extraction process, as discussed in the following sections.

3.2 Kinetics Models and Data Fittings

Multiple kinetic models were employed in this study to comprehensively describe the complex extraction behavior of SFME rather than to identify a single best-fitting equation. Each model represents different assumptions regarding mass transfer mechanisms, including concentration-driven kinetics, heterogeneous surface effects, and diffusion-controlled transport. The use of multiple models enables a mechanistic interpretation of the extraction process by comparing parameter trends and model performance under varying operating conditions. Therefore, the selection of kinetic models in this study was intended to elucidate extraction mechanisms and rate-controlling steps, rather than to achieve purely statistical curve fitting.

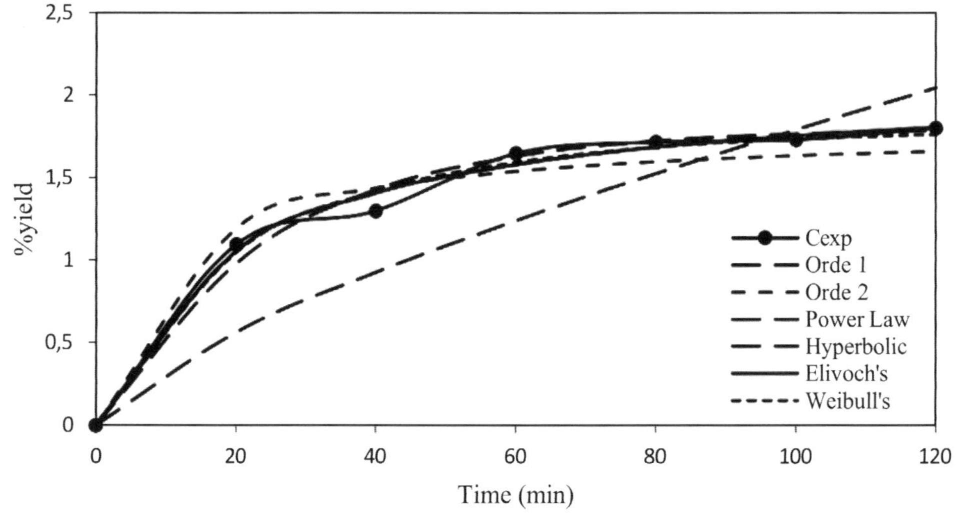

Fig. 3. Comparison of Experimental Data and Model Calculations for SFME (450W, 100gram)

Table 2. Kinetic Model of SFME (450W, 1cm, 100gram)

Order 1		Order 2		Power law		Hyperbolic		Elovich's		Weibull's	
k	0.017	k	0.054	B	0.065	C^1	0.106	Eo	0.106	α	23.64
				n	0.721	C^2	0.050	E_1	0.050	β	0.827
R^2	0.921	R^2	0.620	R^2	0.971	R^2	0.718	R^2	0.718	R^2	0.698
SSE	0.068	SSE	0.128	SSE	0.152	SSE	0.237	SSE	0.237	SSE	0.173

Figure 3 and Table 2 present the time-dependent extraction behavior of SFME at a microwave power of 450 W using 100 g of raw material, together with the corresponding predictions of several kinetic models. The experimental data exhibit a pronounced non-linear increase in extraction yield with time, particularly during the early stage of the process. A rapid rise in yield is observed within the first 20–30 minutes, indicating that a significant portion of the extractable compounds is readily released at the beginning of extraction. This behavior can be attributed to the effect of microwave heating, which induces internal heating and localized pressure gradients within the plant matrix, facilitating structural disruption and enhancing the initial mass transfer of solutes into the extraction medium.

As extraction time increases, the rate of yield enhancement gradually declines, and the extraction curve approaches an asymptotic trend. This change reflects a transition from a fast initial extraction stage to a slower regime dominated by internal diffusion and matrix resistance. At this stage, the remaining compounds are increasingly confined within less accessible regions of the solid matrix, resulting in reduced mass transfer rates despite continued energy input. The progressive flattening of the experimental curve suggests that the extraction system is approaching a quasi-equilibrium condition, where further solute release becomes increasingly limited.

Comparison between the experimental data and the model-generated curves indicates that the extraction process cannot be adequately described by a single linear kinetic expression. While the first-order model reasonably captures the overall trend, noticeable deviations are observed, particularly at short extraction times, whereas the second-order model shows poorer agreement over the full extraction period. Empirical models such as the power law, Weibull, hyperbolic, and Elovich equations better accommodate the non-linear nature of the extraction process, although differences in fitting accuracy remain evident. These observations confirm the presence of multi-stage extraction kinetics and provide a clear rationale for the application of kinetic modeling to further elucidate the underlying extraction mechanism in subsequent sections [1, 10].

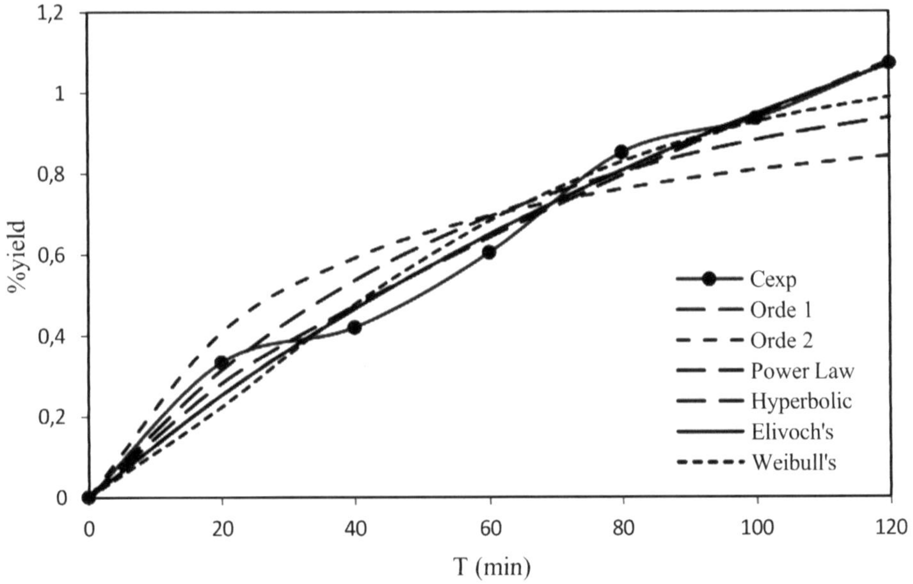

Fig. 4. Comparison of Experimental Data and Model Calculations for SFME (450W, 200gram)

Table 3. Kinetic Model of SFME (450W, 200gram)

Order 1		Order 2		Power law		Hyperbolic		Elovich's		Weibull's	
k	0.018	k	0.029	B n	0.030 0.749	C^1 C^2	0.014 0.004	Eo E_1	0.014 0.004	α β	312.3 1.335
R^2	0.939	R^2	0.851	R^2	0.959	R^2	0.969	R^2	0.969	R^2	0.977
SSE	0.105	SSE	0.059	SSE	0.032	SSE	0.028	SSE	0.028	SSE	0.026

Figure 4 and Table 3 depict the time-dependent extraction behavior of SFME performed at a microwave power of 450 W with a solid loading of 200 g, together with the corresponding predictions of several kinetic models. Similar to the lower solid mass condition, the experimental extraction yield increases progressively with time, although the overall extraction rate is noticeably slower. During the initial extraction period, the yield rises gradually, indicating that the availability of readily extractable compounds is more limited when a higher solid mass is employed. The increased solid loading likely leads to a denser matrix structure and reduced effective exposure of the solid surface, which moderates the early-stage mass transfer despite the same microwave power input.

As extraction proceeds, the yield continues to increase in a nearly linear fashion before gradually approaching an asymptotic trend at longer extraction times. This behavior suggests that the extraction process is increasingly governed by internal diffusion and matrix resistance, which become more pronounced at higher solid mass. The delayed approach to equilibrium observed in Figure 4 reflects the greater diffusion path length and reduced solvent accessibility within the bulk solid matrix. Consequently, the extraction process exhibits a more prolonged kinetic profile compared to the 100 g condition, highlighting the influence of solid loading on the temporal evolution of extraction yield.

The comparison between experimental data and kinetic model predictions shows improved overall agreement for most models under this condition. The first-order and second-order models capture the general extraction trend with relatively high coefficients of determination, although minor deviations are still observed at intermediate extraction times. Empirical models, particularly the power law, hyperbolic, Elovich, and Weibull equations, demonstrate closer conformity with the experimental data over the full time range, indicating their greater flexibility in representing non-linear extraction kinetics at higher solid mass. These results further confirm that SFME under the studied conditions follows a multi-stage extraction mechanism and provide a consistent phenomenological basis for subsequent kinetic parameter analysis.

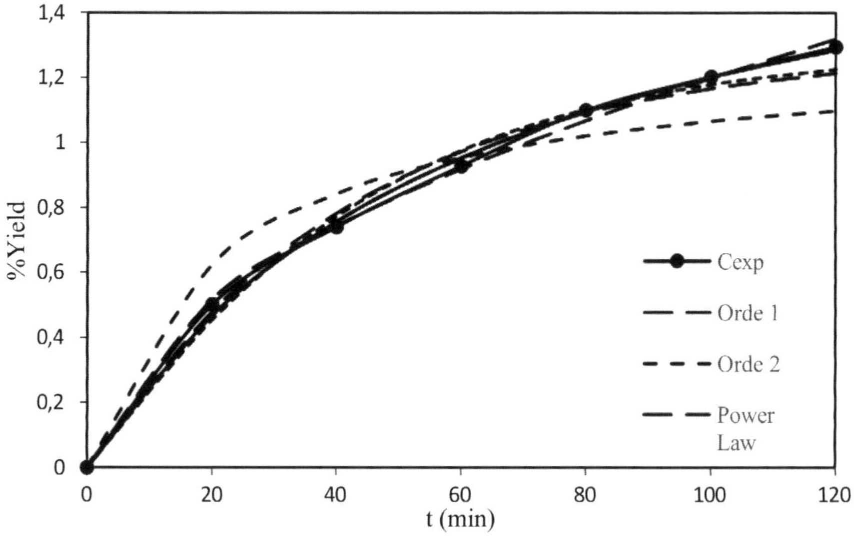

Fig. 5. Comparison of Experimental Data and Model Calculations for SFME (450W, 300gram)

Table 4. Kinetic Model of SFME (450W, 300gram)

Order 1		Order 2		Power law		Hyperbolic		Elovich's		Weibull's	
k	0.018	k	0.036	B	0.108	C^1	0.031	Eo	0.031	α	43.72
				n	0.522	C^2	0.015	E_1	0.015	β	1.070
R^2	0.944	R^2	0.927	R^2	0.942	R^2	0.937	R^2	0.937	R^2	0.913
SSE	0.018	SSE	0.029	SSE	0.238	SSE	0.239	SSE	0.239	SSE	0.026

Figure 5 and Table 4 illustrate the time-dependent extraction behavior of SFME at a microwave power of 450 W with a solid mass of 300 g, along with the corresponding predictions of several kinetic models. The experimental extraction profile shows a gradual and continuous increase in extracted compound mass with time, characterized by a less pronounced initial rise compared to lower solid loadings. This trend indicates that increasing the solid mass further limits the availability of readily extractable compounds at the early stage of extraction. The denser packing of the solid matrix and the increased diffusion path length likely hinder rapid solute release, resulting in a more moderated initial extraction rate under otherwise identical operating conditions.

As the extraction progresses, the yield continues to increase steadily before approaching a plateau at extended extraction times. The absence of a sharp transition between extraction stages suggests that diffusion resistance dominates the mass transfer process throughout most of the extraction period. At higher solid mass, the extraction system requires a longer time to approach equilibrium, reflecting the cumulative effect of matrix compaction and reduced solvent accessibility. The gradual curvature of the experimental data confirms that the extraction process becomes increasingly controlled by internal mass transfer limitations as the solid loading increases.

The comparison between experimental data and model calculations reveals that most kinetic models provide a reasonable representation of the extraction behavior under this condition. The first-order and second-order models show good agreement with the experimental data, as reflected by relatively high coefficients of determination, indicating that concentration-dependent kinetics become more dominant at higher solid mass. Empirical models, including the hyperbolic, Elovich, and Weibull equations, also capture the overall extraction trend, although deviations are still observed in the later extraction stage. These results reinforce the conclusion that SFME at high solid loading follows a multi-stage kinetic process, with diffusion-controlled mechanisms playing a central role, thereby providing a consistent phenomenological basis for subsequent kinetic parameter interpretation.

3.3 Interpretation Kinetics Parameters

The kinetic parameters obtained from the fitting of various models provide deeper insight into the underlying extraction mechanism of SFME at a constant microwave power of 450 W under different solid loadings. The first-order rate constant (k_1) exhibits relatively similar values across all solid masses, ranging from 0.017 to 0.018 min^{-1}. This limited variation suggests that the intrinsic extraction rate driven by concentration gradients is primarily governed by the applied microwave power rather than by solid loading. However, despite comparable k_1 values, the overall extraction profiles differ substantially, indicating that first-order kinetics alone are insufficient to fully describe the extraction process, particularly as solid mass increases and internal diffusion resistance becomes more pronounced.

In contrast, the second-order rate constant (k_2) shows a noticeable dependence on solid loading, with values increasing from 0.029 to 0.054 as the mass decreases. This trend reflects the greater contribution of solute–matrix interactions and concentration-dependent limitations at higher solid masses. The improved fitting performance of the second-order model at 300 g, compared to lower masses, suggests that extraction at high solid loading increasingly follows concentration-controlled behavior, where the rate of solute release becomes sensitive to the remaining solute concentration within the matrix. Nevertheless, the second-order model still fails to adequately capture the early extraction stage, highlighting the presence of additional transport mechanisms beyond simple reaction-order assumptions.

The power law model provides important mechanistic information through its parameters B and n. The exponent n ranges between 0.52 and 0.75 across the investigated solid masses, consistently remaining below unity. This fractional exponent indicates non-Fickian mass transfer behavior, confirming that the extraction process is governed by a combination of rapid initial release and slower diffusion-controlled transport. Notably, the decrease in n with increasing solid mass suggests a shift toward diffusion-dominated kinetics, as the extraction pathway becomes increasingly constrained by longer diffusion distances and reduced solvent accessibility within the bulk matrix. The flexibility of the power law model in capturing this mixed-control behavior explains its consistently high coefficients of determination across all conditions.

The hyperbolic and Elovich models further support the presence of heterogeneous extraction kinetics. The parameters C_1 and E_0, which represent the initial extraction rate, decrease systematically with increasing solid mass, indicating a reduction in readily accessible solute fractions at higher loadings. Simultaneously, the corresponding decay parameters (C_2 and E_1) reflect a rapid decline in extraction rate as extraction progresses, consistent with progressive depletion of easily extractable compounds. The comparable fitting performance of these two models suggests that the extraction surface becomes increasingly heterogeneous over time, with a diminishing number of active sites contributing to solute release.

The Weibull model offers additional insight through its scale (α) and shape (β) parameters. The scale parameter α increases substantially with solid mass, indicating that a longer characteristic time is required to reach a given fraction of total extraction as the matrix becomes more compact. Meanwhile, the shape parameter β remains close to or slightly above unity, reflecting a decelerating extraction rate over time. This behavior is characteristic of systems where the extraction rate decreases continuously due to increasing internal resistance, further confirming the dominance of diffusion-controlled mechanisms at higher solid loading.

Taken together, the interpretation of kinetic parameters across different solid masses demonstrates that SFME at constant microwave power is governed by a multi-stage extraction mechanism. While the applied power establishes the driving force for extraction, solid loading strongly influences the balance between rapid initial release and diffusion-limited transport. The combined analysis of empirical and semi-empirical models reveals that no single kinetic expression can fully describe the extraction behavior across all conditions; instead, the observed kinetics result from the interplay between matrix structure, solute accessibility, and time-dependent mass transfer resistance. These findings provide a robust mechanistic foundation for identifying optimal extraction conditions based on kinetic behavior rather than solely on final extraction yield. The observed two-stage extraction behavior and non-Fickian diffusion characteristics are consistent with previous reports on microwave-based extraction of ginger and other essential oil–bearing plants. Similar values of the diffusion exponent ($n < 1$) have been reported for microwave-assisted extraction of ginger and aromatic herbs, indicating mixed control between rapid solute release and internal diffusion. These similarities suggest that the kinetic behavior observed in this study aligns well with established extraction mechanisms reported in the literature.

3.4 Determination of Optimal Extraction Condition Based on Kinetic Behavior

The determination of the optimal extraction condition was carried out by integrating the kinetic behavior and parameter interpretation obtained from the previous sections, rather than relying solely on final extraction yield. At a constant microwave power of 450 W, variations in solid loading significantly influenced the temporal evolution of extraction and the dominant mass transfer mechanisms. Lower solid mass (100 g) exhibited a rapid initial extraction stage, characterized by higher apparent extraction rates and earlier attainment of a quasi-equilibrium state. This behavior indicates a more efficient utilization of the applied microwave energy, resulting in enhanced solute accessibility and reduced internal diffusion resistance.

At intermediate solid loading (200 g), the extraction process displayed a more balanced kinetic profile. Although the initial extraction rate was lower than that observed at 100 g, the system maintained a relatively sustained extraction rate over a longer time period. The kinetic parameters derived from empirical models showed improved fitting performance under this condition, suggesting a more stable and predictable extraction behavior. This indicates that the extraction process at 200 g benefits from a compromise between solute availability and diffusion resistance, allowing effective mass transfer without excessive limitation from matrix compaction.

In contrast, the highest solid loading (300 g) resulted in a noticeably prolonged extraction process, with diffusion-controlled mechanisms dominating throughout most of the extraction period. Despite achieving a comparable final extraction yield at extended times, the kinetic parameters revealed increased characteristic extraction times and reduced initial extraction rates. From a kinetic perspective, this condition requires longer processing times to approach equilibrium, which may be unfavorable when extraction efficiency and energy utilization are considered. Based on the combined evaluation of extraction rate, kinetic stability, and characteristic extraction time, the optimal condition at 450 W is identified within the lower to intermediate solid loading range. While 100 g provides the fastest extraction kinetics, 200 g offers a more controlled and reproducible extraction profile with favorable kinetic parameters and reduced sensitivity to diffusion limitations. Therefore, the selection of the optimal condition should prioritize kinetic efficiency and process practicality rather than maximum yield alone, reinforcing the importance of kinetic-based evaluation for rational process design.

From a practical perspective, the kinetically favorable condition identified in this study implies reduced extraction time and more efficient energy utilization, which are critical parameters for industrial-scale SFME. The balanced extraction kinetics observed at intermediate solid loading suggest improved process controllability and scalability, supporting the potential application of SFME as an energy-efficient and solvent-free alternative for essential oil production.

3.5 GC–MS Results of Ginger Essential Oil

Table 5. Major volatile compounds of *Zingiber officinale* essential oil

No.	Compound	R.Time	Similarity	% Area	% Component
1	Citral	15.400	89	1.00	3.42
2	Neral	15.505	96	0.67	2.29
3	Geraniol	16.237	96	0.16	0.51
4	α-Terpineol	14.159	95	0.09	0.24
5	Linalool	11.305	97	0.06	0.21
6	Citronellol	15.463	94	0.05	0.17
7	Eucalyptol	9.779	93	0.03	0.10
8	Camphor	12.572	88	0.02	0.07

Gas Chromatography–Mass Spectrometry (GC–MS) analysis revealed that the ginger essential oil obtained by solvent-free microwave extraction (SFME) consisted predominantly of oxygenated monoterpenes with varying relative abundances, as summarized in Table 5. Among the identified compounds, citral was the most abundant constituent, accounting for 3.42% of the total peak area, followed by neral (2.29%), both of which are key aldehydic monoterpenes responsible for the characteristic lemon-like aroma of ginger essential oil. The presence of citral and neral indicates that the SFME process was effective in recovering thermally sensitive volatile compounds without significant degradation.

In addition to aldehydic monoterpenes, several oxygenated monoterpene alcohols were detected at lower concentrations, including geraniol (0.51%), α-terpineol (0.24%), linalool (0.21%), and citronellol (0.17%). These compounds are widely recognized for their contributions to floral, sweet, and fresh aroma notes, as well as for their reported biological activities such as antimicrobial and antioxidant effects. Although present in relatively minor proportions, these monoterpenes play a crucial role in defining the sensory complexity and functional properties of ginger essential oil.

Furthermore, other oxygenated monoterpenes such as eucalyptol (1,8-cineole, 0.10%) and camphor (0.07%) were also identified. These compounds are commonly associated with camphoraceous and cooling aroma characteristics and are frequently reported as minor constituents in ginger essential oil. Their detection suggests that the applied SFME conditions were sufficiently mild to preserve volatile and thermally sensitive compounds that are often susceptible to loss during prolonged conventional heating.

Overall, the GC–MS profile demonstrates that ginger essential oil extracted via SFME exhibits a chemically diverse composition dominated by oxygenated monoterpenes, particularly citral and its isomers. This compositional profile highlights the capability of the SFME technique to selectively recover characteristic ginger volatiles while maintaining their chemical integrity, which is essential for both aroma quality and potential bioactivity.

4. Conclusions

The solvent-free microwave extraction of essential oil from *Zingiber officinale* exhibited a clear multi-stage kinetic behavior, characterized by a rapid initial extraction followed by a diffusion-controlled regime. Kinetic modeling demonstrated that empirical models, particularly the power law and Weibull models, effectively described the non-linear extraction behavior and reflected mixed mass transfer control mechanisms. At a constant microwave power of 450 W, solid loading significantly influenced extraction kinetics, with an intermediate sample mass of 200 g providing the most favorable balance between extraction rate, stability, and characteristic extraction time. These findings confirm that kinetic-based evaluation offers a rational basis for defining optimal SFME operating conditions and supports the development of efficient and sustainable green extraction processes.

References

1. R. Kant, A. Kumar, Review on essential oil extraction from aromatic and medicinal plants: techniques, performance and economic analysis. Sustain. Chem. Pharm. **30**, 100829 (2022). https://doi.org/10.1016/j.scp.2022.100829
2. Q.-Q. Mao, X.-Y. Xu, S.-Y. Cao, R. Y. Gan, H. Corke, T. Beta, H.-B. Li. Bioactive compounds and bioactivities of ginger (*Zingiber officinale* Roscoe). Foods **8**, 185 (2019). https://doi.org/10.3390/foods8060185
3. C. Ravi Kiran, A. K. Chakka, K. P. Padmakumari Amma, A. Nirmala Menon, M. M. Sree Kumar & V. V. Venugopalan, Essential oil composition of fresh ginger cultivars from North-East India. J. Essent. Oil Res. **25**, 380–387 (2013). https://doi.org/10.1080/10412905.2013.796496
4. B. Berka-Zougali, M.-A. Ferhat, A. Hassani, F. Chemat & K. S. Allaf, Comparative study of essential oils extracted from *Myrtus communis* L. leaves using microwaves and

hydrodistillation. Int. J. Mol. Sci. **13**, 4673–4695 (2012). https://doi.org/10.3390/ijms13044673

5. F. Chemat, N. Rombaut, A. Meullemiestre, M. Turk, S. Perino, A.-S. Fabiano-Tixier & M. Abert-Vian, Review of green food processing techniques: preservation, transformation, and extraction. Innov. Food Sci. Emerg. Technol. **41**, 357–377 (2017). https://doi.org/10.1016/j.ifset.2017.04.016
6. H. Haqqyana, A. Altway & M. Mahfud, Kinetic modeling for microwave-assisted green extraction: Effects of power on citronella oil from *Cymbopogon nardus* leaf. Int. J. Appl. Sci. Eng. **19**, 2022147 (2022)
7. Z. Drinić, M. Jovanović, D. Pljevljakušić, N. Čujić-Nikolić, D. Bigović, K. Šavikin, Microwave-assisted extraction of essential oil from ginger (*Zingiber officinale* Rosc.). Lekov. Sirovine **41**, 22–27 (2021). https://doi.org/10.5937/leksir2141022d
8. Y. Qiu, Y. Wang, Y. Li, Solvent-free microwave extraction of essential oils from *Litsea cubeba* (Lour.) Pers. at different harvesting times and their skin-whitening cosmetic potential. Antioxidants **11**, 2389 (2022). https://doi.org/10.3390/antiox11122389
9. E. Widiaswanti, R. Yunitarini, T. Novianti, A. Kartiningsih, Investigasi kajian kinetik pengeringan jahe dalam pembuatan simplisia. J. Serambi Eng. **8**, 4413–4421 (2023)
10. D. Darnengsih, H. Ni'mah, M. Mahfud, A. C. Kartika Fitri, Kinetics model of *Cananga* (*Cananga odorata*) fresh flower extraction using microwave hydrodistillation method. Makara J. Sci. **28**, 287–295 (2024). https://doi.org/10.7454/mss.v28i4.2146
11. Y. Variyana, M. Mahfud, Kinetics study using solvent-free microwave extraction of essential oil from *Allium sativum* L. Key Eng. Mater. **840**, 186–192 (2020). https://doi.org/10.4028/www.scientific.net/KEM.840.186
12. B. Yingngam, A. Brantner, M. Treichler, N. Brugger, A. Navabhatra, P. Nakonrat, Optimization of the eco-friendly solvent-free microwave extraction of *Limnophila aromatica* essential oil. Ind. Crops Prod. **165**, 113443 (2021). https://doi.org/10.1016/j.indcrop.2021.113443

Crystal structure analysis of natural materials in building materials using XRD (X-ray diffraction)

Inas Marwaa Dzakiya[1], *Akida* Mulyaningtyas[1*], and *Adi* Wijayanto[2]

[1]Department of Chemical Engineering, Faculty of Engineering, University of Muhammadiyah Surakarta, Surakarta, Indonesia
[2]Directorate of Nuclear Facility Management, National Research and Innovation Agency (BRIN), Tangerang Selatan, Indonesia

Abstract. Natural building materials such as cement, sand, clay bricks, and lightweight bricks are widely used in Indonesia and play an important role in construction and material engineering applications. Understanding their crystallographic characteristics is essential, as crystal structure influences the physical and mechanical properties of materials. This study aims to identify the dominant crystalline phases and crystal systems of common building materials collected from several regions in Indonesia. X-ray diffraction (XRD) analysis was employed, and qualitative phase identification was conducted using Match! 4.0 software, with crystal structure visualization performed using Mercury 3.8. The results indicate that cement, clay brick, and lightweight brick samples are predominantly associated with phases belonging to the monoclinic crystal system, while sand samples are mainly characterized by an orthorhombic crystal system. The phase identification is qualitative in nature and based on diffraction pattern matching with standard reference databases. This study provides preliminary insight into the crystallographic characteristics of commonly used building materials in Indonesia and may serve as a basis for further quantitative investigations to better relate crystal structure to material performance and potential applications.

1 Introduction

Building materials are an essential component of construction systems, comprising both primary and auxiliary materials that provide structural integrity and specific functional properties in buildings [1]. In construction engineering, the selection and performance of materials are strongly influenced by their physical, chemical, and mineralogical characteristics. Consequently, material science research in the construction sector focuses on understanding the properties, composition, durability, and technological behavior of building materials to ensure safety, efficiency, and long-term performance [2].

Among the most commonly used building materials are cement, sand, clay bricks, and lightweight bricks, which are widely applied in residential, commercial, and infrastructure projects [3-7]. These materials are generally composed of naturally occurring mineral phases

* Corresponding author: akida.mulyaningtyas@ums.ac.id

such as quartz, feldspar, calcite, mullite, and calcium silicate compounds, whose presence and relative abundance affect mechanical strength, thermal stability, and chemical resistance. Therefore, identifying the mineralogical composition of building materials is a critical step in evaluating their suitability for various engineering applications.

X-ray diffraction (XRD) is one of the most widely used techniques for the identification of crystalline phases and crystal structures in solid materials. XRD enables the qualitative analysis of mineral phases through diffraction pattern matching with standard reference databases and provides insight into the crystallographic systems present in a material. In the context of building materials, XRD is particularly useful for characterizing cementitious materials, aggregates, and ceramic-based products, as it allows the identification of dominant mineral phases and their structural characteristics [8].

In addition to conventional construction applications, building materials are also considered for use in radioactive waste management systems, where long-term structural stability and chemical compatibility are essential. Understanding the crystallographic characteristics of these materials is important because crystal structure can influence properties such as porosity, phase stability, and resistance to radiation-induced degradation. Preliminary crystallographic characterization using XRD can therefore provide foundational information for assessing material behavior in radioactive environments.

Based on these considerations, this study aims to identify the dominant crystalline phases and crystal systems present in commonly used natural building materials through XRD analysis. The analysis is focused on qualitative phase identification and crystallographic classification using diffraction pattern matching. The results are intended to serve as an initial database of crystallographic information for natural building materials, which may support future studies employing quantitative XRD methods to more rigorously correlate crystal structure with material performance in radioactive waste management applications.

2 Methodology

Analysis was performed using X-Ray Diffraction (XRD) to identify the crystalline phase in the samples. Diffraction patterns were analyzed with Match! 4.0 software and crystal structure visualization performed using Mercury 3.8. [8]. The analysis results are presented in the form of graphs, intensity tables, and crystal structure visualization for each sample.

2.1 Material

The tools used in this research were a sieve, mortar, spoon, 5 mL sample container, mask, small label, ballpoint pen, mask gloves, protective glasses, hammer, tray, X-Ray Diffraction tool, plastic mat, and pestle. The materials used in this research consisted of several building materials such as cement, sand, bricks and lightweight bricks originating from the Jakarta, Bogor Depok, Tangerang, Bekasi, Serang, Jepara and Kudus areas.

2.2 Procedures

The following is how the work carried out in the research is that each building material is ground using mortar and then sieved using a 200mesh sieve according to the standard size for XRD (X-Ray Diffraction) analysis. Then the results of the 200mesh sieve are put into a small 5 mL sample container which has been given the identity of the material and area of the building material. After the sample preparation time is complete, the sample is analyzed using an X-ray diffractometer. The resulting diffraction patterns are then analyzed to identify the crystal phases present in each building material sample. The results of the XRD (X-Ray

Diffraction) analysis that were obtained were then processed using Match!4.0 software, then the crystal phase content in each sample that had been tested was obtained and conclusions were drawn.

2.3 Data analysis

Qualitative and quantitative analysis will be carried out on data obtained from the results of XRD (X-Ray Diffraction) analysis. Qualitative analysis involves identifying the crystal phase and quantitative analysis involves measuring crystal structure parameters such as grain size and diffraction patterns. After obtaining the research data, the data results from the XRD analysis tool were then analyzed using the Match!4.0 software. Match!4.0 software will match the diffraction pattern of a sample with a reference database using the "Search-Match" menu and then display the matching results in the form of a list of candidate phases. From the matching results, the 3 highest phase data which are close to the diffraction peak are taken, then a conclusion is obtained in the form of the phase content in each sample. Next, an analysis of the shape of the crystal system was carried out using Mercury 3.8 Software to determine the shape of the crystal structure of the identified phases of each sample. Relative intensity values represent matching scores from the search–match process and do not reflect actual phase fractions.

3 Result and discussion

The X-ray diffraction (XRD) results for each investigated sample are presented in the form of diffractogram patterns obtained from cement, sand, brick, and lightweight brick materials collected from various regions. All measurements were conducted at the Physics Laboratory of the National Research and Innovation Agency (BRIN), Serpong, on June 21, 2024, using standard XRD procedures. The diffractograms reflect the diffraction behavior of crystalline phases present in the samples and provide qualitative information regarding peak positions, relative intensities, and overall diffraction patterns. Phase identification was performed through pattern matching with reference databases using Match! 4.0 software, while crystal structure visualization was carried out using Mercury 3.8.

The interpretation of the XRD results in this study is limited to qualitative phase identification and crystallographic system classification. The analysis focuses on identifying dominant phases and their associated crystal systems based on the presence of characteristic diffraction peaks, without conducting quantitative phase analysis or refinement. As such, the results are intended to provide an initial mineralogical and crystallographic overview of common building materials rather than a comprehensive assessment of their structural or mechanical performance. Any discussion of material behavior is therefore confined to general crystallographic characteristics and supported by established literature, while recognizing that further quantitative analysis and complementary testing would be required to establish robust structure–property relationships.

3.1 X-ray diffraction analysis using the search and match method

The diffraction pattern of the SMN – TGR (Tangerang cement) test samples identified that the dominant phases found in the three highest peaks are calcium silicate (Ca_3O_5Si), portlandite ($Ca(OH)_2$), and calcite ($CaCO_3$) (Table 1-4).

Table 1. Search and match results for Tangerang cement samples.

Location	Identified Phase	Relative Intensity (%)	Crystal System
Tangerang	Calcium silicate (Ca_3O_5Si)	96.7	Monoclinic
	Portlandite ($Ca(OH)_2$)	3.2	Hexagonal
	Calcite ($CaCO_3$)	0.1	Orthorhombic

The diffraction pattern of the PSR – JKT (Jakarta sand) test sample identified that the dominant phases are quartz (SiO_2), albite ($NaAlSi_3O_8$), and microcline ($KAlSi_3O_8$).

Table 2. Search and match results for Jakarta sand samples.

Location	Identified Phase	Relative Intensity (%)	Crystal System
Jakarta	Quartz (SiO_2)	82.4	Orthorhombic
	Albite ($NaAlSi_3O_8$)	11.7	Monoclinic
	Microcline ($KAlSi_3O_8$)	5.9	Monoclinic

The diffraction pattern of the BB – KDS (Kudus brick) test sample identified that the dominant phases are aluminosilicate phase, quartz (SiO_2), and mullite ($Al_6Si_2O_{13}$).

Table 3. Search and match results for Kudus brick samples.

Location	Identified Phase	Relative Intensity (%)	Crystal System
Kudus	Aluminosilicate phase	96.5	Monoclinic
	Quartz (SiO_2)	3.1	Monoclinic
	Mullite ($Al_6Si_2O_{13}$)	0.4	Monoclinic

The diffraction pattern of the HBL – BGR (Bogor lightweight brick) test sample identified that the dominant phases are calcium silicate hydrate, quartz (SiO_2), and calcite ($CaCO_3$).

Table 4. Search and match results for Bogor lightweight brick samples.

Location	Identified Phase	Relative Intensity (%)	Crystal System
Bogor	Calcium silicate hydrate	85.5	Hexagonal
	Quartz (SiO_2)	7.4	Monoclinic
	Calcite ($CaCO_3$)	7.1	Orthorhombic

3.2 Crystal system analysis

The results from the search and match analysis with the highest relative intensity for each sample material were further analyzed to determine the dominant crystal systems.

For the Tangerang cement sample, the crystal system with the highest relative intensity is monoclinic. Hexagonal and orthorhombic systems were also detected, though with lower intensity. For the Jakarta sand sample, the orthorhombic crystal system has the highest relative intensity and is dominant. In the Kudus brick samples, the monoclinic crystal system has the highest relative intensity and is the dominant system. Finally, based on the analysis of the lightweight brick sample, the dominant crystal system is Monoclinic, despite the hexagonal system showing the highest relative intensity for a single phase (Figure 1 and 2).

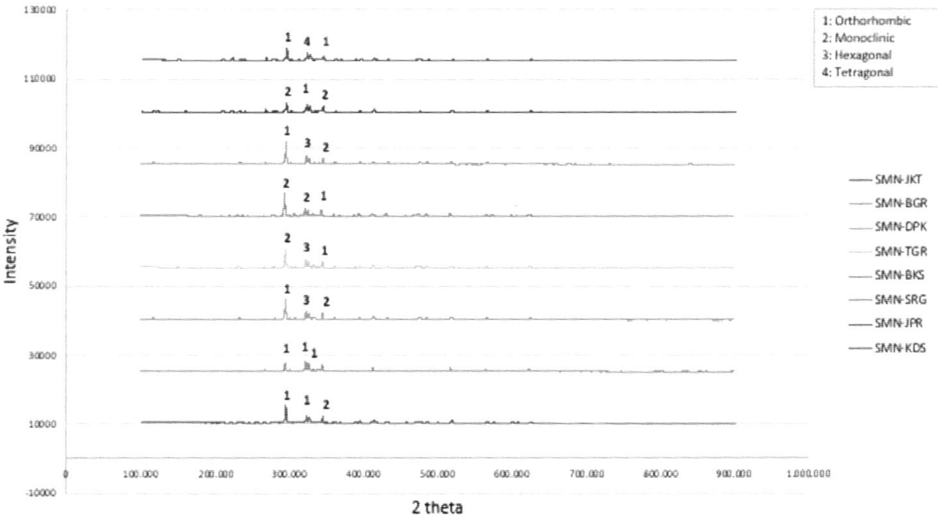

Fig. 1. Results of crystal system analysis of cement samples from various regions.

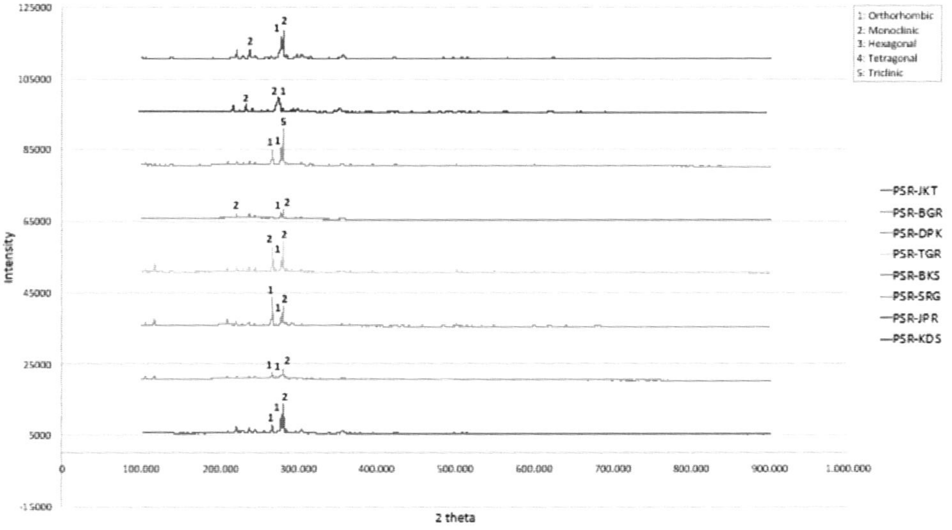

Fig. 2. Results of crystal system analysis in sand samples from various regions.

3.3 Analysis of diffraction patterns

Analysis of the crystal structure of building materials is an important step in understanding their physical and chemical properties. The X-ray diffraction (XRD) pattern is used to analyze the crystallographic properties of the material. The diffraction pattern shows the intensity of X-rays at various angles (2θ), providing information about the interplanar distances within the crystal structure. The peaks on the graph correspond to specific crystallographic plane orientations. This identification helps determine the crystal phases present, which is crucial for understanding the material's properties such as strength, durability, and thermal stability. By understanding the diffraction patterns, researchers can design and select appropriate materials for specific construction applications (Figure 3).

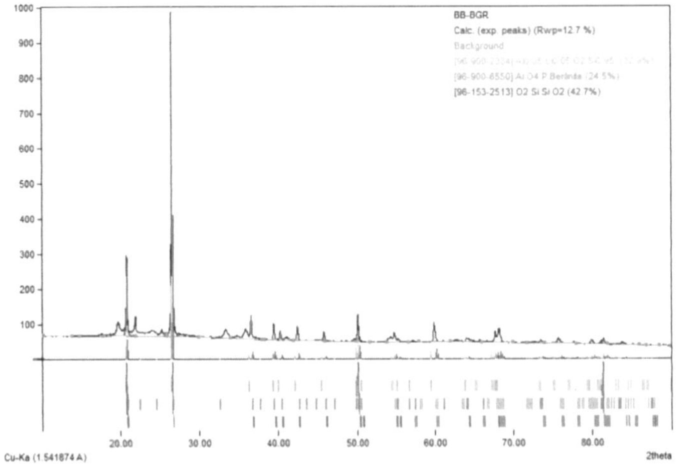

Fig. 3. Diffraction patterns related to crystallographic properties.

3.4 XRD database development

Providing a comprehensive database for XRD data is a crucial step in the analysis and development of building materials. Such a database functions as a central repository for diffraction patterns, allowing researchers to make comparisons and analyses of the building materials they study. This would make the identification and verification process for building materials more efficient, as results can be quickly compared with existing patterns in the database. In the long term, a complete and well-structured XRD database will support advances in building materials technology and encourage collaboration between academia and industry.

3.5 Interpretation and discussion

The identification of dominant crystal systems in the investigated building materials provides descriptive information on their crystallographic characteristics rather than direct evidence of their mechanical or functional performance. Although monoclinic and orthorhombic crystal systems were observed as dominant in several samples, it should be emphasized that crystal system classification alone does not determine material strength, durability, or long-term stability. Mechanical properties of construction materials are influenced by a complex interplay of factors, including phase composition, microstructure, porosity, grain size distribution, and processing history. Therefore, the present results should be interpreted as a qualitative crystallographic overview rather than as a direct explanation of material performance.

The dominance of a particular crystal system in cementitious and ceramic-based materials should be understood within the context of commonly occurring mineral phases. For example, phases such as quartz, feldspar, mullite, alite, and belite are frequently reported in cement, sand, and brick materials and may crystallize in different crystal systems depending on their chemical composition and thermal history. However, the presence of these systems does not imply a uniform or homogeneous microstructure, nor does it directly correlate with macroscopic properties without further quantitative analysis. Consequently, the current study does not attempt to establish a direct causal relationship between crystal system dominance and engineering performance.

The XRD analysis conducted in this study was limited to qualitative phase identification based on diffraction pattern matching. No quantitative phase analysis, such as Rietveld refinement, crystallite size estimation, or lattice parameter determination, was performed. As a result, parameters that are often linked to mechanical strength and thermal stability—such as phase abundance, peak broadening, and lattice distortion—could not be evaluated. This limitation restricts the interpretation of the data and underscores the need for complementary quantitative techniques in future studies to strengthen structure–property correlations.

With respect to the potential application of building materials in radioactive waste management, the present findings should be regarded as preliminary and conceptual. While crystallographic characterization is an important initial step in assessing material suitability, the current study does not provide direct experimental evidence related to radiation shielding effectiveness, radionuclide immobilization, or long-term radiation resistance. These properties depend on additional factors such as material density, chemical binding mechanisms, radiation-induced phase stability, and microstructural evolution under irradiation, which were not addressed in this work.

The relevance of the identified crystal structures to radioactive waste management is therefore limited to providing baseline mineralogical information. Such information may support subsequent investigations that integrate crystallographic data with radiation interaction studies, leaching tests, or long-term durability assessments. Without these complementary analyses, any conclusions regarding performance in radioactive environments must be considered speculative and should be interpreted with caution.

Overall, this study contributes to the existing literature by documenting the qualitative crystallographic characteristics of commonly used natural building materials from different regions. Although the approach is primarily descriptive, the results can serve as a reference dataset for future research employing more advanced analytical frameworks. To enhance scientific contribution and applicability, further work should incorporate quantitative XRD analysis, mechanical and thermal testing, and a clearer experimental linkage to specific application domains, including but not limited to radioactive waste management.

4 Conclusion

This study presents a qualitative X-ray diffraction analysis of commonly used building materials, including cement, sand, bricks, and lightweight bricks, to identify their dominant crystalline phases and crystal systems. The results indicate that cement, brick, and lightweight brick samples are primarily associated with phases belonging to the monoclinic crystal system, while sand samples are dominated by phases with an orthorhombic crystal system. These findings provide descriptive insight into the crystallographic characteristics of the investigated materials; however, the dominance of a particular crystal system should not be interpreted as a direct determinant of mechanical strength, thermal stability, or long-term durability. Material performance is governed by multiple factors such as phase composition, microstructure, porosity, and processing conditions, which were not quantitatively analyzed in this study. Therefore, the present results should be regarded as preliminary crystallographic characterization that may serve as a reference for future investigations. Further research incorporating quantitative XRD analysis and complementary mechanical, thermal, and durability testing is required to establish reliable correlations between crystal structure and material performance in construction-related applications.

References

1. D. Kurniawan, V. Kuswanto, A.H. Gunawan, Analisis dan perancangan sistem informasi penjualan bahan bangunan berbasis web pada toko bangunan daerah Tigaraksa menggunakan metode user acceptance testing. J. Algor. **4**, 58–74 (2023). https://doi.org/10.31253/algor.v4i2.1526
2. K. Asaduzzaman, F. Mannan, M. U. Khandaker, M. S. Farook, A. Elkezza, Y. B. M. Amin, S. Sharma, H. B. A. Kassim, Assessment of natural radioactivity levels and potential radiological risks of common building materials used in Bangladeshi dwellings. PLoS One. **10**, 1–16 (2015). https://doi.org/10.1371/journal.pone.0140667
3. R. Fitriyanti, M. Fatimura, Aplikasi produksi bersih pada industri semen. J. Redoks. **4**, 10–15 (2019). https://doi.org/10.31851/redoks.v4i1.3070
4. G. Marino, Y.D. Setiyarto, Penggunaan tanah liat untuk mengurangi jumlah semen pada beton geopolimer. CRANE: Civ. Eng. Res. J. **1**, 82–88 (2020) . https://doi.org/10.34010/crane.v1i2.4186
5. M. Udi, B. Sugito, Y. Bernadi, Pengolahan limbah radioaktif resin bekas dengan metode sementasi, in Proceedings Hasil Penelitian dan Kegiatan Tahun 2018, PTLR-BATAN, 119–125 (2019)
6. H. Prayuda, E. A. Setyawan, F. Saleh, Analisis sifat fisik dan mekanik batu bata merah di Yogyakarta. J. Ris. Rekayasa Sipil. **1**, 94–104 (2018). https://doi.org/10.20961/jrrs.v1i2.20658
7. G. Rori, D. R. O. Walangitan, R. L. Inkiriwang, Analisis perbandingan biaya material pekerjaan pasangan dinding bata merah dengan bata ringan. J. Sipil Statik. **8**, 311–318 (2020)
8. A.D. Rahmanto, S. Wulandari, Pengaruh kandungan mineral tanah pasir terhadap nilai CBR tanah, J. Jalan-Jembatan. **37**, 28–35 (2020)

Fabrication of reusable and low-cost sers substrate for explosives detection

Syifa Hilya Tsaniya[1], *Elis* Agustiana[1], *Nina Siti* Aminah[1], *Herman*[2], *Anita* Alni[3], *Isnaeni*[4], *Vivi* Fauzia[5], *Husin* Alatas[6], *Sigit Puji* Santosa[7,8] , and *Mitra* Djamal[1*]

[1]Instrumentation and Computational Physics, Institut Teknologi Bandung, Indonesia
[2]Physics of Magnetism and Photonics, Institut Teknologi Bandung, Indonesia
[3]Organic Chemistry, Institut Teknologi Bandung, Indonesia
[4]Research Center for Photonics, National Research and Innovation Agency, Indonesia
[5]Physics, Universitas Indonesia, Indonesia
[6]Physics, Institut Pertanian Bogor, Indonesia
[7]PT. Pindad, Indonesia
[8]Light-weight Structures, Institut Teknologi Bandung, Indonesia

Abstract. Surface-enhanced Raman Spectroscopy (SERS) is a highly sensitive and powerful analytical technique for molecular identification and analysis and is used in various fields. SERS uses the plasmonic effect on the surface of specific substrates, such as grating plasmonic, to enhance the Raman signal of the sample. However, the fabrication of grating structures as substrates is challenging and often comes with high costs. This research presents an approach to address this challenge by developing a low-cost SERS substrate using DVDs as a grating-based substrate coated with gold thin film. The fabricated substrate will be used as a SERS substrate to detect explosives: trinitrotoluene (TNT), Tetryl, Research Department Explosive (RDX), and pentaerythritol tetranitrate (PETN). According to our research, the concentration of explosives that can be detected is up to 0.01M for TNT, PETN, and Tetryl and 0.1M for RDX. Furthermore, a reusability test was conducted for the SERS substrate from DVD by cleaning the substrate using an acetone solution. Based on the results of the reusability test, the substrate can be reused up to three times. The periodicity of grating used is 713 nm, the thickness of the gold thin film used is 50 nm, and the depth of grating is 29.43 nm.

1 Introduction

An explosion is an event that occurs due to pressure caused by a sudden increase in volume and the release of a large amount of energy. Explosions can happen due to chemical reactions. For example, an explosion caused by trinitrotoluene (TNT) decomposition or nuclear reactions, such as the Chernobyl nuclear tragedy in 1986 [1]. Terrorist using explosive as a weapon to endanger national security and often targets civilians. The increase in terrorism using explosives emphasizes the significance of Explosive Trace Detection (ETD) due to the

*Corresponding autors: mitra@itb.ac.id

growing utilization of Home-Made Explosives (HMEs). Therefore, there is a need to improve existing Explosive Trace Detection (ETD) technology [2].

Raman spectroscopy is an analytical method to analyze complex mixtures and molecules quantitatively [3]. Raman spectroscopy provides detailed chemical properties and molecular information, even for samples with similar chemical properties and structures that are difficult to distinguish [4]. It is widely used in various fields, such as medical applications, food ingredients analysis, narcotics detection and many other areas. The signals of Raman spectroscopy are very weak due to the nature of the Raman scattering process itself [5]. SERS is a development of Raman spectroscopy that significantly enhances the weak Raman signal of molecule [5].

SERS is a highly sensitive and powerful molecular identification and analysis technique used in various fields. SERS uses the plasmonic effect on the surface of specific substrates, such as grating plasmonic, to enhance the Raman signal of the sample [6]. However, the fabrication of grating structures as a substrate is difficult and often comes with high costs, and currently, few reported SERS substrates can be reused [7].

This paper presents an approach to address this challenge by fabricating a reusable and low-cost SERS substrate using DVD as a grating-based substrate coated with gold thin film. The produced substrate will be used for the detection of explosives; Trinitrotoluene (TNT), Tetryl, Research Department Explosive (RDX), and Pentaerythritol Tetranitrate (PETN).

2 Materials and Methods

The fabricated SERS substrate (Figure 1) is made from DVD-RW plus GT-PRO Multi Speed 4x with a capacity of 4.7 GB. The DVD outer layer must be separated to obtain a grating layer by using a knife razor tool. Subsequently, the separated DVD layer is cut into 1 cm x 1 cm size.

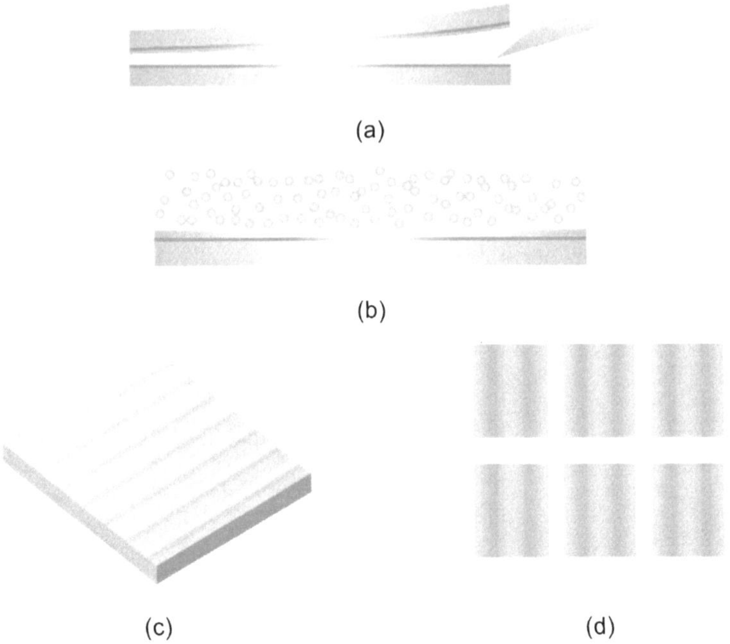

Fig. 1. Preparation steps to fabricate SERS substrates. (a) separation, (b) sputtering of gold thin film, (c) fabricated SERS substrate, (d) coated substrate.

The substrate was then coated with gold thin film using the Ion Beam Sputtering (IBS). The substrate was placed in an MC1000 Ion Sputter Coater, the parameters used for this coating process are 10 mA plasma current and a coating rate of 5 nm per minute. The sputtering process took 10 minutes to obtain a layer of gold thin film with a thickness of 50 nm.

Raman characterization of the SERS substrate is performed by measuring the Raman spectra of explosive materials using a Raman spectrometer. The Raman spectrometer used is The MacroRAMTM Raman Spectrometer with a laser wavelength of 785 nm. The laser power used is 10% (4.5 mW) and an exposure time of 5 seconds (1 exposure per second). We were using AFM (Atomic Force Microscopy) and SEM (Scanning electron Microscopy) to observe the substrate structure.

The samples used are 0.1 M, 0.01 M, and 0.001 M of TNT, PETN, RDX, and Tetryl. The samples used are explosives in a solid phase dissolved in acetone. The sample was prepared by drop-and-dry technique at room temperature.

For the reusability test, the substrate is cleaned by using acetone, and the Raman measurements were performed on the substrates to ensure no residual sample remained on the substrate surface. After that, the explosives sample is dropped onto the substrate to check whether the substrate can be used repeatedly or not.

3 Result and Discussion

3.1 Limit of Detection Test

Figure 2 shows the Raman spectra of TNT, Tetryl, PETN, and RDX on SERS substrate. The concentrations of TNT, Tetryl, PETN, and RDX used are 0.1 M (blue line), 0.01 M (red line), and 0.001 M (black line). As shown in Figure 2, when the concentration of the samples decreases to 0.01 M and 0.001 M, the intensity of the peaks in the Raman spectrum decreases or becomes undetectable because the number of particles that vibrated also reduces along with a decrease in the sample's concentration. For TNT, PETN, and Tetryl, as the concentration is decreased, the peak intensities decrease, and some peaks become undetectable. However, for the RDX sample, no Raman spectrum was observed when the concentration of the sample was decreased. For a concentration of 0.001 M, no Raman spectrum is obtained for any of the samples.

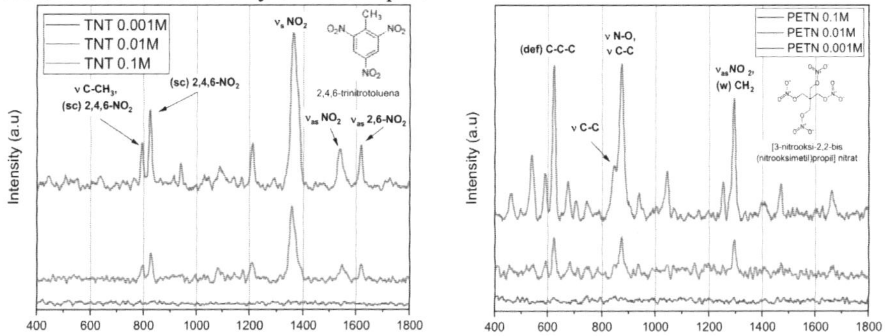

(a) (b)

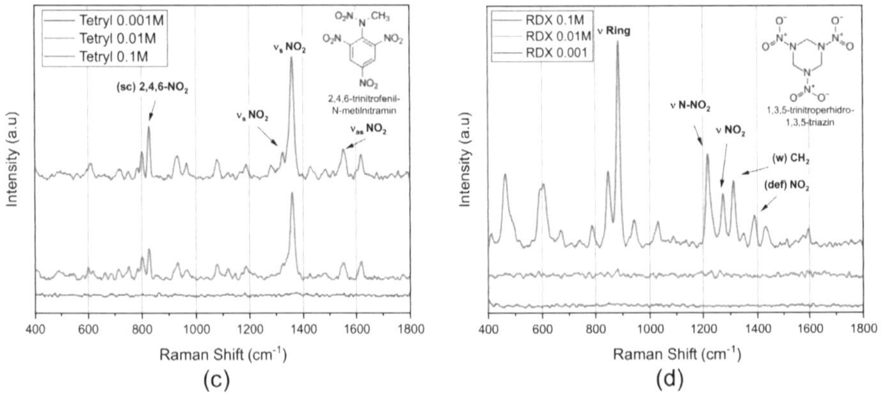

Fig. 2. The Raman spectra of explosives (a) TNT, (b) PETN, (c) Tetryl, and (d) RDX are shown with varying concentrations of 0.1M, 0.01M, and 0.001 M.

As can be seen in blue line Figure 2 (a), the strong peak of TNT appears at Raman shift of 793 cm^{-1}, 825 cm^{-1}, and 1362 cm^{-1}. Raman shifts and vibrational modes for the Raman spectrum of TNT can be seen in Table 1. When the concentration of TNT decreases to 0.01 M (red line in Figure 2 (a)), the intensity of the peak decreases, and the peak at 940 cm^{-1} is not detected. When the concentration is decreased to 0.001 M (black line in Figure 2 (a)), no Raman spectrum is detected.

Table 1. Assignment of Vibrational Mode of TNT Raman Spectra [8,9].

Raman Shift (cm^{-1})		Vibrational Mode
Ref	Exp	
1617	1619.75	$2{,}6 - NO_2$ Asymmetric stretch
1536	1537.94	NO_2 Asymmetric stretch
1362	1362.32	NO_2 Symmetric stretch
1212, 1089	1212.39, 1089.88	$C - H$ (ring) In-plane bend
940	942.173	$C - H$ (ring) out-plane bend
825	825.105	$2{,}4{,}6 - NO_2$ Scissoring
793	795.565	$C - CH_3$ Stretching, $2{,}4{,}6 - NO_2$ Scissoring

PETN molecule consists of carbon atom bonds, nitrogen and oxygen atom bonds forming nitrate groups, and nitrate groups with carbon atom bonds. From PETN Raman spectra result (Figure 2 (b) blue line), the strong peaks appear at the Raman shift of 622 cm^{-1}, 844 cm^{-1}, and 872 cm^{-1}. Raman shift and vibrational modes for the Raman spectrum of PETN can be seen in Table 2. When the concentration of PETN decreases to 0.01 M (red line in Figure 2 (b)), the intensity of the peak decreases, and the peak at 1044 cm^{-1}, 1251 cm^{-1}, and 1469 cm^{-1} are difficult to distinguish from noise. When the concentration is decreased to 0.001 M (black line in Figure 2 (b)), no Raman spectrum is detected.

Table 2. Assignment of Vibrational Mode of PETN Raman Spectra [8, 10, 11].

Raman Shift (cm^{-1})		Vibrational Mode
Ref	Exp	
1469	1469.64	CH_2 Scissoring
1292	1294.21	NO_2 Asymmetric stretch, CH_2 wagging
1251	1251.56	$C-H$ Bending stretch
1044	1044.14	$C-C$ Bending
872	874.094	$N-O$ Stretching, $C-C$ Stretching
844	842.776	NO_2 Scissoring
676	678.304	$N-O$ Stretching, $C-C$ Stretching
622	624.07	$C-C-C$ Deformation, $C-C$ Stretching
587	591.759	$O-NO_2$ Rock, $C-C$ Bend
538	538.981	C_5 Skeletal, CH_2 Wagging, $N-O$ Stretching

The molecular adsorption behavior on the substrate surface plays a critical role in determining the Raman cross-section, where any reduction in molecular adsorption leads to a corresponding attenuation in the signal intensity. This phenomenon is particularly evident in the SERS spectrum of PETN, specifically at the Raman shift of 1294.21 cm^{-1}. While standard library references[12] typically identify this peak as the most dominant for PETN, our experimental measurements show a lower relative intensity. This discrepancy is attributed to the specific adsorption orientation and the unique interfacial interactions between the PETN molecules and the gold-coated DVD grating, which likely differ from the conditions found in bulk phase or standard library spectra. Consequently, these variations in peak intensity provide a clearer physical interpretation of how the molecular alignment on the nanostructured surface influences the resulting Raman enhancement.

From the Tetryl Raman spectra result (Figure 2 (c) blue line), the strong peaks appear at the Raman shift of 827 cm^{-1}, 1325 cm^{-1}, 1358 cm^{-1}, and 1552 cm^{-1}. Based on the figure, the Raman spectra of TNT and Tetryl have similarities, but there is a shoulder peak at the Raman shift of 1325 cm^{-1}, and no peak appears at 1325 cm^{-1} in Raman spectra of TNT. This is because Tetryl and TNT are nitroaromatic compounds with different functional groups attached to their aromatic rings [13, 14]. Raman shifts and vibrational modes for the Raman spectrum of Tetryl can be seen in Table 3. When the concentration of Tetryl is decreased to 0.01 M (red line in Figure 2 (c)), the intensity of the peak is decreased, and the shoulder peak is not clear. When the concentration is decreased to 0.001 M (black line in Figure 2 (c)), no Raman spectrum is detected.

Table 3. Assignment of Vibrational Mode of Tetryl Raman Spectra [9].

Raman Shift (cm^{-1})		Vibrational Mode
Ref	Exp	
1617	1619.75	$2,6-NO_2$ Asymmetric stretch
1552	1554.02	NO_2 Asymmetric stretch
1358	1360.49	NO_2 Symmetric stretch
1325	1323.74	NO_2 Symmetric stretch
1078	1078.47	$C-H$ (ring) In-plane bend

930	934.422	$C - H$ (ring) out-plane bend
827	829.035	2,4,6 $- NO_2$ Scissoring
801	801	$C - CH_3$ Stretching, 2,4,6 $- NO_2$ Scissoring

In the RDX Raman spectra (Figure 2 (d) blue line), the strong peaks appear at the Raman shift of 885 cm^{-1}, 1218 cm^{-1}, 1273 cm^{-1}, 1310 cm^{-1}, and 1387 cm^{-1}. Raman shift, and vibrational modes for the Raman spectrum of RDX can be seen in Table 4. The peak at 1218 cm-1 is a characteristic of the vibrational mode of the nitramine group [10]. When the concentration of RDX is decreased to 0.01 M (red line in Figure 2 (d)) and 0.001 M (black line in Figure 2 (d)), no Raman spectrum is detected.

Table 4. Assignment of Vibrational Mode of RDX Raman Spectra [9, 10].

Raman Shift (cm^{-1})		Vibrational Mode
Ref	Exp	
1596	1596.72	NO_2 Asymmetric stretch
1431	1433.44	$CH_2 - N$ Deformation
1387	1389.76	NO_2 Asymmetric stretch
1310	1389.76	NO_2 Symmetric vibration
1273	1275.7	NO_2 Stretching
1218	1218	$N - NO_2$ Stretching
944	944.109	$N - O$ Deformation
885	885.807	$N - N$ Symmetric ring breathing
848	850	CH_2 rocking, $C - N - C$ Deformation
787	789.643	$N - O$ Deformation
605	605.991	NO_2 Wagging

3.2 Reusability Test of SERS Substrate

To evaluate the reusability of SERS substrates, a reusability test was conducted by cleaning the used substrates with acetone. After cleaning, Raman measurements were performed on the substrates to ensure no residual sample remained on the substrate surface. As observed in Figures 3, all peaks in the Raman spectrum reappeared upon reusing the substrate, albeit with a decrease in peak intensity. This phenomenon arises from enhancements in low sample concentrations that depend on the detected molecular locations. SERS enhancements rely on hot spots and the distribution of sample molecules [15]. Therefore, Raman measurement results may vary from one measurement point to another. The SERS substrate can be reused up to three times.

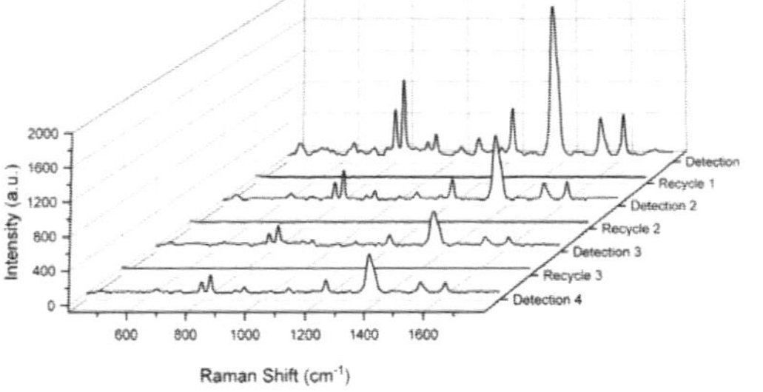

(a) Spectrum Raman of TNT 0.1 M

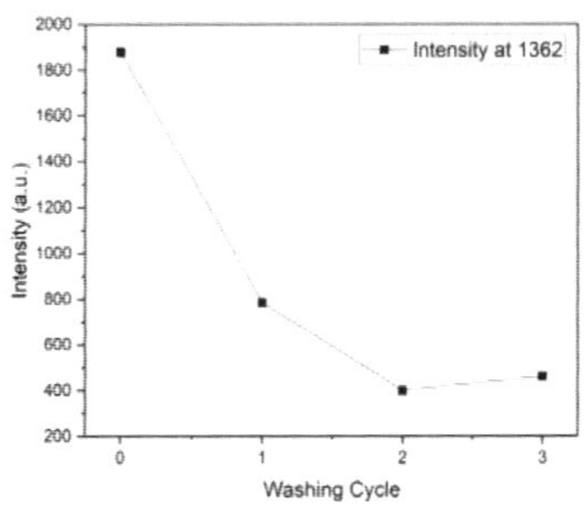

(b) Intensity at 1362 cm^{-1} of Spectrum Raman of TNT 0.1 M

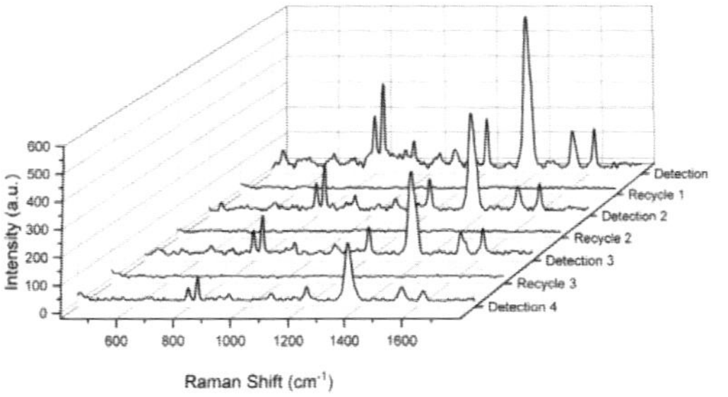

(c) Spectrum Raman of TNT 0.01 M

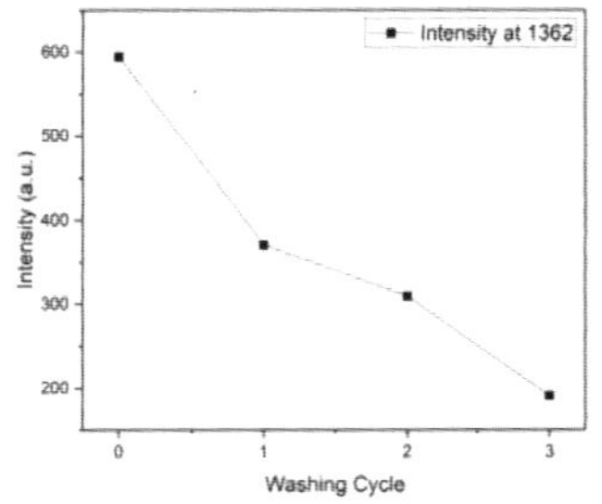

(d) Intensity at 1362 cm^{-1} of Spectrum Raman of TNT 0.01 M

Fig. 3. Result of reusability of SERS substrate with 3 cycles. (a) and (c) represent the Raman spectra of measurement results. Recycle is the condition when the sample is dropped to the substrate, and detection is the condition after substrate has been cleaned. (b) and (d) represent the Raman spectra intensity at specific Raman shift for each sample.

3.3 SEM and AFM Characterization for SERS Substrate

Scanning Electron Microscope (SEM) (Figure 4) and Atomic Force Microscope (AFM) (Figure 5) were performed on the fabricated SERS substrate to observe the grating structure of the substrate. Based on the SEM result, the grating structure has a period of 713 nm, and the thickness of the coated gold thin film is 50 nm. Based on the AFM result, the depth of grating is 29.43 nm.

(a)

(b)

Fig. 4. SEM images showing grating structure of a gold-coated DVD substrate for (a) unused substrate and (b) triple-use substrate.

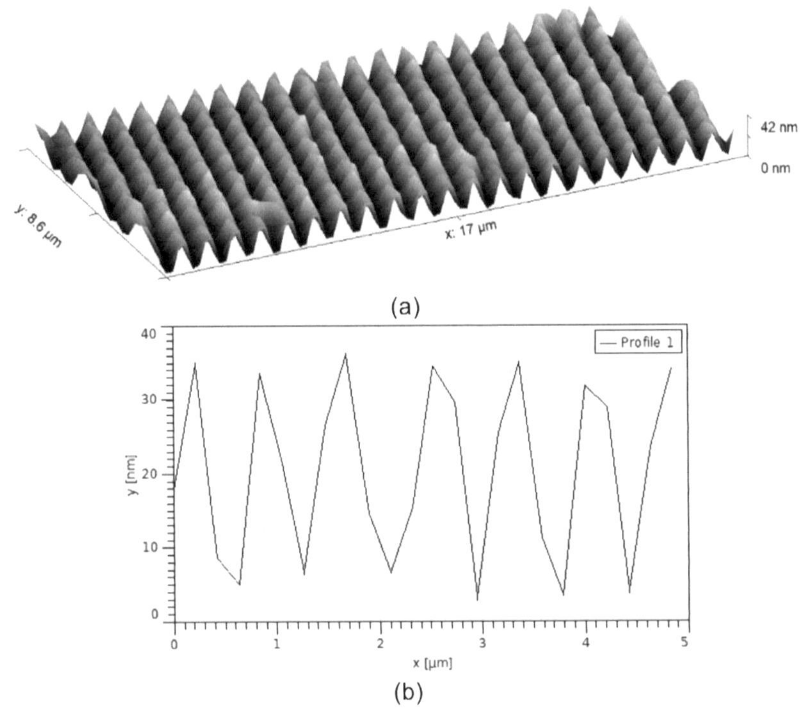

Fig. 5. AFM topographic image of DVD substrate showing the grating structure of DVD layered with gold thin film (a) and AFM profile measurements perpendicular to the grating to show the depth of the grating (b).

4 Conclusions

The fabrication of SERS substrates using DVDs coated with gold thin film has been successfully achieved. The periodicity of the grating used is 713 nm, the thickness of the gold thin film used is 50 nm, and the grating depth is 29.43 nm. The detectable concentration of explosive materials is up to 0.01 M for TNT, PETN, and Tetryl, while for RDX, it is 0.1 M. The SERS substrate can be reused up to three times after the substrate has been cleaned.

This research is fully funded by the Indonesian Ministry of Education, Culture, Research, and Technology under 2022 Indonesian Collaborative Research Program managed by Institut Teknologi Bandung. The authors would like to express gratitude to PT PINDAD Indonesia for support and assistance in the completion of this research project.

References

1. Bell, S., Forensic Chemistry (Third Edition), (CRC Press, Florida, 2022)
2. K. C. To, S. Ben-Jaber, I. P. Parkin, Recent developments in the field of explosive trace detection. *ACS Nano* **14**, 10804–10833 (2020). https://doi.org/10.1021/acsnano.0c01579
3. Khandpur, R. S., Handbook of Analytical Instruments (3rd ed.), (Tata McGraw-Hill, Uttar Pradesh 201309, 2015)
4. K. C. Doty, C. K. Muro, J. Bueno, L. Halámková, I. K. Lednev, What can Raman spectroscopy do for criminalistics? *J. Raman Spectrosc.* **47**, 39–50 (2016). https://doi.org/10.1002/jrs.4826

5. Smith, E., & Dent, G., Modern Raman Spectroscopy-A Practical Approach, (John Wiley and Sons, New York, 2019)
6. D. Graham, M. Moskovits, Z.-Q. Tian, SERS – facts, figures and the future. *Chem. Soc. Rev.* **46**, 3864–3865 (2017). https://doi.org/10.1039/C7CS90060K
7. Y. Hu, C. Wu, S. Huang, X. Luo, R. Yuan, X. Yang, A novel SERS substrate with high reusability for sensitive detection of miRNA 21. *Talanta* **228**, 122240 (2021). https://doi.org/10.1016/j.talanta.2021.122240
8. S. Elbasuney, A. F. El-Sherif, Complete spectroscopic picture of concealed explosives: laser induced Raman versus infrared. *TrAC Trends Anal. Chem.* **85**, 34–41 (2016). https://doi.org/10.1016/j.trac.2016.04.023
9. L. C. Pacheco-Londoño, W. Ortiz-Rivera, O. M. Primera-Pedrozo, S. P. Hernández-Rivera, Vibrational spectroscopy standoff detection of explosives. *Anal. Bioanal. Chem.* **395**, 323–335 (2009). https://doi.org/10.1007/s00216-009-2954-y
10. Y. A. Gruzdkov, Y. M. Gupta, Vibrational properties and structure of pentaerythritol tetranitrate. *J. Phys. Chem. A* **105**, 6197–6202 (2001). https://doi.org/10.1021/jp004425j
11. J. B. Oleske, B. T. Smith, J. Barber, J. C. Weatherall, Identifying Raman and infrared vibrational motions of erythritol tetranitrate. *Appl. Spectrosc.* **69**, 1397–1402 (2015). https://doi.org/10.1366/14-07684
12. A. Tripathi, E. D. Emmons, P. G. Wilcox, J. A. Guicheteau, D. K. Emge, S. D. Christesen, A. W. Fountain III, Semi-automated detection of trace explosives in fingerprints on strongly interfering surfaces with Raman chemical imaging. *Appl. Spectrosc.* **65**, 611–619 (2011). https://doi.org/10.1366/10-06214
13. D. Lin-Vien, N. B. Colthup, W. G. Fateley, J. G. Grasselli, *The handbook of infrared and Raman characteristic frequencies of organic molecules*. London: Academic Press (1991).
14. A. Sakamoto, M. Tasumi, Symmetry of the benzene ring and its normal vibrations: The "breathing" mode is not always a normal vibration of a benzene ring. *J. Raman Spectrosc.* **52**, 2282–2291 (2021). https://doi.org/10.1002/jrs.6131
15. S. Cortijo-Campos, R. Ramírez-Jiménez, A. de Andrés, *Raman and fluorescence enhancement approaches in graphene-based platforms for optical sensing and imaging*. *Nanomaterials* **11**, 644 (2021). https://doi.org/10.3390/nano11030644

Synthesis of ZnO and SnO_2 catalysts using roselle flower extract for biodiesel production from waste cooking oil

Maisari Utami[1,2*], *Izzaty Fauziya* Salwa[1], and *Marisa*[1]

[1]Department of Chemistry, Faculty of Mathematics and Natural Sciences, Universitas Islam Indonesia, Yogyakarta 55584, Indonesia
[2]Nanomaterials and Sustainable Chemistry Research Centre, Universitas Islam Indonesia, Yogyakarta 55584, Indonesia

Abstract. The development of biodiesel from waste cooking oil using green-synthesized heterogeneous catalysts offers a sustainable alternative to fossil fuels. In this study, ZnO and SnO_2 catalysts were synthesized using roselle flower (*Hibiscus sabdariffa* L.) extract and applied to transesterification. Characterization revealed that SnO_2 exhibited a surface area of 51.52 m^2/g and a pore volume of 0.080 cm^3/g, nearly 12 times higher than ZnO (4.11 m^2/g; 0.009 cm^3/g), resulting in superior catalytic activity. FTIR analysis of biodiesel catalyzed by ZnO showed ester-specific peaks at 1115 cm^{-1} (C–O) and 1022 cm^{-1} (C–O–C), while biodiesel from SnO_2 displayed a strong ester carbonyl band at 1744 cm^{-1}, confirming FAME formation. GC–MS results indicated that ZnO-catalyzed biodiesel consisted mainly of methyl oleate (60.23%), methyl palmitate (34.83%), methyl stearate (2.27%), and methyl myristate (0.34%), with a total FAME content of 97.34%. In contrast, SnO_2-catalyzed biodiesel contained methyl oleate (55.85%), methyl palmitate (37.12%), methyl stearate (4.50%), and methyl myristate (0.64%), yielding nearly 100% FAME. The dominance of methyl oleate and palmitate confirmed the production of high-quality biodiesel with balanced oxidative stability and combustion properties. Therefore, roselle-extract-based SnO_2 demonstrated greater potential than ZnO as an eco-friendly catalyst for biodiesel production from waste cooking oil.

1 Introduction

Currently, the use of cleaner and more sustainable energy is a global focus due to the depletion of fossil fuel resources, global warming, and rising energy costs [1]. The development of abundant and environmentally friendly energy sources is a key solution to climate change, energy security, and economic stability. Biodiesel, derived from vegetable oils or animal fats, is a promising renewable fuel that can be used directly in diesel engines and is environmentally friendly, biodegradable, and low in toxicity.

* Corresponding author: maisariutami@uii.ac.id

Recent studies highlight biodiesel from waste cooking oil (WCO) as a potential circular economy approach to reduce greenhouse gas emissions and valorize waste [2].

Transesterification is a catalytic process in which triglycerides from animal fats or vegetable oils react with alcohols, typically methanol, to yield methyl esters as the primary product and glycerol as a by-product [3]. Transesterification, also referred to as alcoholysis, is a dissociation process analogous to hydrolysis, in which one alcohol is substituted by another within an ester molecule [4]. The reaction rate is mainly influenced by temperature, reaction time, catalyst, and alcohol concentration. Typically, it is carried out at 30–65 °C, near the boiling point of methanol, where higher temperatures shorten reaction time while lower temperatures may improve conversion efficiency [5].

In the transesterification process, catalysts play a crucial role in accelerating the reaction and increasing conversion efficiency. Catalysts can be homogeneous, such as NaOH and KOH, or heterogeneous, such as CaO, ZnO, and SnO_2. Heterogeneous catalysts offer several advantages, including easier separation from the product, reusability, and improved environmental compatibility. ZnO and SnO_2 are increasingly employed due to their stability, high catalytic activity, and resistance to fouling. Mechanistically, metal oxide catalysts function as base sites that accelerate the breakdown of triglycerides into methyl esters and glycerol [6]. For instance, a SnO_2 catalyst impregnated with ZnO nanowires can achieve FAME conversions of up to 92% and can be reused up to five times without regeneration [7].

Green synthesis approaches using plant extracts as reducing agents and stabilizers in catalyst synthesis have received increasing attention, as bioactive plant compounds, such as flavonoids, phenols, and anthocyanins, can facilitate the formation of uniform and catalytically active materials [8]. One potential candidate is the roselle flower (*Hibiscus sabdariffa* L.), which is rich in bioactive compounds and has proven effective as a reducing agent in the synthesis of ZnO and SnO_2 catalysts [9].

Although numerous studies have employed metal oxide catalysts, few have utilized roselle flower extract for WCO-based biodiesel production. Therefore, this study aims to evaluate the effectiveness of roselle extract-based ZnO and SnO_2 catalysts in producing biodiesel from WCO, compare the methyl ester yields obtained from each catalyst, and determine the optimal reaction conditions for maximum conversion.

2 Materials and Methods

2.1 Materials

The materials used in this research were methanol p.a (CH_3OH) from Merck, waste cooking oil, the precursor $SnCl_2.H_2O$ (Tin (II) chloride monohydrate) branded as stannous chloride dihydrate, the precursor $ZnO(NO_2)_3.6H_2O$ (Zinc nitrat hexahydrate) branded as, and filter paper.

2.2 Extraction of Roselle Flower (*Hibiscus sabdariffa* L.)

The ground dried roselle flowers were weighed according to the catalyst requirements, namely 1 g for ZnO and 8 g for SnO_2. Each sample was then dissolved in 100 mL of deionized water and continuously stirred for 1 hour at 60°C to extract the active compounds that function as a bioreducing agent. After the extraction process was complete, the solution was filtered to remove the solids, yielding a clear roselle extract ready to be used in the catalyst synthesis stage.

2.3 Synthesis of Tin (IV) Oxide (SnO_2) Material

The hydrothermal synthesis of SnO_2 was initiated by weighing 0.021 g of $SnCl_2 \cdot 2H_2O$ and dissolving it in 100 mL of distilled water, followed by the addition of 50 mL of roselle flower extract. The mixture was heated at 60 °C for 4 hours under continuous stirring with a magnetic stirrer. The obtained solution was subsequently transferred into an autoclave and maintained at 60 °C for 12 hours. After the reaction, the mixture was filtered to separate the precipitate from the filtrate. The precipitate was collected, placed in a crucible, and oven-dried at 60 °C until completely dry, then calcined at 400 °C for 2 hours in a furnace. Finally, the resulting SnO_2 material was ground into a fine powder and stored in a sealed container. The final product was designated as SnO_2.

2.4 Synthesis of Zinc Oxide (ZnO) Material

The synthesis of ZnO material was carried out by dissolving 2 g of Zn $(NO_3)_2 \cdot 6H_2O$ in 42 mL of roselle extract. The solution was stirred with a magnetic stirrer at 60 °C for 1 hour until completely dissolved, followed by heating in a water bath at the same temperature until it formed a paste-like consistency. The obtained sample was then oven-dried at 60 °C until fully dry. Finally, the dried sample was calcined at 400 °C for 1 hour and subsequently ground into a fine white powder.

2.5 Biodiesel Production Process

In the biodiesel production process, 10 mL of waste cooking oil was first prepared and then mixed with 100 mL of analytical-grade methanol and 0.2 g of metal oxide catalyst (SnO_2/ZnO). The mixture was transferred into a reactor, and the reaction was carried out at 60 °C for 2 hours under continuous stirring with a magnetic stirrer. After completion of the reaction, the mixture was separated using a separatory funnel to divide the upper and lower layers. The upper layer was subsequently filtered using standard filter paper to remove residual catalyst, yielding a purer biodiesel product

3 Results and Discussion

3.1 Materials Characterization

3.1.1 Fourier transmission-infrared spectroscopy (FT-IR)

Fourier Transform Infrared (FTIR) spectroscopy is a technique used to identify the functional groups present in a compound. FTIR works by measuring molecular vibrations and rotations that are influenced by infrared radiation at specific wavelengths. The FTIR results for ZnO show a broad band at 3427 cm^{-1} (O–H stretching) and 1630 cm^{-1} (H–O–H bending), indicating the presence of hydroxyl groups and adsorbed water molecules, along with additional bands at 1385 cm^{-1} (carbonate C-O), 1116 cm^{-1} (C–O), and 870 cm^{-1} (=C–H), which suggest organic residues from the synthesis process (Figure 1). A characteristic band at 514 cm^{-1} confirms the presence of Zn–O bonds, consistent with the theoretical lattice vibration range of Zn–O (430–520 cm^{-1}) reported in the literature [10]. For SnO_2, the spectrum exhibits an O–H band at 3434 cm^{-1} and an H–O–H bending band at 1637 cm^{-1}, followed by a band at 1081 cm^{-1} associated with Sn–OH or C–O, as well as an intense band at 561 cm^{-1} characteristic of Sn–O vibrations, in agreement with the theoretical range of Sn–O bonds at 550–630 cm^{-1} [11].

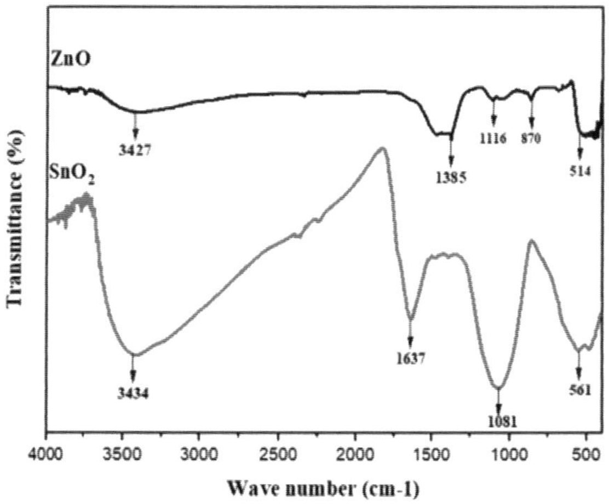

Fig. 1. FTIR spectra of ZnO dan SnO_2.

3.1.2 X-Ray Diffraction (XRD)

X-ray Diffraction (XRD) is a fundamental technique used to identify the crystal structure of materials by analyzing their diffraction patterns. The main parameters such as diffraction angle (2θ), Miller indices (hkl), full width at half maximum (FWHM), and d-spacing are important to determine the crystal phase, degree of crystallinity, and crystallite size. When X-rays strike a material, they are diffracted by the atoms within it. Each crystalline material produces a unique diffraction pattern, much like a fingerprint. The resulting pattern can be compared with a standard crystal pattern database, commonly referred to as JCPDS (Joint Committee on Powder Diffraction Standards). Good crystallinity is indicated by sharp peaks, where the sharper the peaks, the better the crystallinity of the material.

For synthesized SnO_2, the diffraction peaks correspond to the tetragonal rutile phase with JCPDS card no. 41-1445. The most intense diffraction peak is observed at 2θ = 26.589° on the (110) plane with an FWHM of 0.127, confirming strong preferential orientation and good crystallinity. Additional peaks are observed at (101) at 33.888° (FWHM 0.391), (201) at 40.508° (FWHM 0.158), and (211) at 52.704° (FWHM 0.467) (Figure 2). Narrow FWHM values, especially at (110) and (201), indicate larger crystallite size and higher crystallinity, whereas broader FWHM values at (101) and (211) suggest smaller crystallites or strain within the lattice. These findings are in line with Liu et al. (2022), who reported that hydrothermal SnO_2 typically shows a dominant (110) reflection with narrow FWHM, while the (211) peak often appears broader due to heterogeneous crystallite size distribution [12].

Meanwhile, ZnO is identified as a hexagonal wurtzite structure with JCPDS card no. 36-1451. Its diffraction peaks are observed at (100), (002), (101), and higher planes with FWHM values ranging between 0.236-0.387. For instance, at 2θ = 31.68° (hkl 100) with FWHM 0.236, the crystallite size is calculated to be 34.91 nm. At higher angles such as 68.98° with FWHM 0.387, the crystallite size decreases to 24.87 nm. The average crystallite size of ZnO is about 29.04 nm, consistent with nanoscale dimensions and high crystallinity. This result agrees with Zhang et al. (2023), who reported crystallite sizes of 25–35 nm for modified ZnO nanostructures [13].

Comparatively, SnO_2 exhibits its strongest diffraction peak at the (110) plane with a sharp and narrow FWHM, while ZnO shows multiple characteristic peaks (100, 002, 101) with relatively larger average crystallite sizes. This indicates that SnO_2 may have smaller crystallites on certain planes but stronger preferential orientation, whereas ZnO has more uniform crystallinity. Such differences in crystallite size and crystallographic orientation significantly influence the optical absorption and photocatalytic activity of both materials.

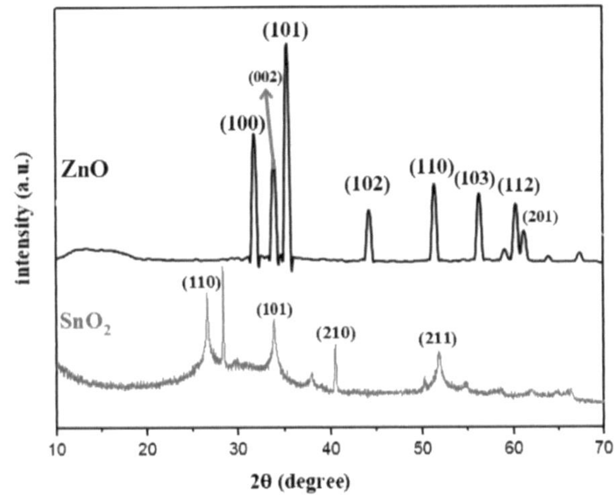

Fig. 2. XRD diffactogram of ZnO and SnO_2.

3.1.3 Gas Sorption Analysis (GSA)

Gas sorption analysis (GSA) is used to determine the surface area, pore volume, and pore size of a material. The analysis is carried out by measuring the amount of gas adsorbed onto the surface of a solid at specific pressures and temperatures [14]. The Brunauer Emmett Teller (BET) equation is used to determine the specific surface area of a material based on gas adsorption data. In its application, the volume of gas adsorbed at various relative pressures is analyzed to obtain the monolayer adsorption capacity (*Vm*), which represents the amount of gas molecules covering the material surface in a single layer. This *Vm* value is then used as the basis for calculating the specific surface area of the material, reflecting the number of active sites available on the surface.

BET equation:

$$\frac{1}{V\left(\frac{P_0}{P}-1\right)} = \frac{C-1}{V_m C}\left(\frac{P}{P_0}\right) + \frac{1}{V_m C} \quad (1)$$

where:
V is the volume of gas adsorbed at a given pressure,
P is the gas pressure,
P_0 is the saturation vapor pressure of the gas,
V_m is the monolayer adsorption capacity, and
C is the BET constant related to the interaction energy between the adsorbed gas and the material surface.

Three types of gases are commonly used in the analysis: helium for micropores (<2 nm), nitrogen for mesopores (2–5 nm), and argon for macropores (>5 nm). The results showed that ZnO exhibited a specific surface area of 4.11 m^2/g, pore volume of 0.00924 cm^3/g, and an average pore size of 2.13 nm. This relatively low surface area suggests that ZnO tends to agglomerate and possesses a dense morphology with limited porosity, resulting in fewer available active sites and reduced photocatalytic and adsorption activity. In contrast, SnO_2 demonstrated significantly better textural properties with a specific surface area of 51.52 m^2/g, pore volume of 0.08015 cm^3/g, and an average pore size of 3.12 nm (Figure 3 and Table 1). This value is nearly twelve times higher than that of ZnO, indicating the presence of a mesoporous structure that facilitates easier molecular access to active sites and enhances both catalytic and adsorption performance.

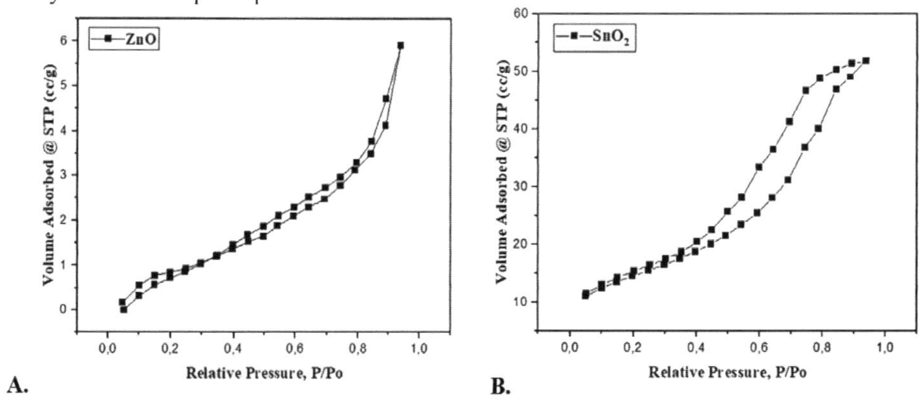

Fig. 3. The adsorption and desorption isotherm curves from the GSA (a) ZnO and (b) SnO_2.

In conclusion, the GSA results highlight a stark contrast between ZnO and SnO_2. ZnO shows low surface area and limited porosity, far below the values reported in recent studies, making it less suitable for photocatalytic applications unless further synthesis optimization is carried out. In contrast, SnO_2 exhibits higher surface area and mesoporosity, aligning well with recent literature, thus offering superior potential for photocatalysis and adsorption applications compared to ZnO.

Table 1. The quantitative GSA results of ZnO and SnO_2.

Description	**ZnO**	**SnO_2**
Specific surface area (BET)	4.10952 m^2/g	51.5183 m^2/g
Total pore volume	0.00924157 cm^3/g	0.080153 cm^3/g
Average pore size	2.12638 nm	3.11782 nm

3.1.4 Scanning Electron Microscopy-Energy Dispersive X-ray Spectroscopy (SEM-EDS)

Scanning Electron Microscopy-Energy Dispersive X-ray Spectroscopy (SEM-EDS) is a crucial characterization technique that provides comprehensive information on both the surface morphology and the elemental composition of a material. SEM generates high-resolution images through the interaction of an electron beam with the sample surface, enabling the observation of particle shape, size, and distribution [15]. The combination of these techniques is essential for evaluating the success of inorganic material synthesis, such as ZnO and SnO_2, in terms of both morphological uniformity and elemental purity.

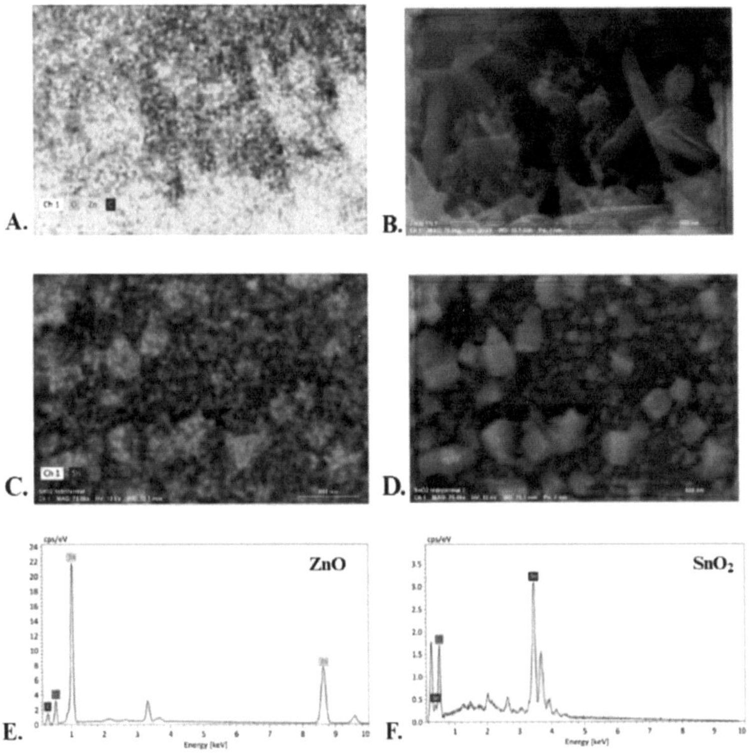

Fig. 4. (a) EDS ZnO elemental mapping, (b) SEM image ZnO, (c) EDS SnO_2 elemental mapping, (d) SEM image SnO_2, EDS spectra of (e) ZnO, and (f) SnO_2.

The characterization results for ZnO reveal that elemental mapping (Figure 4A) shows a homogeneous distribution of Zn and O, indicating the uniform formation of the ZnO phase across the material surface. The SEM image (Figure 4B) demonstrates a morphology dominated by rod-like structures and irregularly shaped particles that tend to agglomerate, which is typically associated with the anisotropic growth behavior of ZnO crystals. The corresponding EDS spectrum (Figure 4E) confirms the presence of Zn and O with mass compositions of 59.05% and 9.48%, respectively. A minor carbon peak was also detected, which is most likely attributed to the conductive carbon coating used during SEM sample preparation rather than being part of the actual ZnO composition.

For SnO_2, elemental mapping (Figure 4C) shows a uniform distribution of Sn and O without significant contamination from other elements, suggesting the successful formation of the SnO_2 phase. The SEM micrograph (Figure 4D) reveals cube-like particles with relatively uniform sizes and dense distributions, indicating a more isotropic crystal growth compared to ZnO. The EDS spectrum of SnO_2 (Figure 4) exhibits dominant peaks of Sn and O, with mass compositions of 83% and 17%, respectively (Table 2). This ratio closely matches the theoretical stoichiometry of SnO_2 (1:2 ratio of Sn to O), thereby confirming the high purity of the synthesized material.

Table 2. Elemental composition of ZnO and SnO_2.

Sample	Mass Concentration (%)		
	Zn	Sn	O
ZnO	59.05	-	9.48
SnO_2	-	83.00	17.00

Overall, the SEM-EDS analysis demonstrates that both ZnO and SnO_2 were successfully synthesized with distinctive morphologies, homogeneous elemental distributions, and mass ratios consistent with theoretical values. The morphological differences ZnO exhibiting rod-like structures and SnO_2 displaying cube-like particles can be attributed to differences in crystal growth mechanisms and synthesis conditions. Furthermore, the elemental composition results reinforce the effectiveness of the applied synthesis method in producing high-quality ZnO and SnO_2 materials. These findings suggest that the synthesized materials hold significant potential for applications in various fields, particularly as photocatalysts for clean energy reactions and wastewater treatment.

3.2 Biodiesel

The biodiesel synthesis proceeded via a transesterification reaction between triglycerides and methanol using SnO_2 and ZnO as heterogeneous catalysts under mild thermal conditions (60 °C) without light irradiation (Figure 5). SnO_2 and ZnO functioned as solid acid–base catalysts, where Lewis acidic sites (Sn^{4+} and Zn^{2+}) facilitated the activation of the carbonyl (C=O) groups of triglycerides, while basic surface oxygen species promoted the adsorption and activation of methanol into reactive methoxide species.

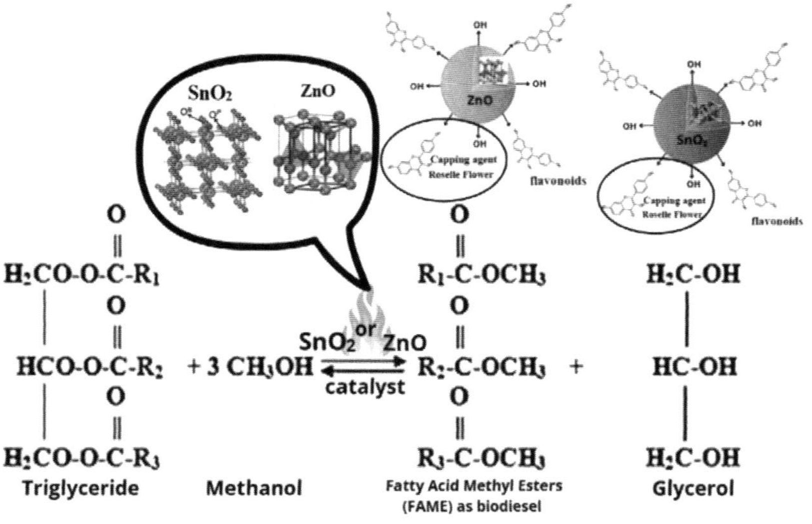

Fig. 5. Illustration of biodiesel production using SnO_2 and ZnO.

The generated methoxide subsequently performed a nucleophilic attack on the electrophilic carbonyl carbon of the triglyceride ester groups, forming tetrahedral intermediates that decomposed into fatty acid methyl esters (FAME) as biodiesel and glycerol as the by-product. The catalytic process was governed solely by thermal energy and catalyst surface interactions, with no involvement of photoinduced electron–hole pairs, confirming the purely heterogeneous catalytic role of SnO_2 and ZnO in the transesterification reaction.

3.3 Biodiesel Characterization

3.3.1 *Fourier transmission-infrared spectroscopy (FT-IR)*

The FTIR spectrum of ZnO catalyst (Figure 6) shows an absorption peak at 3312 cm^{-1}, which indicates the presence of –OH groups from moisture or residual alcohol, while the bands at

2951 cm^{-1} and 2832 cm^{-1} correspond to asymmetric and symmetric stretching vibrations of C–H from methyl and methylene groups in the aliphatic chain. The absorption at 1451 cm^{-1} indicates C–H bending vibrations, whereas the strong band at 1115 cm^{-1} is related to C–O stretching from ester groups, along with a sharp peak at 1022 cm^{-1} that further confirms the presence of C–O–C bonds in esters, and the band at 630 cm^{-1} is associated with aliphatic C–H bending vibrations, thus confirming that the use of ZnO catalyst produces biodiesel with ester functional group characteristics.

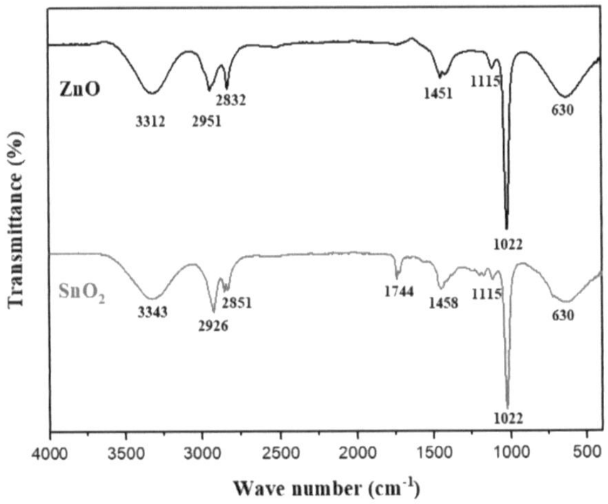

Fig. 6. FTIR spectra results of SnO_2 HT Vis, SnO_2 HT UV, SnO_2 SG Vis, and SnO_2 SG UV.

Meanwhile, the FTIR spectrum of SnO_2 catalyst exhibits an absorption band at 3343 cm^{-1} corresponding to –OH groups, followed by bands at 2926 cm^{-1} and 2851 cm^{-1} indicating C–H stretching vibrations from hydrocarbon chains, as well as a characteristic band at 1744 cm^{-1} that confirms the presence of ester carbonyl (C=O) groups as the main indicator of biodiesel formation. The band at 1458 cm^{-1} is attributed to C–H deformation, while the peaks at 1115 cm^{-1} and 1022 cm^{-1} correspond to C–O stretching vibrations of esters, and the absorption at 630 cm^{-1} is associated with aliphatic C–H bonds, thereby confirming that SnO_2 is also effective in assisting the transesterification process, with the ester carbonyl peak serving as strong evidence for the formation of fatty acid methyl ester biodiesel.

3.3.2 Gas Chromatography–Mass Spectrometry (GC–MS)

Gas Chromatography–Mass Spectrometry (GC–MS) is an analytical technique used to separate, identify, and quantify chemical compounds with high accuracy. GC–MS analysis produces chromatograms showing analyte separation based on retention time and fragmentograms displaying mass fragmentation patterns that provide information on molecular structure.

GC–MS analysis of the catalyst ZnO (Figure 7) without light sample revealed that the biodiesel is dominated by four major compounds with a total content reaching 97.34%. The most abundant compound is methyl oleate (RT 16.836 min, 60.23%, SI = 96), which plays a crucial role in maintaining a balance between oxidative stability and cold flow properties of biodiesel. The second largest component is methyl palmitate (RT 15.057 min, 34.83%, SI = 96), which enhances thermal stability and flash point, although it may reduce cold flow performance.

In addition, methyl stearate (RT 17.061 min, 2.27%, SI = 96) was also detected, contributing to increased viscosity and cetane number of the biodiesel. The last key component is methyl myristate (RT 12.994 min, 0.34%, SI = 97), which, despite being present in a small amount, shows the highest Similarity Index, confirming its strong identification.

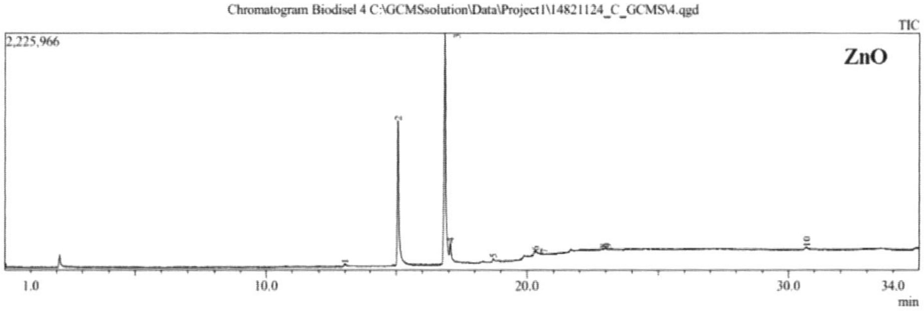

Fig. 7. GC-MS spectra result of ZnO.

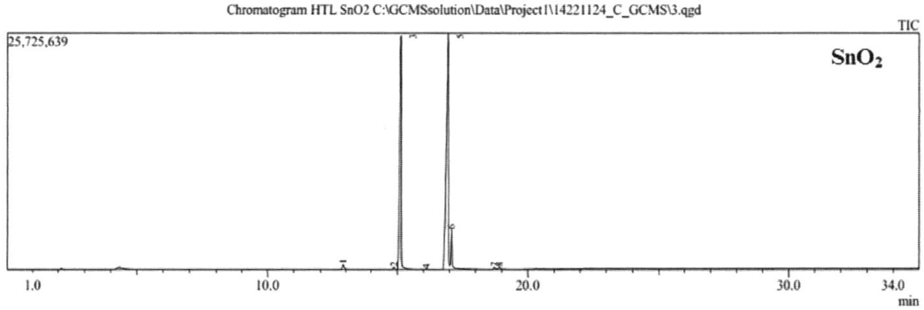

Fig. 8. GC-MS spectra result of SnO_2.

The dominance of methyl oleate and methyl palmitate indicates that biodiesel produced using ZnO catalyst without light has a typical vegetable oil profile, reflecting a balance between saturated and unsaturated fatty acids. This composition confirms that the resulting biodiesel is of high quality, with a high FAME content and chemical characteristics that support its application as a renewable alternative fuel.

GC–MS analysis of the catalyst SnO_2 (Figure 8) without light sample revealed that the biodiesel is dominated by four main compounds with a total content approaching 100%. The most abundant compound is methyl oleate (RT 16.925 min, 55.85%, SI = 96), a methyl ester of oleic acid that plays an important role in providing oxidative stability while maintaining the cold flow properties of biodiesel (**Table 3**). The second major component is methyl palmitate (RT 15.120 min, 37.12%, SI = 96), a methyl ester of palmitic acid that enhances thermal stability and flash point, although it tends to reduce cold flow properties. The next compound is methyl stearate (RT 17.071 min, 4.50%, SI = 96), which contributes to increasing viscosity and improving the cetane number of biodiesel. Lastly, methyl myristate (RT 12.902 min, 0.64%, SI = 97) is present in a small amount but shows the highest identification match. The dominance of methyl oleate and methyl palmitate indicates that the resulting biodiesel achieves a balance between oxidative stability and combustion properties, classifying it as high-quality biodiesel suitable for alternative fuel applications.

Table 3. GC-MS analysis summary.

Catalyst (Without Light)	RT (min)	Compound (Methyl Ester)	%Area	SI (Similarity Index)
SnO_2	16.925	Methyl oleate ($C_{19}H_{36}O_2$)	55.85%	96
	15.120	Methyl palmitate	37.12%	96

		($C_{17}H_{34}O_2$)		
	17.071	Methyl stearate ($C_{19}H_{38}O_2$)	4.50%	96
	12.902	Methyl myristate ($C_{15}H_{30}O_2$)	0.64%	97
FAME			**100%**	–
ZnO	16.836	Methyl oleate ($C_{19}H_{36}O_2$)	60.23%	96
	15.057	Methyl palmitate ($C_{17}H_{34}O_2$)	34.83%	96
	17.061	Methyl stearate ($C_{19}H_{38}O_2$)	2.27%	96
	12.994	Methyl myristate ($C_{15}H_{30}O_2$)	0.34%	97
FAME			**97.34%**	–

4 Conclusions

This study demonstrated that ZnO and SnO_2 catalysts synthesized using roselle flower extract can effectively produce high-quality biodiesel from waste cooking oil. FTIR analysis confirmed the presence of ester groups in biodiesel, with ZnO showing peaks at 1115 cm^{-1} and 1022 cm^{-1}, while SnO_2 exhibited a strong ester carbonyl band at 1744 cm^{-1}, providing clearer evidence of FAME formation. GC–MS analysis revealed that ZnO-catalyzed biodiesel contained 97.34% FAME, dominated by methyl oleate (60.23%) and methyl palmitate (34.83%). In contrast, SnO_2-catalyzed biodiesel achieved nearly 100% FAME, consisting of methyl oleate (55.85%), methyl palmitate (37.12%), with additional contributions from methyl stearate (4.50%) and methyl myristate (0.64%). These results highlight that SnO_2, with a surface area of 51.52 m^2/g, provided significantly higher catalytic performance than ZnO (4.11 m^2/g). Therefore, roselle-extract-based SnO_2 is a promising and eco-friendly heterogeneous catalyst for efficient biodiesel production from waste cooking oil.

References

1. L. R. Kumar, S. K. Yellapu, X. Zhang and R. D. Tyagi, Energy balance for biodiesel production processes using microbial oil and scum. Bioresour. Technol. **272**, 379–388 (2019) https://doi.org/10.1016/j.biortech.2018.10.071
2. D. Singh, D. Sharma, S. L. Soni, S. Sharma, P. Kumar Sharma and A. Jhalani, A review on feedstocks, production processes, and yield for different generations of biodiesel. Fuel. **262**, 116553 (2020) https://doi.org/10.1016/j.fuel.2019.116553
3. B. Panchal, H. M. Kalaji, S. Deshmukh, M. Sharma and W. R. Strobel, Synthesis of methyl esters from silk cotton tree seed kernel oil using dimethyl carbonate and KOH catalysis. Eur. J. Sustain. Dev. Res. **2**, 20 (2018) https://doi.org/10.20897/ejosdr/84899
4. V. Gadore, S. R. Mishra, N. Yadav, G. Yadav and M. Ahmaruzzaman, Metal oxide-based heterogeneous catalysts for biodiesel production. Next Sustain. **2**, 100012 (2023) https://doi.org/10.1016/j.nxsust.2023.100012
5. D.F. Dall'Oglio, M.A.S. Garcia, J.L. Fiorio, W.C. de Abreu, L.N.S. Pereira, A. Braga, E.M. de Moura, A. Guldhe, F. Bux and C.V.R. de Moura, Reusable heterogeneous SnO_2/ZnO catalyst for biodiesel production from acidified/acid oils. J. Braz. Chem. Soc. **32**, 182–193 (2021) https://dx.doi.org/10.21577/0103-5053.20200167
6. C.A. Soto-Robles, P.A. Luque, C.M. Gómez-Gutiérrez, O. Nava, A.R. Vilchis-Nestor, E. Lugo-Medina, R. Ranjithkumar and A. Castro-Beltrán, Study on the effect of the

concentration of Hibiscus sabdariffa extract on the green synthesis of ZnO nanoparticles. Results Phys. **15**, 102807 (2019) https://doi.org/10.1016/j.rinp.2019.102807
7. M. O. Guerrero-Pérez and G. S. Patience, Experimental methods in chemical engineering : Fourier transform infrared spectroscopy—FTIR. Can. J. Chem. Eng. **98**, 25–33 (2020) https://doi.org/10.1002/cjce.23664
8. W. Zhang, F. Liu, F. Liu, C. Huang, H. Zheng, Q. Zhang, Y. Zheng and J. Gao, Microstructural evolution and cracking behavior of Hastelloy X superalloy fabricated by laser directed energy deposition. J. Alloys Comp. **5**, 164179 (2022) https://doi.org/10.1016/j.jallcom.2022.164179
9. E. Q. González, E. L. Medina, R. V. Q. Robles, H. E. G. Gálvez, Y. A. B. Lopez, E. V. Viveros, F. A. Molina, A. R. V. Nestor, P. A. L. Morales. A study of the optical and structural properties of SnO_2 nanoparticles synthesized with Tilia cordata applied in methylene blue degradation. Symmetry, **14(11)**, 2231 (2022). https://doi.org/10.3390/sym14112231
10. S. Fatimah, R. Ragadhita, D. F. Al Husaeni, and A. B. D. Nandiyanto, How to calculate crystallite size from X-ray diffraction (XRD) using Scherrer method. ASEAN J. Sci. Eng. **2**, 65–76 (2022) https://doi.org/10.17509/ajse.v2i1.37647
11. N. Aggarwal, H. Kaur, N. Kumar, J. Kaur, S. Malhotra, A. Sharma, S. Tripathi, R. S. Panwar, P. Patial. Evaluation of photocatalytic efficacy of biosynthesized tetragonal SnO_2 nanoparticles. Results Chem., **5**, 100803 (2023). https://doi.org/10.1016/j.rechem.2023.100803
12. Y. Sivarajan, K. A. Ishak, J.S. Yaacob. Green synthesis of tin oxide (SnO_2) nanoparticles using ginger extracts for photocatalytic degradation of organic dyes in wastewater. Appl. Biochem. Biotechnol. **197(9)**, 5668–5693 (2025) https://doi.org/10.1007/s12010-025-05293-2
13. D. Aviliani, F. I. P. Sari, and R. G. Mahardika, The effectiveness of Pandan sea (Pandanus tectorius) cellulose active charcoal in reducing BOD, COD, TSS levels in liquid waste tapioca. Stannum J. Sains dan Terap. Kim. **4**, 52–59 (2022). https://doi.org/10.33019/jstk.v4i2.3326
14. M. Thommes, K. Kaneko, A.V. Neimark, J.P. Olivier, F. Rodriguez-Reinoso, J. Rouquerol and K.S.W. Sing, Physisorption of gases, with special reference to the evaluation of surface area and pore size distribution (IUPAC technical report). Pure App. Chem. **87**, 1051-1069 (2015). https://doi.org/10.1515/pac-2014-1117
15. J.I. Goldstein, D.E. Newbury, J.R. Michael, N.W.M. Ritchie, J.H.J. Scott and D.C. Joy, Scanning electron microscopy and X-ray microanalysis (Spinger, 2018)

Characterization of Pixell Arabica and Robusta coffee varieties: physicochemical properties and antioxidant

Tantry Eko Putri Mariastuty[1], *Kokom* Komariyah[2*], *Wahyu Eko* Widodo[3], *Astuti*[1], *Gigih* Atmaji[3], *Wahyu Bahari* Setianto[1], *Nenie* Yustiningsih[3], *Edi Priyo* Pramono[1], *Henky* Henanto[3], *Ni* Wayan T[4], *Andini* Sundowo[5], *Siti* Machmudah[6], and *Lully Natharina* Prasetyani[1].

[1]Research Center for Process Technology, Research Organization for Energy and Manufacture, National Research and Innovation Agency, Indonesia
[2]Research Centre for Behavioral and Circural Economics, Reasearch Organization for Governancse, Economy, and Public Welfare, National Research and Innovation Agency, Indonesia
[3]Research Center for Equipment Manufacturing Technology, Research Organization for Energy and Manufacture, National Research and Innovation Agency, Indonesia
[4]Research Center for Process Food Technology and Processing, Research Organization for Agriculture and Food, National Research and Innovation Agency, Indonesia
[5]Center for Research on Raw Materials for Medicines and Traditional Medicines, National Research and Innovation Agency, Indonesia
[6]Chemical Engineering Department, Sepuluh November Institute of Technology, Indonesia

Abstract. Coffee is one of the most widely consumed beverages globally, with Arabica (Coffee arabica) and Robusta (Coffee canephora) being the two main commercial species. This research aims to investigate the physicochemical characteristics of Pixell Arabica and Robusta coffee varieties by analyzing their physicochemical profile. The results indicated that Pixell Arabica coffee exhibited lighter color values, lower moisture and ash contents, softer texture, lower fat levels, and reduced friction values compared to Robusta. Pixell Arabica showed a reduced total fat content (11.57%) and higher fracturability, but Pixell Robusta displayed an elevated lipid content (17.26%), increased hardness, and heightened static friction, especially on aluminum surfaces (0.42 ± 0.26). Additionally, Arabica samples showed higher antioxidant activity and total phenolic content than Robusta samples. Bioactive compounds were extracted by traditional and assisted techniques, while antioxidant activity was measured using the DPPH assay, with measurements of total phenolic and flavonoid content. The Soxhlet procedure yielded the highest antioxidant content (98.199 mg ascorbate/g sample), followed by maceration (75.619 mg/g) and ultrasonic extraction at 30 °C (48.212 mg/g), and ultrasonic extraction at 50 °C yielded the least (47.656 mg/g). These findings offer significant insights into the selection of coffee types and extraction methods to enhance the recovery of bioactive compounds and processing efficiency

* Corresponding author: koko001@brin.go.id

1 Introduction

Coffee ranks as one of the most extensively used beverages worldwide, possessing considerable economic and cultural significance. The two most commercially significant species, Coffee arabica (Arabica) and Coffee canephora (Robusta), exhibit notable differences in their chemical makeup, encompassing moisture, ash, and bioactive component levels. Green coffee beans are also called unroasted coffee beans. The chemical composition of green coffee beans plays a vital role in determining the final product's aroma [1] Green coffee beans have been reported to exhibit numerous biological functions, such as obesity reduction, type II diabetes prevention, mitigation of inflammation, enhancement of cardiovascular parameters, reduced risk of chronic hepatic disorders, as well as neuroprotective, antitumor, antioxidant, anti-inflammatory, and antimicrobial effects [2]

In recent years, advancements in selective breeding and cultivation practices have resulted in the development of localized coffee varieties such as Pixell Arabica and Pixell Robusta. These variants are cultivated in specific agroecological regions with the goal of enhancing both yield and cup quality. However, limited scientific data are available regarding their physicochemical characteristics and functional properties, particularly their antioxidant activity, which is closely linked to consumer health benefits [3]. Meanwhile, the antioxidant capacity of coffee, primarily derived from phenolic compounds such as chlorogenic acids, has been associated with potential protective effects against oxidative stress-related diseases [4].

Extraction methods significantly influence the yield and content of bioactive chemicals from coffee and its by-products. Studies have shown that solve choice, extraction time, and temperature are critical factors affecting the rcovery of phenolic compounds and antioxidants from spent coffee grounds. For coffee husks, a mixture of water and ethanol (1 :1) using coventional water bath extraction proved most effective for otaining bioactive compounds, particularly chlorogenic acid [5]. In green coffee beans, soxhlet and ultrasound-assisted extractions yield the highest bioactive compound content, with chlorogenic acid being the most abundant [6].

This study focused on clarifying the correlation between extraction type, extraction conditions, and the end product extracts, as well as to ascertain the physical properties of the obtained green coffee bean extracts.

2 Methods

The tools used include erlenmeyer, beaker glass, funnel, soxhlet, ultrasonic cell crusher biobase Elisa Biotek Synergy HTX Multi-Mode Reader at a wavelength of 750 nm. The used compounds are the 2,2-diphenyl-1-picrylhydrazyl (DPPH) and methanol, both sourced from Sigma-Aldrich. The ingredients include two varieties of Arabica and Robusta green coffee bean samples gathered from the Temanggung district of Central Java during May and June of 2024.

2.1 Sample preparations

The green bean pulverized with a grinder to a size of 40 mesh. Samples were stored in aluminum foil valves at ambient temperature until analysis. Extraction methods generally fall into two main categories: maceration and Soxhlet extraction, which are examples of conventional methods, and water bath and ultrasound-assisted extraction procedures use water (100%) with a sample-to-water ratio of 1:10.

2.2 Determination Physicochemical Properties

2.2.1 Proximate

A small sample was dried in an oven at 103°C for 24 hours for determining the moisture content values. The proximate composition of roasted beans was analyzed by Indonesia Standar National (SNI) [7]. The material was burned at 550°C in an electric furnace to determine the ash content.The crude fat content was measured use the Soxhlet and N-hexane extraction solvent.

2.2.2 Co-efficient of static friction

To determine the coefficient of friction, a sample was placed into a PVC cylinder (diameter 10 cm, height 5 cm) positioned on the friction surface. To keep the barrel from coming into contact with the friction surface, the PVC cylinder was then raised by 3 mm. The friction surface was continuously raised until the barrel began to descend. The angle (θ) between the elevated friction surface and the horizontal plane is referred to as the static angle of friction between the sample and the friction surface. Static friction was measured using four different material surface types: 1 mm stainless steel, 1 mm aluminum, 3 mm acrylic, and 18 mm plywood.

The coefficient of static friction obtained from the equation (1) [8] [9].

$$\mu = \tan \theta \qquad (1)$$

2.2.3 Quantitative Measurement of the Bean Weight and Dimensions

The weight of the green beans was measured using a digital scale Mettler Toledo. Dimensions in the main axes were measured using digital calliper. According to indications dimensions correspond to L (length), W (width), and D (depth) were measured and their averages taken from 100 beans each selected randomly [10].

2.2.4 Determination of color

The color of the roasted coffee was determined using an Agera Spectrophotometer/Colorimeter (Hunter Lab), it provided direct measurements of the index values L* (brightness), a* (red or green intensity), and b* (yellow or blue intensity). The hue (h*) and chroma (c*) values are calculated from the values of a* and b*[11]

$$h* = arctan \frac{b*}{a*} \qquad (2)$$

$$c* = \sqrt{a* + b*} \qquad (3)$$

2.2.5 Measurement of Hardness and fracturability

The hardness texture was measured using the TA X Plus Texture Analyser from Stable Micro Systems. The compression test mode, test speed of 0.1 mm/sec, distance of 2.5 mm, and probe type P/5 were the TA's operational settings. Samples were randomly chosen and positioned horizontally on the platform of the texture analyzer instrument. The peak force recorded during the test before the sample's fracture has been designated as the sample's fracture force. Deformation is determined by the fracture force. The values of ten samples were recorded [8].

2.2.6 Determination of antioxidant activity

Sample preparation and solvent extraction. 0.5 grams of ground coffee is put into a 50 ml volumetric flask then add methanol, followed by an addition of methanol to the required mark. The stock solution is concentrated to 1 mg/mL, equivalent to 1000 ppm. Prepare a dilution series of 10, 20, 40, 60, 80, and 100 ppm in 10 mL volumetric flasks, dissolving with methanol.
Sample Testing Place the samples in a microplate. Add 25 μL of sample from each dilution with 150 mM DPPH. Incubate in a dark conditions for 30 minutes, then assess the absorbance using an ELISA Biotek SYNERGY HTX Multi-Mode Reader at a wavelength of 517 nm [12]

$$\% \text{ Inhibition} = (A0 - A) / A0 \times 100\% \quad (4)$$

Note:

- A0 = absorbance of control
- A = absorbance of sample

3 Results and discussion

3.1 Physicochemical Properties

3.1.1 Proximate

The moisture content of Pixell coffee beans is essential, as it affects the flavor, aroma, and overall quality of the coffee. Nevertheless, the moisture levels of the Arabica and Robusta samples exhibited minimal variation. The moisture level of the Robusta samples ranged from 8.45% to 8.54%, while the Arabica samples presented a range of 8.65% to 8.76%. The moisture contents of the Arabica and Robusta green beans met the Indonesian standard (SNI 01-2907-2008), which requires a maximum moisture content of 12% on green beans. Table 1 presents the proximate composition of Pixell Arabica and Robusta green coffee beans.

Table 1. Proximate composition of Pixell Arabica and Robusta green coffee beans

Parameter	Pixell Arabica	Pixell Robusta
Moisture content (%)	8.65 – 8.76	8.45 – 8.54
Ash content (%)	5.34 – 5.37	4.79 – 4.83
Total fat content (%)	11.57	17.26

Measurements of the ash contents of Arabica and Robusta green coffee beans revealed that the Arabica samples had an ash content ranging from 5.34% to 5.37%, while the Robusta samples had an ash level among 4.79% to 4.83%. It's possible that a high ash concentration signifies a high mineral content [1]

The total fat content of coffee beans may affect the overall quality and flavor of coffee. Lipids enrich the aroma and flavor profile of coffee. Therefore, a higher lipid content might produce a more intense and complex aroma and flavor. However, higher lipid content raises susceptibility to rancidity, reducing the flavor and aroma of coffee. Pixels of Arabica showed the lowest total fat levels at 11.57%, while Robusta showed the highest total fat content at 17.26%.

Robusta coffee beans are known for their robust and strong flavor, mainly attributed to their higher lipid content. The results show that the amount of lipids can change a lot between different types of coffee and can be affected by things like how they are grown and processed, including how long they are dried [1, 13].

3.1.2 Co-efficient of static friction

Experimental findings for various agricultural products show that the coefficient of internal friction of a material exceeds its coefficient of friction against common construction materials. For a non-free-flowing substance like parchment coffee, it is expected that the material will move smoothly over other surfaces (e.g., stainless steel, glass, aluminum), resulting in a smoother contact surface. The coefficient of friction decreases significantly due to free or surface moisture from the parchment coffee. Subsequently, it is only slightly influenced by further reductions in moisture content [9].

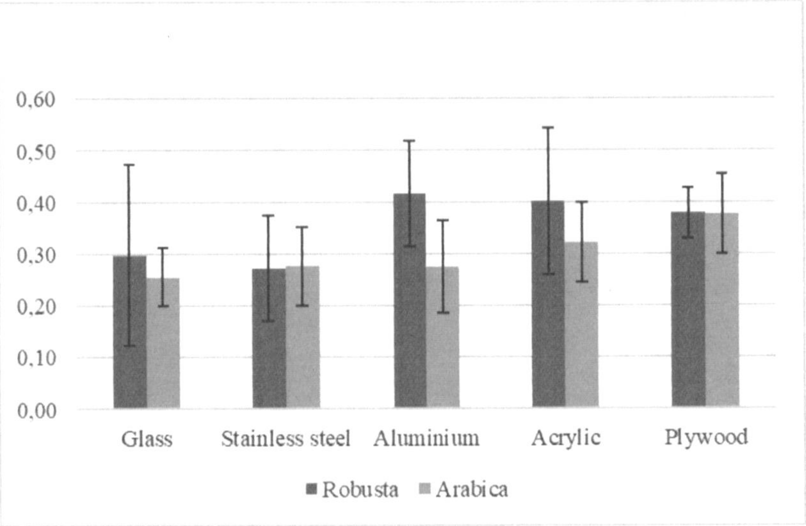

Fig. 1. Co-efficient of static friction of Pixel Robusta and Arabica

Figure 1 shows the strongest static friction for pixel Robusta was found on aluminum (0.42±0.26) and on plywood for Arabica (0.38±0.08); this friction coefficient helps in designing agricultural equipment, bins, silos, and other storage and handling structures [8].

3.1.3 Weight and Dimensions

The mean values of experimentally found weights (g), dimensions (mm) along main bean axes are listed in Table 2. Mean values represent the average of 100 individual measurements in each lot of coffee beans. As can be seen in Table 1, the dimensions of Pixell Robusta coffee were found to be larger than those of Arabica. This is consistent with the observation that the static coefficient of friction for Robusta is higher than that of Arabica.

Table 2. Weight and dimensions of Pixel Robusta and Arabica

Coffe var.	Massa (g)	Length (mm)	Width (mm)	Depth (mm)
Robusta	0.226±0.063	8.229±0.971	5.587±1.026	7.909±1.303
Arabica	0.149±0.036	7.404±1.056	4.699±1.200	7.408±1.144

The geometric diameter correlated with the length, width, and depth features. The surface area correlated with length, width, depth, and geometric diameter. Sphericity has been correlated with length, geometric diameter, and surface area. Porosity correlated with particle density, bulk density, and sphericity. The aspect ratio correlates with the length feature, dimension, geometric diameter, surface area, and sphericity [8].

3.1.2 Color Analysis

Regarding the L* indexes values observed between the two varieties were similar for both Robusta and Arabica coffees. Arabica samples showed a significantly lighter color compared with Robusta. Robusta coffee samples showed higher a* and b* values compared to Arabica coffee. The index values of h* at Robusta are lower than those at Arabica, revealing that Robusta coffee samples present higher chroma values compared with Arabica. The data are presented in Table 3.

Table 3. Color analysis of Pixel Robusta and Arabica.

Coffee var.	L*	a*	b*	Hue (h*)	Chroma indexes (c*)
Robusta	43.832	7.722	18.840	1.182	13.281
Arabica	44.198	3.953	18.310	1.358	11.132

The color of coffee beans ranges from light brown to dark brown as a result of the pyrolysis of organic compounds and the formation of melanoidins [11]

3.1.4 Hardness and fracturability

The hardness of Pixell Robusta coffee exhibited a significantly higher value compared to Arabica. Conversely, its fracturability showed a lower value than that of Pixell Arabica coffee. This is related to the larger dimensions of Pixell Robusta coffee compared to Arabica. Fracturability is a characteristic attribute of materials that show a high degree of hardness combined with a low degree of cohesiveness. The design for equipment, such as the crusher, requires the use of fracture force. The fracture force will determine the maximum strength that could be applied to a material without compromising the structural integrity of the processed product [8]. The minimum, maximum, mean, and standard deviation values for hardness and fracturability are presented in Table 4.

Table 4. Texture analysis Hardness and Fracturability

Coffee var.	Minimum (kg-force)	Maximum (kg-force)	Mean (kg-force)	Std. Deviation
Robusta				
Hardness	30.652	33.288	31.929	1.319
Fracturability	5.220	7.359	6.394	1.085
Arabica				
Hardness	12.356	12.856	12.636	0.255

Fracturability	15.745	19.770	17.388	2.112

3.1.5 Determination of antioxidant activity

Extraction was performed using three methods: (1) maceration with 100% distilled water for 24 hours at room temperature; (2) Soxhlet extraction using hexane as the solvent for 6 hours; and (3) ultrasonic-assisted extraction at time intervals of 5 and 15 minutes with temperature settings of 30°C and 50 °C. Figure 2 presents the data obtained antioxidant activity from the extraction results.

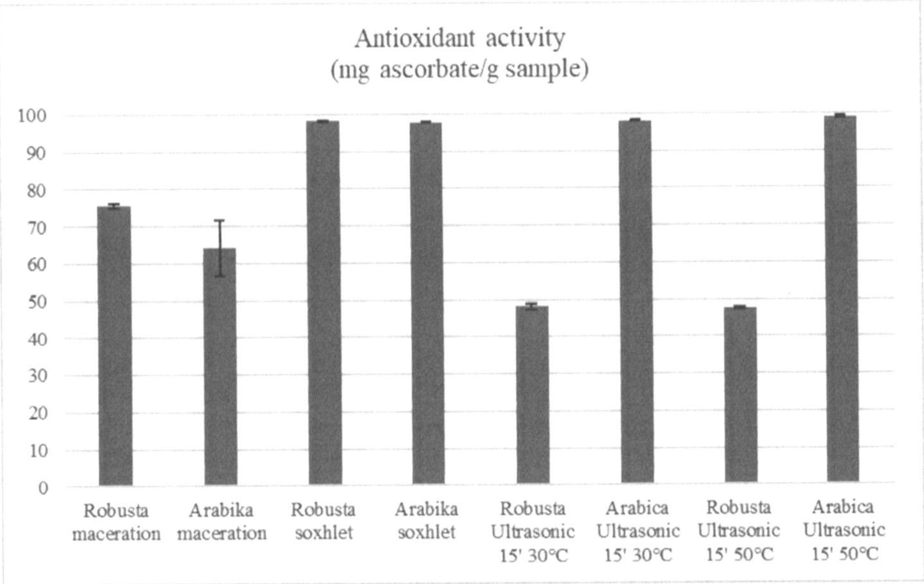

Fig. 2. Results of antioxidan activity

Figure 2 confirmed that using hexane as a solvent improved the extraction of compounds with antioxidant activity, but both the conventional maceration technique with a 1:1 mixture of water and ethanol and the ultrasonic method yielded favorable results; however, there were no significant differences. While evaluating the chemical composition, including total phenolic compounds, total flavonoids, condensed tannins, and the phenolic profile, a significant quantity of bioactive compounds was subsequently identified [11].

The antioxidant extraction process from Pixell Arabica coffee using the ultrasonic method for 15 minutes at 50 °C yielded the highest antioxidant activity (99.054 mg ascorbate/g sample). This was followed by ultrasonic extraction at 30 °C for 15 minutes (98.167 mg ascorbate/g sample, soxhlet extraction (97.843 mg ascorbate / g sample), and finally, maceration (64.281 mg ascorbate / g sample). These results indicate that ultrasonic-assisted extraction, particularly at higher temperatures, is more effective in enhancing the release of antioxidant compounds from coffee beans.

However, a different pattern was observed for Pixell Robusta coffee, where the highest antioxidant yield was obtained through the soxhlet method (98.199 mg ascorbate / g sampel), followed by maceration (75.619 mg ascorbate / g sample), and then ultrasonic extraction for 15 minutes at 30 °C (48.212 mg ascorbate / g sample). The lowest antioxidant activity was recorded from ultrasonic extraction for 15 minutes at 50 °C (47.656 mg ascorbate / g sample). The highest antioxidant activity was observed in the extract of pixel coffee Arabica. This coffee extract also showed the highest content of total phenols and

flavonoids. The lowest antioxidant capacity was measured in the extract pixel coffee Robusta [4].
According to Thouri et al. [15], the polarity of the solvent used during extraction significantly affects the level of extracted polyphenols. The most effective extraction of polyphenols takes place with polar solvents due to their superior effectiveness in the solvation process, attributed to the interactions (hydrogen bonds) between the polar sites of the antioxidant compounds and the solvent.
Following implementing the DPPH assay, no substantial differences were noted between the samples treated with extraction conditions of 15 minutes at 30°C and 50°C. However, samples exposed to longer durations and higher temperature values showed a significant decrease in levels [6]. The most abundant phenolic compounds of coffee are CGAs. The main hydrophilic antioxidant compounds in coffee are 3-, 4-, and 5-monocaffeoylquinic acids and 3,4-, 3,5-, and 4,5-diCQAs. Also, little amounts of caffeic acid, ferulic acid, p-coumaric acid, 4-hydroxybenzoic acid, and sinapic acid can be found in coffee [14]. Antioxidant compounds are subject to destruction when subjected to elevated temperatures for extended durations. This happens when antioxidant molecules can't give electrons to radical substances anymore to make them neutral. More total phenols and flavonoids increase the ability of antioxidants to give away electrons, which helps stop free radicals from forming. The main compounds that act as antioxidants are phenolic compounds, or flavonoids [5, 12].

4 Conclusion

This study shows major differences in physicochemical properties and antioxidant activity between Pixell Arabica and Robusta coffee varieties. Pixell Arabica exhibited the lowest total fat content (11.57%), while Pixell Robusta had the highest (17.26%). The static coefficient of friction varied by surface, with Robusta showing the highest value on aluminum (0.42 ± 0.26) and Arabica on plywood (0.38 ± 0.08). Morphological characteristics such as surface area, sphericity, and porosity were found to correlate strongly with dimensional and density-related parameters. Variations in bean color, from light to dark brown, are attributed to the pyrolysis of organic compounds and melanoidin formation, particularly prominent in the larger-sized Robusta beans.
The mechanical properties showed that Robusta had higher hardness, whereas Arabica demonstrated higher fracturability, characteristics essential for equipment design in coffee processing. In terms of antioxidant extraction, the Soxhlet method yielded the highest antioxidant content (98.199 mg ascorbate/g sample), followed by maceration (75.619 mg/g), and ultrasonic extraction at 30 °C (48.212 mg/g). The lowest antioxidant yield was obtained from ultrasonic extraction at 50 °C (47.656 mg/g). These findings provide valuable insights into the selection of coffee varieties and extraction methods for optimizing bioactive compound recovery and processing efficiency.

References

1. Zainuri, et al., The chemical characteristics of Arabica and Robusta green coffee beans from Geopark Rinjani, Indonesia. Biotropia. **30**, 318–328 (2023). https://doi.org/10.11598/BTB.2023.30.3.1940
2. V. Pimpley, S. Patil, K. Srinivasan, N. Desai and P.S. Murthy, The chemistry of chlorogenic acid from green coffee and its role in attenuation of obesity and diabetes. Prep. Biochem. Biotechnol. **50**, 969-978 (2020). https://doi.org/10.1080/10826068.2020.1786699

3. M. Daglia, Polyphenols as antimicrobial agents. Curr. Opin. Biotechnol. **23**, 174–181 (2012). https://doi.org/10.1016/j.copbio.2011.08.007
4. J. Pokorná, P.R. Venskutonis, V. Kraujalyte, P. Kraujalis, P. Dvořák, B. Tremlová, V. Kopřiva and M. Ošťádalová, Comparison of different methods of antioxidant activity evaluation of green and roast C. arabica and C. robusta coffee beans. Acta Aliment. **44**, 454–460 (2015). https://doi.org/10.1556/066.2015.44.0017
5. M.O. Silva, J.N.B. Honfoga, L.L. Medeiros, M.S. Madruga and T.K.A. Bezerra, Obtaining bioactive compounds from the coffee husk (Coffea arabica L.) using different extraction methods. Molecules. **26**, 46 (2021). https://doi.org/10.3390/molecules26010046
6. O. Gligor, S. Clichici, R. Moldovan, D. Muntean, A.-M. Vlase, G.C. Nadăş, I.A. Matei, G.A. Filip, L. Vlase and G. Crişan, The effect of extraction methods on phytochemicals and biological activities of green coffee beans extracts. Plants. **12**, 712 (2023). https://doi.org/10.3390/plants12040712
7. Badan Standardisasi Nasional, SNI 2907-2008 Biji kopi (2008)
8. S.I. Kuala, D.D. Hidayat, C.E.W. Anggara and R. Saparita, Characterization and evaluation of physical and mechanical properties of unhulled Arabica coffee bean. IOP Conf. Ser.: Earth Environ. Sci. **251**, 012040 (2019). https://doi.org/10.1088/1755-1315/251/1/012040
9. B.N. Ghosh, Effect of moisture content on the static coefficient of friction of parchment coffee beans. *J. Agric. Eng. Res.* **13**, 249-253 (1968)
10. L. Severa, J. Buchar and Š. Nedomová, Shape and size variability of roasted Arabica coffee beans. Int. J. Food Prop. **15**, 426–437 (2012). https://doi.org/10.1080/10942912.2010.487967
11. V.V. Freitas, L.L.R. Borges, G.A.D. Castro, M.H. dos Santos, M.C.T.R. Vidigal, S.A. Fernandes and P.C. Stringheta, Impact of different roasting conditions on the chemical composition, antioxidant activities, and color of Coffea canephora and Coffea arabica L. samples. Heliyon. **9**, e19580 (2023). https://doi.org/10.1016/j.heliyon.2023.e19580
12. Astuti, S. Machmudah, S. Winardi, L.P. Manalu, G. Atmaji, W.E. Widodo, L.K. Hartono, K. Komariyah, M.N. Alfa and I. Kurniasari, The effect of roasting Toraja Arabica coffee with a fluidized roaster on the antioxidant and proximate content. IOP Conf. Ser.: Earth Environ. Sci. **1182**, 012076 (2023). https://doi.org/10.1088/1755-1315/1182/1/012076
13. X. Liu, Y. Fei, W. Wang, S. Lei, C. Cheng and Z. Xing, Physicochemical difference of coffee beans with different species, production areas and roasting degrees. Bever. Plant Res. **2**, 7 (2022) https://doi.org/10.48130/BPR-2022-0007
14. M.C. Cid and M.P. de Peña, Coffee : Analysis and composition. In Encyclopedia of Food and Health (Elsevier Inc., 2015), 225–231 https://doi.org/10.1016/B978-0-12-384947-2.00185-9
15. A. Thouri, H. Chahdoura, A. El Arem, A. O. Hichri, R. B. Hassin and L. Achour, Effect of solvents extraction on phytochemical components and biological activities of Tunisian date seeds (var. Korkobbi and Arechti). BMC Complement. Altern. Med. **17**, 248 (2017). https://doi.org/10.1186/s12906-017-1751-y

The effect of volume fraction on hybrid composites with fiber and trunk powder coconut filler

Cindia Ramadhan[1], *Zulnazri* Zulnazri[2], and *Ahmad* Fikri[2*]

[1]Material Enginnering, Universitas Malikussaleh, 24355 Aceh, Indonesia
[2] CoE TechnoPlast, Universitas Malikussaleh, 24355 Aceh, Indonesia

Abstract. Hybrid composites are composites composed of multiple matrices and fibers, where the types of fibers include both aligned and randomly oriented fibers. The objective of this study is to investigate the effect of volume fraction variations on the mechanical properties of composites reinforced with a mixture of coconut fiber and coconut trunk powder, in order to obtain a hybrid composite with optimal mechanical characteristics.The materials used for the fabrication of the hybrid composite include coconut fiber, coconut trunk powder, polyester resin, and catalyst. The volume fraction variations of fiber and powder were set at 25%:5%, 20%:10%, 15%:15%, 10%:20%, and 5%:25%, with a constant resin content of 70%. Specimen fabrication and testing procedures followed ASTM D3039 for tensile testing and ASTM D256-03 for impact testing. The tests conducted in this study include tensile test, impact test, and Scanning Electron Microscopy (SEM) analysis. The results showed that the highest impact and tensile strength wre 0.0087 J/mm^2 and 14,06 MPa for the composition of 70% resin, 25% fiber, and 5% powder. The morphology of the hybrid composite shows a closed surface structure and a fiber diameter of 10 μm. The bonding between the filler and the fiber is formed.

1 Introduction

Composite materials have recently a shift from the use of synthetic fiber reinforcements toward natural fiber-reinforced composites. The demand for environmentally friendly and recyclable composite materials is increasing [1]. One of the most promising composite materials for industrial applications is a composite reinforced with fillers, either in the form of natural or synthetic fibers [2]. A composite is a material formed by combining two or more constituents through a non-homogeneous mixing process, resulting in mechanical properties that are superior to those of the individual components [3]. Composites consist of two main

* Corresponding author: ahmadfikri@unimal.ac.id

parts: the matrix, which functions as the binder or protective phase, and the filler, which serves as the reinforcement [4].

Natural and synthetic fiber-reinforced composites are among the most extensively researched and developed types of composites, considered as potential alternatives to metals. To foster innovation in composite materials, advancements in advanced materials—such as hybrid composites—are essential. Hybrid composites are materials composed of multiple types of matrices and fibers, including both aligned and randomly oriented fibers [5]. In this study, the fibers used for hybrid composite fabrication are natural fibers, specifically coconut coir fiber and coconut trunk powder as fillers.

Coconut fiber is a lignocellulosic fiber that can serve as an alternative raw material due to its abundance and favorable mechanical properties [6]. It is durable, highly resistant to abrasion, and not easily broken, making it a promising natural filler for composite materials [7, 8]. In addition to its easy availability, coconut coir fiber is composed of approximately 44% cellulose, 45% lignin, 3% pectin and associated compounds, and 5% moisture [9]. The high lignin content contributes to its hardness and stiffness. Coconut trunk powder is a natural powder used as a filler in hybrid composites, consisting of organic compounds such as cellulose, hemicellulose, lignin, pentosans, and others [10]. The use of coconut coir fiber and coconut trunk powder offers a viable alternative as fillers in hybrid composites.

This study aims to evaluate the effect of adding coconut coir fiber and coconut trunk powder as fillers in hybrid composite materials. The influence of these fillers is assessed through impact and tensile testing. Additionally, morphological characterization using Scanning Electron Microscopy (SEM) is conducted to examine the bonding between the matrix and the filler.

Method

The hybrid composite was carried out by mixing resin polyester, catalyst, fiber, and coconut shell powder according to the specified volume fraction variations. The mixture was stirred thoroughly and poured into a mold that had been pre-lubricated with wax gel. The mixture was then pressed to ensure uniform distribution of all components and left to dry. The volume fraction variations used were: 70% resin with combinations of fiber and coconut shell powder as follows — 5% + 25% (KS5B25), 10% + 20% (KS10B20), 15% + 15% (KS15B15), 20% + 10% (KS20B10), and 25% + 5% (KS25B5). Characterization of the hybrid composite was conducted through impact testing, tensile strength testing, and Scanning Electron Microscopy (SEM) analysis to evaluate its mechanical and morphological properties. Testing procedures followed ASTM D3039 for tensile testing and ASTM D256-03 for impact testing.

Result and Discussion

3.1 Impact strength analysis

Impact strength analysis was carried out to determine the resistance of the hybrid composite to sudden impact or shock loads [11]. The results of this analysis indicate the toughness of the hybrid composite material. Table 1 presents the energy absorption and impact strength of hybrid composites reinforced with coconut fiber and coconut shell powder. Figure1A and 1B show the increase in absorbed energy and impact strength in the hybrid composite material.

Table 1. Energy Absorbed and Impact Strength of hybrid composite

Sample	Angle of Pendulum (°)		Energy Absorbed (J)	Impact Strength (J/mm2)
	Start	Finish		
KS5B25	160	155	0,317	0,0024
KS10B20	160	154	0,388	0,003
KS15B15	160	152	0,539	0,0042
KS20B10	160	151	0,618	0,0048
KS25B5	160	150	0,699	0,0055

The energy absorbed by the hybrid composite material increased with the addition of coconut fiber and the reduction of coconut shell powder content. The lowest absorbed energy was recorded for sample KS5B25, at 0.317 J, while the highest was observed in sample KS25B5, at 0.699 J. A similar trend was observed in the impact strength of the hybrid composite materials. The lowest impact strength was exhibited by sample KS5B25, at 0.0024 J/mm^2, whereas the highest impact strength was observed in sample KS25B5, at 0.0055 J/mm^2.

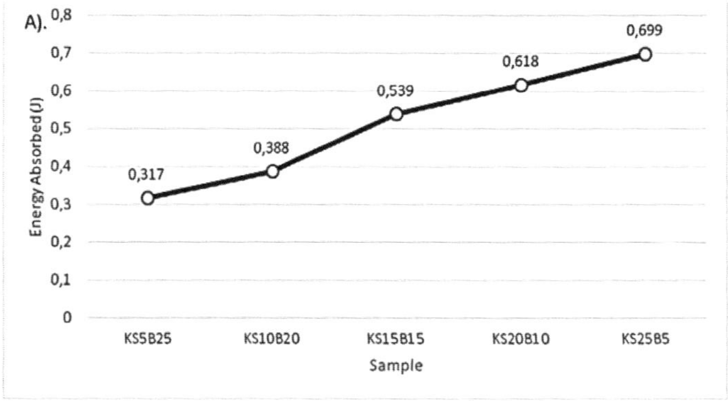

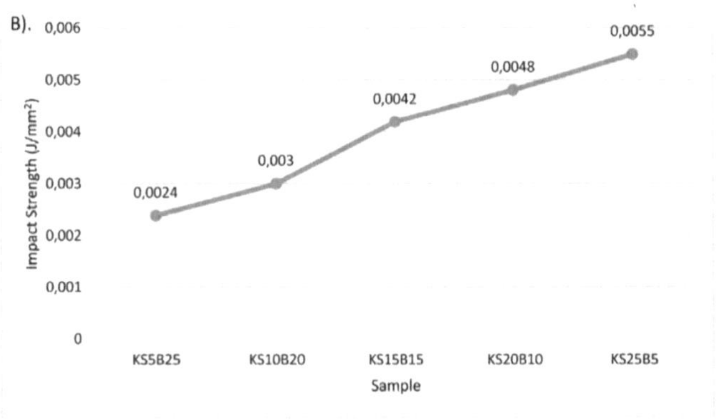

Fig. 1. Energy abLsorbed and impact strength showing of hybrid composite

The addition of coconut fiber in the hybrid composite significantly influenced the increase in impact strength. Fiber length plays a crucial role in the hybrid composite material [12]. During the impact test, the energy from the pendulum is transmitted through the filler in the form of coconut fiber. A higher content of fiber filler enhances the material's ability to absorb energy, thereby increasing its impact strength.

3.2 Tensile strength analysis

Tensile strength analysis was conducted to evaluate the behavior of the hybrid composite material under tensile loading [13]. The analysis provided data on maximum load, tensile strength, elongation, and elastic modulus, as presented in Figure 2. The hybrid composite was subjected to loading perpendicular to the sample surface. Table 2 further summarizes the maximum load, tensile strength, elongation, and elastic modulus of the hybrid composites.

Table 2. Tensile strength, elongation and modulus elasticity of hybrid composite

Sample	Max Force (kgN)	Tensile Strength (MPa)	Elongation (%)	Modulus Elastisity (MPa)
KS5B25	861,71	10,25	1,08	949,07
KS10B20	1.007,99	12,14	1,66	731,32
KS15B15	1.020,26	12,89	1,48	870,94
KS20B10	1.141,11	13,58	1,55	876,12
KS25B5	1.181,79	14,06	0,9	1.420,20

The data on maximum load and tensile strength exhibit similar trends. Both values increased with the addition of coconut stem fibers and a reduction in coconut shell powder as the hybrid composite filler. The lowest maximum load was recorded for sample KS5B25 at 861.71 kgN, while the highest was for KS25B5 at 1,181.79 kgN. The lowest tensile strength was found in sample KS5B25 at 10.25 MPa, and the highest in KS25B5 at 14.06 MPa.

The elongation of the hybrid composites exhibited a different trend compared to maximum load and tensile strength. Elongation tended to decrease as the maximum load increased. Distinct elongation trends were observed in samples KS5B25 and KS20B10, with elongation values of 1.08% and 1.55%, respectively. Additionally, the elastic modulus initially showed a decreasing trend, followed by an increase in line with the rising maximum load applied to the samples. A different pattern was observed for sample KS5B25, which had an elastic modulus of 949.07 MPa.

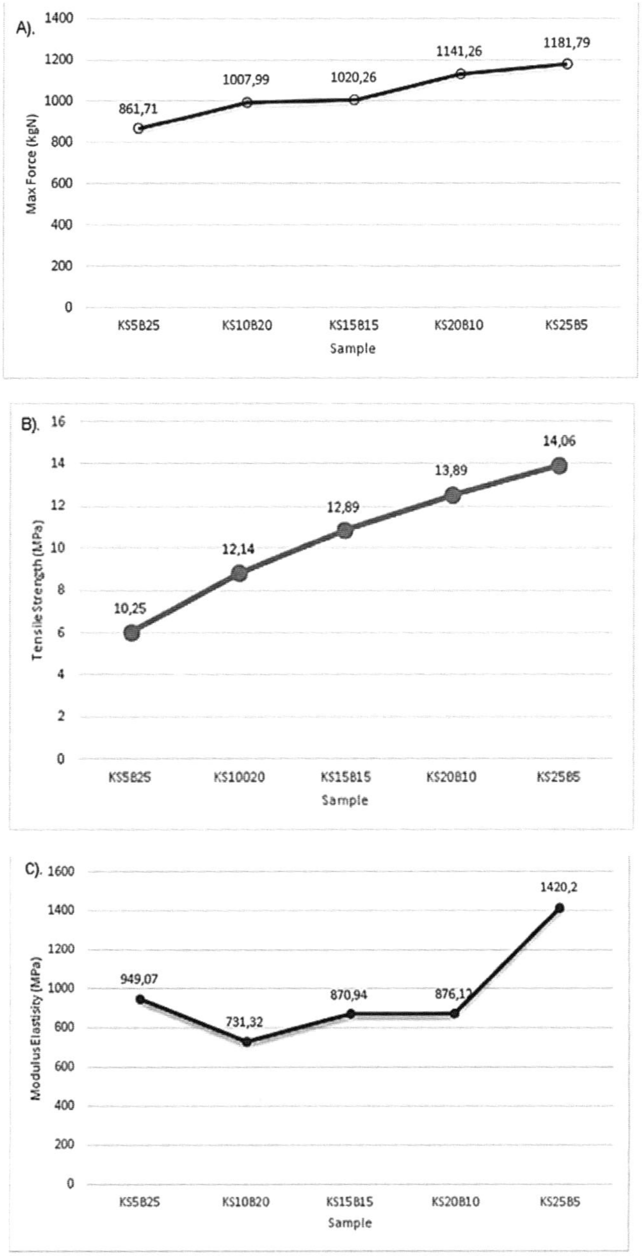

Fig. 2. Max force (A), tensile strength (B), and modulus elasticity (C) of hybrid composite

The increase in maximum load and tensile strength of the hybrid composite indicates strong interfacial adhesion between the filler and the matrix [14]. When tensile loads are applied perpendicularly, the stress is distributed through the polyester resin matrix and the fillers, which include coconut fibers and coconut shell powder. Enhanced interfacial bonding between matrix and filler leads to an increase in internal stress distribution, resulting in higher maximum load and tensile strength values.

The differing elongation behavior in samples KS5B25 and KS20B10 suggests non-uniformity in filler dimensions [15]. Filler geometry affects the stiffness of the material, as reflected in variations in elongation. The individual particles of the coconut shell powder and fibers contribute to the particle–matrix interaction. Furthermore, the variation in elastic modulus observed in sample KS5B25 may be attributed to poor interfacial bonding between the particles and the matrix, resulting in less efficient stress transfer across the interface [15].

3.3 Morphological analysis

Figure 3 shows the SEM results illustrating the surface morphology of the KS25B5 sample. The KS25B5 sample exhibits the highest energy absorption, impact strength, tensile strength, tensile resistance, and elastic modulus. Scanning Electron Microscopy (SEM) was employed to observe the surface morphology of the hybrid composite material [14]. The characterization was conducted using secondary electrons at an accelerating voltage of 15 kV with a magnification of ×23.

The SEM results reveal a relatively closed surface of the hybrid composite, with the presence of fibers, coconut shell powder, and pores on the surface. The fibers appear as long filaments with a diameter of approximately 10 µm. Similarly, the coconut shell powder shows particle dimensions of around 10 µm. The fiber diameter was determined by comparing the fiber diameter with the reference line (white line) located in the lower right corner of Figure 3. Some bonding between the matrix and the filler is observed, as indicated by the absence of voids at the matrix–filler interface [14].

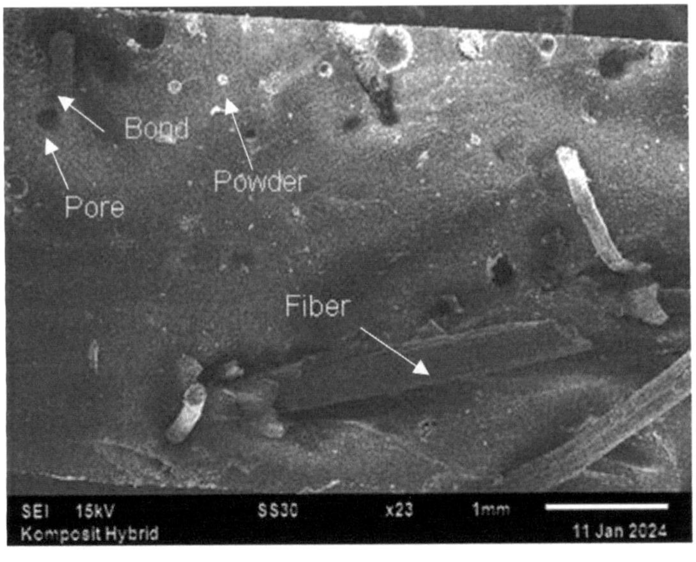

Fig. 3. Scanning electron microscopy (SEM) showing surface of hybrid composite

Conclusion

The variation in volume fractions of coconut fiber and coconut trunk powder significantly affects the mechanical properties of the polyester hybrid composite. The highest impact strength and tensile strength were observed at a volume fraction of 25% fiber and 5% powder. As the volume fraction of the powder increases, both impact strength and tensile strength tend to decrease. Hybrid composite indicates strong interfacial adhesion between the filler and the matrix

References

1. Mitra BC. Environment friendly composite materials: Biocomposites and green composites. Def. Sci. J. **64,** 244–261 (2014). https://doi.org/10.14429/dsj.64.7323
2. K.N. Keya, N.A. Kona, F.A. Koly, K.M. Maraz, Md. N. Islam. Natural fiber reinforced polymer composites: history, types, advantages, and applications. Mater. Eng. Res. **1**, 69–87 (2019). https://doi.org/10.25082/MER.2019.02.006
3. Y. Chen, F. Liou. Additive Manufacturing of Metal Functionally Graded Materials: A Review, in Proceedings of the 29th Annual International Solid Freeform Fabrication Symposium. Austin, TX-USA, August 13-15 (2018)
4. A.K. Mazitova, I.I. Zaripov, G.K. Aminova, M.V. Ovod, N.L. Suntsova. Fillers for polymer composite materials. Nanotech. Construction. **14,** 294–299 (2022). https://doi.org/10.15828/2075-8545-2022-14-4-294-299
5. M. Mohammed, J.K. Oleiwi, A.M. Mohammed, A.J.M. Jawad, A.F. O.T. Adam, B.O. Betar, S.C.B. Gopinath. A Review on the Advancement of Renewable Natural Fiber Hybrid Composites: Prospects, Challenges, and Industrial Applications. J. Renew. Mater. **12**, 1237–1290 (2024). https://doi.org/10.32604/jrm.2024.051201
6. F.Vieira, H. E. P. Santana, M. Jesus, F. Mata, P.Pires, M.Vaz-Velho, D. P. Silva, D. S. Ruzene. Comparative Study of Pretreatments on Coconut Fiber for Efficient Isolation of Lignocellulosic Fractions. Sustain. **16,** 4784 (2024). https://doi.org/10.3390/su16114784
7. M. Milosevic, P. Valášek, A. Ruggiero. Tribology of natural fibers composite materials: An overview. Lubricants. **8**, 42 (2020). https://doi.org/10.3390/lubricants8040042
8. F. R. B. Martinelli, F. R .C. Ribeiro, M. T. Marvila, S. N. Monteiro, F.da. C. G. Filho, A. R. G. de. Azevedo. A Review of the Use of Coconut Fiber in Cement Composites. Polymers. **15**, 1309 (2023). https://doi.org/10.3390/polym15051309
9. K. M. F. Hasan, P. G. Horváth, M. Baka, T. Alpár. A state-of-the-art review on coir fiber-reinforced biocomposites. RSC Adv. **11**, 10548-10571 (2021). https://doi.org/10.1039/D1RA00231G
10. N. Sienkiewicz, M. Dominic, J. Parameswaranpillai. Natural Fillers as Potential Modifying Agents for Epoxy Composition: A Review. Polymers. **14**, 265 (2022). https://doi.org/10.3390/polym14020265
11. C. Stephen, B. Shivamurthy, A.-H. I. Mourad, R. Selvam, M. Mohan. Experimental and finite element study on high-velocity impact resistance and energy absorption of hybrid and non-hybrid fabric reinforced polymer composites. J. Mater. Res. Technol. **18**, 5406–5418 (2022). https://doi.org/10.1016/j.jmrt.2022.05.007

12. R. Muthuraj, M. Misra, F. Defersha, A. K. Mohanty. Influence of processing parameters on the impact strength of biocomposites: A statistical approach. Compos Part A Appl. Sci. Manuf. **83**, 120–129 (2016) https://doi.org/10.1016/j.compositesa.2015.09.003
13. R.T. D. Prabhakaran, T. L. Andersen, C.M. Markusse, B. Madsen, H. Lilholt. General rights Tensile and compression properties of hybrid composites-A comparative study. In Proceedings of the 19th International Conference on Composite Materials (iccm19), Canada, 28 Jul 2013-2 Aug (2013), 1029-1035
14. A. Lekrine, A. Belaadi, I. Dembri, M. Jawaid, A. S. Ismail, D. Ghernaout. Fiber treatment impact on the thermal behavior of biomass/palm-fibers polylactic-acid hybrid biocomposites. Mater. Chem. Phys. **338**, 130651 (2025). https://doi.org/10.1016/j.matchemphys.2025.130651
15. R.E. Njokua, A.E. Okona, T.C. Ikpakia. Effect of Variation of Particle Size and Weight Fraction on The Tensile Strength and Modulus of Periwingkle Shell Reinforced Polyester Composite. Niger. J. Technol. **30,** 87-93 (2011) https://doi.org/10.4314/njt.302.32

Antioxidant assay and component identification using lc-hrms of ecoenzyme products from *sandoricum koetjape* and *pometia pinnata* fruit peel waste

Hepy Terima Telaumbanua[1], *Joy Sisokhi Putra* Lase[1], *Chatrina Yohana* Sihombing[1], and *Candra Yulius* Tahya[1,2*]

[1]Biology Education Study Program, Faculty of Education, Universitas Pelita Harapan, Tangerang, Indonesia
[2]D4 Medical Laboratory Technology Program, Faculty of Health Sciences, Universitas Pelita Harapan, Tangerang, Indonesia

Abstract. Kecapi (Sandoricum koetjape) and matoa (Pometia pinnata) fruit peel, which are often considered as waste, have the potential as a source of natural antioxidants. This study examines the antioxidant potential of kecapi fruit peel and matoa fruit peel in ecoenzyme production using the DPPH method and reducing power using the FRAP method. The chemicals content in the ecoenzyme samples were also identified using LC-HRMS. The EcoE-2 sample had the highest total phenolic content of 1613.962 ppm eq. gallic acid and also the highest total flavonoid content of 66.653 ppm eq. quercetin. The antioxidant power using the DPPH method for the EcoE-2 product (crude solution) was also greater than EcoE-1, the IC_{50} value is 2633.0 ppm, the IC_{50} for the EcoE-2 paste (without water/ethanol) was 119.3 ppm. For the reduction power with the FRAP method, data obtained that EcoE-2 also has a better reduction power than EcoE-1. It can be concluded that the higher content of matoa fruit peel will increase the antioxidant power of this ecoenzyme product. There are 118 putative chemical substances identified from the EcoE-2 sample, including several flavonoids and phenolic derivatives such as catechin, coumarin derivatives, and benzofuran compounds, supporting the observed antioxidant activity.

1 Introduction

Fruit is a vital source of food and nutrition for humans, and is an essential necessity utilized daily by people, both for direct consumption and processed into various products. This utilization process often leaves waste in the form of peels and unused fruit scraps, which become garbage. Disposing of waste carelessly or without adequate processing can negatively impact environmental cleanliness and potentially endanger health. Furthermore, this waste can also pollute the environment. In general, organic waste management still faces

*Corresponding author: candra.tahya@uph.edu

significant challenges because it is often directly disposed of in trash bins around residential areas, awaiting transport to final disposal sites (TPA). In some areas, piles of organic waste are often found on roadsides, riverbanks, and gutters. This can cause various problems, such as greenhouse gas emissions and methane gas produced from the decomposition process of organic substances. Furthermore, leachate produced from waste has the potential to pollute the environment because environmental conditions become increasingly dirty and produce unpleasant odors [1]. Blocked waterways become breeding grounds for insects such as mosquitoes and flies, which can impact residents' health. The government has implemented various programs to clean up, but it seems that the public is not very concerned because they feel that the remaining fruit and vegetable peels are useless, even though this paradigm is wrong. To support government programs in preserving healthy environment and maintaining public health, efforts are being made to reduce the problem of organic waste which has become a common problem encountered around us, both in rural and urban areas. Therefore, fruit waste and peels need to be processed into more useful products with economic value, one of which is the production of ecoenzymes [2] in this study. Hopefully, this research will also contribute to building public understanding of organic waste and strengthening public interest in processing and obtaining economic and health benefits from ecoenzyme (EcoE) products.

The Kecapi tree (Sandoricum koetjape) is a large, lush tree, reaching 15–30 meters tall, with branches and smooth, straight young leaves. The kecapi tree's trunk can reach 90 cm in diameter. It can be found in primitive and secondary forests at altitudes of 1,200 m or more and is also found in lowland forests [3]. The Kecapi tree has many benefits, including treating roundworms, lowering fever, treating bloating and diarrhea, strengthening the mother's body after childbirth [4], and possessing antioxidant properties [5]. The Matoa plant (Pometia pinnata) is a woody plant from the Sapidantheaceae family found in Southeast Asia and the Pacific. The Matua tree is also recognized as one of Indonesia's endemic plants, particularly in the Papua region [6]. Parts of this plant, including the leaves, fruit, bark, fruit skin, and roots, can be used medicinally. Phytochemical analysis showed that ethanol extract of matoa bark contains secondary metabolites such as alkaloids, flavonoids, glycosides, saponins, tannins, and terpenoids. Phenolic compounds such as flavonoids, tannins, and saponins exhibit high antioxidant activity [7].

Ecoenzymes (EcoE) are complex organic substances consisting of polypeptide, enzymes, organic acids, and mineral salts, which have function in structural components of cell, or as catalyst. EcoE is a fermentation product obtained from fruit and vegetable waste and brown sugar or molasses. During the fermentation process, carbohydrates derived from molasses are converted into volatile acids, while organic acids and enzymes are extracted from fruit peels to form an enzyme solution [8]. The process of making EcoE is quite simple and can be done at home using organic waste from households. EcoE does not require complex or extensive media. You only need to use used plastic bottles or regular plastic or glass containers. The biggest challenge in making EcoE is the time required, which is approximately 3 months from the manufacturing process until it is ready for use [9]. EcoE produced from fruit peels has a wide range of applications in everyday life. Generally, EcoE can be used in the food industry as a natural and environmentally friendly alternative to chemical catalysts. Overall, the utilization of EcoE from fruit peels is a sustainable and economical solution to address various real problems, such as reducing waste and environmental pollution, as well as increasing health or economic benefits.

Antioxidants are substances needed by the body and generally function to inhibit fat oxidation. Free radicals are formed as a byproduct of the energy production process in the body. Antioxidants are included in a group of chemicals that protect biological systems from the potential harmful effects caused by oxidation reactions, through various mechanisms to prevent damage caused by these radicals [10]. Antioxidants play a very important role in maintaining human health, especially in preventing damage due to oxidation. The oxidative process involves chemical reactions in the body that involve the transfer of electrons, which in turn produce free radicals. Antioxidants are easily oxidized, so free radicals will oxidize these antioxidants, which function to protect other molecules in the cell from damage caused by oxidation by free radicals [11].

Liquid chromatography-high-resolution mass spectrometry, commonly known as LC-HRMS, is a sophisticated instrumental analytical technique frequently applied in various scientific fields, including analytical chemistry, biology, pharmacy, and medicine. One of the main advantages of LC-HRMS is its high sensitivity and ability to analyze very complex samples. Thanks to these capabilities, LC-HRMS has become a crucial tool in modern science, helping researchers and practitioners answer various scientific and medical questions more precisely and accurately. Liquid chromatography-mass spectrometry is an analytical method that integrates two main techniques to separate and analyze the components contained in a sample.

The purpose of this study was to examine the potential of fermented fruit peel waste from plants that thrive and are endemic to Indonesia, namely kecapi and matoa, to become antioxidant agent. This study was conducted to determine the total phenolic and flavonoid values, antioxidant activity, and chemical content in EcoE products using a combination of organic waste raw materials that have never been used before as far as our current literature study.

2 Methods

2.1 Tools and Materials

The materials used in this study were kecapi and matoa fruit peel, distilled water, palm sugar, DPPH (1,1-diphenyl-2-picrylhydrazyl) (Merck), ethanol p.a (Merck), Folin-Ciocalteu solution (Merck), gallic acid (Merck), Na_2CO_3 (Merck), aluminum chloride (Merck), quercetin, $FeCl_3$ (Merck), phosphate buffer solution pH 6.6 (Merck). Fe (III) $6H_2O$, 2,4,6-Tris(2-pyridyl)-1,3,5-triazine (TPTZ), HCl 37%, distilled water, sodium acetate, glacial acetic acid, gallic acid, methanol. The tools used are laboratory glassware such as test tubes, Erlenmeyer flasks, petri dishes, beakers in several brands and non-glass equipment such as spatulas, magnetic rods, micropipettes, tips, 2 ml microtubes, 15 mL centrifuge tubes, ovens, hotplates, microplate readers, UV-Vis spectrophotometers, and LC-HRMS (Liquid chromatography - Orbitrap high-resolution mass spectrometry - Thermo Scientific™ Q Exactive™ Hybrid Quadrupole Orbitrap™ High Resolution Mass Spectrometer). LC-HRMS testing is carried out according to the standard procedures of PT. INVILAB biotechnological analysis.

In the production of ecoenzymes, variations were made in the weight ratio of kecapi fruit peel to matoa fruit peel. EcoE-1 is a variation in the amount of matoa fruit peel 400 g and kecapi fruit peel 200 g. EcoE-2 is a variation in the amount of matoa fruit peel 200 g and kecapi fruit peel 400 g. The dependent variable in this study is the ability of ecoenzymes from kecapi fruit peel and matoa in inhibiting the formation of free radicals with the DPPH method and the ability to reduce Fe^{3+}. Concentration variations were made to determine the IC_{50} value.

2.2 Procedures

2.2.1 Ecoenzyme Production

The matoa and kecapi fruit peels are cut into small pieces and placed in a fermenter container. 3 liters of distilled water and 0.5 kg of refined palm sugar are added. The fermenter container is made from a plastic gallon jug and equipped with a hose and bottle to channel the gas formed during the fermentation process to prevent it from exploding or breaking. The fermenter container is stored in a safe, dark place, out of direct sunlight. The ecoenzyme is stirred weekly while monitoring any changes. After 10 weeks of fermentation, the ecoenzyme is harvested by filtering the liquid from the fruit peel pulp. This process prepares the ecoenzyme for further processing. The EcoE pulp is spread in the students' gardens as fertilizer.

2.2.2 Total Phenolic Acid Test

Total phenolic content was calculated using a spectroscopic method with Folin-Ciocalteu reagent. The total phenolic content was measured as gallic acid equivalents (mgGAE) per gram of dry sample. Gallic acid was prepared in a series of concentrations of 20, 40, 60, 80, and 100 ppm. A 0.3 ml gallic acid solution was mixed with 1.5 mL of 10% Folin-Ciocalteu reagent and 1.2 mL of 7.5% Na_2CO_3. The solution was then vortexed and stored in the dark at room temperature for 30 minutes. A blank solution was also prepared, and its absorbance was recorded using a spectrometer at 765 nm [12]. For the triplet experiment, 1 mL of EcoE solution was diluted 20-fold with ethanol before reaction, and its absorbance was measured to match the standard curve used, and calculations were adjusted accordingly.

2.2.3 Total Flavonoid Test

The total flavonoid content test was conducted using the aluminum chloride technique. This test quantifies the total flavonoid content of the eco-enzyme. A 2 mL sample of quercetin was taken and mixed with 2 mL of 2% aluminum chloride before vortexing. A calibration curve was created using serial concentrations of quercetin. The absorbance of the mixture was measured using a spectrophotometer at 415 nm. The number of flavonoids was expressed as mg QE/g sample [12]. Each analysis was performed three times. A 1 mL Eco-E solution was diluted 10-fold with ethanol before reaction, and its absorbance was measured to match the standard curve used, and calculations were adjusted accordingly.

2.2.4 LC-HRMS Analysis

The test sample was prepared by dissolving 0.1 mL of the sample into a 10 mL volumetric flask, then adding ethanol solvent and homogenizing after ultrasonication for 30 minutes. Next, a 0.22 μm GHP/PTFE filter membrane was inserted into the instrument. The LC setup used a ThermoScientific™ Accucore™ Phenyl-Hexyl Column 100 mm × 2.1 mm ID × 2.6 μm. Eluent A – water containing 0.1% formic acid and Eluent B – methanol containing 0.1% formic acid. Flow rate – 0.3 mL/min. MS Settings: Operation Mode: ESI (-) / ESI (+), acquisition range: 66.7–1000 m/z.

2.2.5 Ecoenzyme Antioxidant Assay

The free radical scavenging activity of EcoE and the standard quercetin was determined using the DPPH radical scavenging method [13]. The IC50 value of EcoE was determined using varying concentrations. For antioxidants, the IC50 value of quercetin was determined using varying concentrations. Pipette 80 μL of sample and standard solutions into 96-well plates, starting from the highest concentration to the lowest. Each concentration was added to three wells (triplicate). Empty three wells for blank media (the blank media contained 80 μL of solvent and added 80 μL of 0.1 mM DPPH solution). Transfer the 0.1 mM DPPH solution to a plastic container (sufficient volume). Add 80 μL of 0.1 mM DPPH solution using a multichannel pipette to the wells of the plate containing the standard and sample. Cover the microplate with aluminium foil and incubate for 30 minutes at 25°C (dark). After

30 minutes, the solution mixture was read on a microplate reader at a wavelength of 492 nm for 30 seconds with a vibration shake medium [14].

$$\% \text{ inhibition} = \frac{\text{Abs Blank - Abs sample}}{\text{Abs Blank}} \text{ x } 100\% \qquad (1)$$

2.2.6 Ecoenzyme Reducing Power Assay

The Ferric Reducing Antioxidant Power Assay (FRAP) was investigated using a modified method by Tang et al. [13]. Pipette 20 μL of sample and standard gallic acid solutions into 96-well plates, starting from the highest concentration to the lowest. Each concentration was added to three wells (triplicate). Transfer the FRAP DYE working solution to the reservoir (sufficient volume). Add 280 μL of the solution using a multichannel pipette to the wells containing the standard and sample. Note: The solution was added in the dark. Cover the well plate with aluminum foil and incubate for 10 minutes at room temperature (dark conditions). After 10 minutes, the solution mixture was read on a microplate reader at a wavelength of 492 nm for 30 seconds with a vibration shake medium. The FRAP DYE solution consisted of sodium acetate solution, TPTZ solution, and Fe(III) solution in a 10:1:1 ratio.

2.2.7 Research Data Analysis

The data analysis results were processed using MS Excel to obtain the average value and standard deviation for each measurement. Chemical content data was analyzed using LC-HRMS using the Compound Discoverer 3.2 dan Sirius 6.2.2 application.

3 Results and Discussion

3.1 Analysis of Total Flavonoids and Total Phenolics EcoE

The total flavonoid content test (Figure 1) is an analytical method to calculate the total amount of flavonoids in ecoenzymes using the aluminium chloride technique.

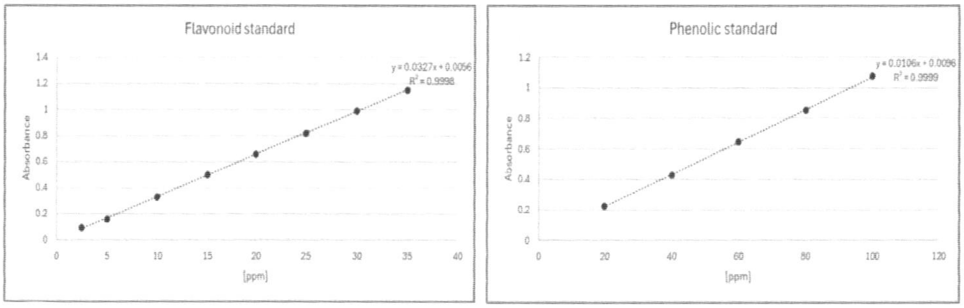

Fig.1. Standard curve of total flavonoid (quercetin standard) and total phenolic (gallic acid standard) tests.

The test results for total flavonoids and total phenolics of EcoE-1 and 2 are shown in Tables 1 and 2.

Table 1. Results of total flavonoid tests of EcoE-1 and 2 with quercetin as standards

Sample	Absorbance			Avg	Conc. (ppm eq. Quercetin)	Average Conc. (ppm)	SD
	Abs 1	Abs 2	Abs 3				
EcoE-1	0.138	0.240	0.102	0.160	47.217		

Sample	Absorbance			Avg	Conc. (ppm eq. Quercetin)	Average Conc. (ppm)	SD
	Abs 1	Abs 2	Abs 3				
EcoE-1	0.132	0.234	0.102	0.156	45.994	46.673	0.622
EcoE-1	0.136	0.238	0.102	0.159	46.809		
EcoE-2	0.255	0.333	0.078	0.222	66.177		
EcoE-2	0.260	0.338	0.078	0.225	67.197	66.653	0.513
EcoE-2	0.257	0.335	0.078	0.223	66.585		

Note: diluted 10 times (1 mL EcoE + 9 mL ethanol) before testing. Avg = Average

Table 2. Results of total phenolic tests of EcoE-1 and 2 with gallic acid as standards

Sample	Absorbance			Avg	Conc. (ppm eq. galic acid)	Average Conc. ppm	SD
	Abs 1	Abs 2	Abs 3				
EcoE-1	0.762	0.768	0.777	0.769	1432.830		
EcoE-1	0.759	0.768	0.771	0.766	1427.170	1428.428	3.928
EcoE-1	0.758	0.767	0.77	0.765	1425.283		
EcoE-2	0.872	0.86	0.852	0.861	1606.415		
EcoE-2	0.861	0.875	0.869	0.868	1619.623	1613.962	6.803
EcoE-2	0.866	0.856	0.875	0.866	1615.849		

Note: diluted 20 times (0.5 mL EcoE + 9.5 mL ethanol) before testing. Avg = Average

Based on the results of the total phenolic content, it was seen that both of ecoenzymes had high total phenolic content, i.e. EcoE-1 = 1428.428 ppm equivalent of gallic acid and EcoE-2 = 1613.962 ppm equivalent of gallic acid, but the total flavonoid content was small which means flovonoid content is low. But it is important to aware that Folin–Ciocalteu assay is non-specific because it responds to any compound capable of reducing the reagent, not only phenolic compounds. In fermentation-derived samples, this limitation becomes critical because fermentation generates and concentrates many non-phenolic reducing substances—such as reducing sugars, organic acids, amino acids, peptides, vitamins, and other microbial or plants metabolites—that can strongly contribute to the Folin–Ciocalteu signal.

3.2 Analysis of EcoE Antioxidant Test DPPH Method

The results of the EcoE-1 and 2 antioxidant tests, as well as quercetin as a standard, are shown in Tables 3 and 4 below. The EcoE solution used for this antioxidant test was the result of fermentation (called crude solution) that had been filtered with a 0.22 µm microfilter at various volumes and diluted with ethanol. The concentration was measured based on the specific density of the fermented EcoE solution. The specific density was determined using the pycnometer method (triplicate).

Table 3. Results of the DPPH method antioxidant activity test for EcoE-1 (crude solution).

Vol. EcoE (mL)	Conc. EcoE (ppm)	Abs1	Abs2	Abs3	Avg	SD	% Inhibition	IC_{50} (ppm)
0	0.00	0.881	0.854	0.891	0.875	0.019	0	
0.2	1998.04	0.535	0.629	0.557	0.574	0.049	34.46	
0.4	3996.08	0.39	0.38	0.335	0.368	0.029	57.92	3902.1

0.6	5994.12	0.196	0.216	0.217	0.210	0.012	76.05	
0.8	7992.16	0.126	0.135	0.117	0.126	0.009	85.60	

Note: EcoE used for antioxidant testing was diluted 1:10 with ethanol p.a. Abs = Absorbance. Avg = Average

Table 4. Results of the DPPH method antioxidant activity test for EcoE-2 (crude solution).

Vol. EcoE (mL)	**Conc. EcoE** (ppm)	**Abs1**	**Abs2**	**Abs3**	**Avg**	**SD**	**% Inhibition**	**IC$_{50}$** (ppm)
0	0.00	1.105	1.013	1.067	1.062	0.046	0	2633.0
0.2	1006.18	0.899	0.859	0.844	0.867	0.028	18.3	
0.4	2012.35	0.583	0.621	0.598	0.601	0.019	43.4	
0.6	3018.53	0.466	0.436	0.423	0.442	0.022	58.4	
0.8	4024.70	0.294	0.296	0.293	0.294	0.002	72.3	

Note: EcoE used for antioxidant testing was diluted 1:20 with ethanol p.a. Abs = Absorbance, Avg = Average

For the quercetin antioxidant test, standard quercetin solids were used, dissolved in ethanol according to the concentration variations used. The results of the antioxidant test are shown in Table 5.

Table 5. Results of the DPPH method antioxidant activity test for quercetin solution

Quercetin Conc. (ppm)	**Abs1**	**Abs2**	**Abs3**	**Avg**	**SD**	**% Inhibition**	**IC$_{50}$** (ppm)
0.00	1.105	1.013	1.067	1.062	0.046	0	32.7
16.67	0.722	0.758	0.727	0.736	0.020	30.7	
33.33	0.464	0.482	0.467	0.471	0.010	55.6	
50.00	0.252	0.234	0.267	0.251	0.017	76.4	
66.67	0.097	0.069	0.088	0.085	0.014	92.0	

Note: Abs = Absorbance, , Avg = Average

Antioxidant activity of EcoE-1 and 2 crude solution is low as we see from the IC$_{50}$ in ppm, but if we see the volume of EcoE crude solution that is needed to inhibit DPPH radical is quite low compared to the total volume of ecoenzyme we have produced. For EcoE only about 0.4-0.5 mL EcoE capable to inhibit 50% of DPPH radical activity, while we can produce easily 2000 mL of ecoenzyme, it means that volume capacity of ecoenzyme as antioxidant is bigger, because ecoenzyme is cheaper, easy to produce, and having better economics benefit if to applied in large scale antioxidant applications.

We can then evaluate the antioxidant of ecoenzyme after all the solvent i.e., water have been evaporated leave only paste of ecoenzyme. The EcoE product was freeze-dried to evaporate all water and small number of ethanol solvents. The freeze-drying method was chosen to ensure that the active chemical components contained in EcoE still remain in the EcoE paste. The EcoE paste was collected and then measured for antioxidant activity using the DPPH method and for reducing activity using the FRAP method. The results are shown in Tables 6 and 7. This antioxidant activity test was conducted to determine the antioxidant and reducing activity of EcoE-1 and 2 products and can be compared to the antioxidant activity test of the EcoE solution still contained in solvents from fermentation. Calculation of antioxidant and FRAP reducing power tests using this paste will provide more accurate view about the activity of the active compound and it can be seen in the IC$_{50}$ value. The IC$_{50}$ value for DPPH inhibition of EcoE-1 is 168.63 ppm and EcoE-2 is 119.26 ppm.

Table 6. DPPH antioxidant activity test of EcoE-1 paste (water/EtOH solvent free)

Conc. (ppm)	**% Inhibition** 1	2	3	**Avg**	**SD**	**IC$_{50}$** (ppm)
200	53.071	55.038	58.637	55.58	2.82	
50	17.850	14.587	14.875	15.77	1.81	
25	7.246	7.582	8.109	7.65	0.44	168.6
12.5	3.83	3.791	3.263	3.63	0.32	
6.25	0.672	5.182	3.503	3.12	2.28	

Note: SD = Standard Deviation, Avg = Average

Table 7. DPPH antioxidant activity test of EcoE-2 paste (water/EtOH solvent free)

Conc. (ppm)	**% Inhibition** 1	2	3	**Avg**	**SD**	**IC$_{50}$** (ppm)
200	70.777	71.737	72.265	71.59	0.75	
100	42.658	42.083	43.762	42.83	0.85	
25	7.870	7.054	9.261	8.06	1.12	119.3
12.5	4.367	1.919	5.374	3.89	1.78	
6.25	1.775	0.960	2.447	1.09	1.80	

Note: SD = Standard Deviation, Avg = Average

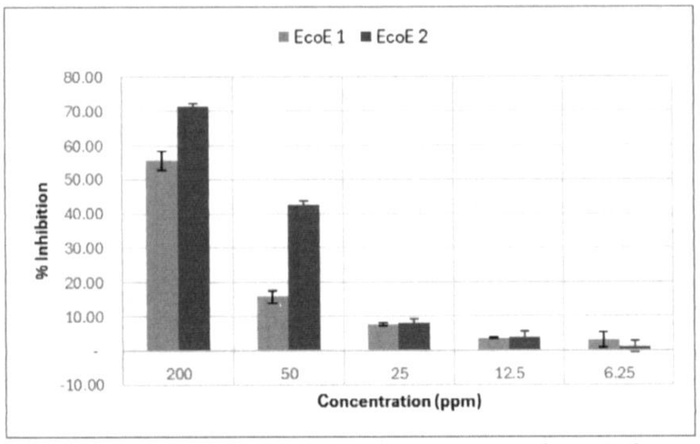

Fig. 2. The percentage of inhibition of radical DPPH by EcoE-1 and EcoE-2.

EcoE-2 seems to have better antioxidant activity compared to EcoE-1, for both form of solution or paste (solvent free) (Figure 2). Even though based on the pair T-test 2 tail calculation on Excel program, we observed that the p-value is > 0.05, which means that the antioxidant activities of both EcoE paste are not significantly different. The EcoE paste antioxidant power is approximately 4-5 times weaker than standard pure quercetin (IC$_{50}$ = 32.65 ppm).

A FRAP test was also performed on Fe(III) to determine the ability of EcoE to release electrons and reduce iron ions. This ability was compared against a gallic acid standard. The reducing power data for EcoE-1 and 2 are shown in Figure 3. Table 8 shows the reducing power for gallic acid.

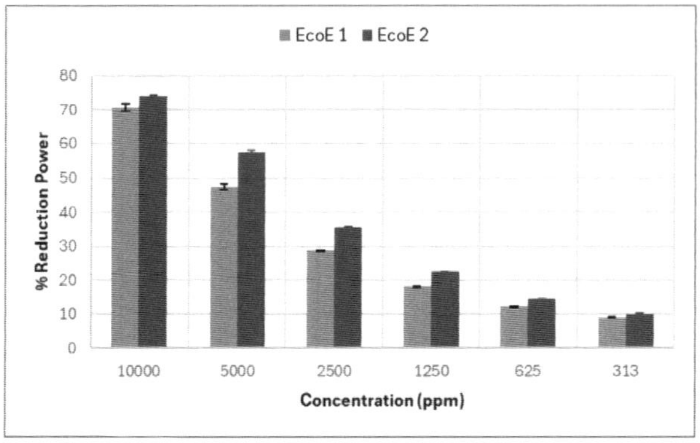

Fig. 3. The percentage of reduction power by EcoE-1 and EcoE-2.

Based on the pair T-test 2 tail calculation on Excel program, we observed that the p-value is < 0.05 i.e. 0.02 which means that the reduction power of both EcoE is significantly different. The EcoE-1 (RP_{50} = 5473.2 ppm) has weaker reduction power than the EcoE-2 (RP_{50} = 3969.9 ppm). Both of EcoE paste still have weaker reduction power in comparison to gallic acid (RP_{50} = 92.4 ppm). The finding support assumption that EcoE-2 which has more flavonoid content and phenolic content will show better antioxidant and reduction power compared to EcoE-1.

Table 8. Results of the FRAP reduction power test of gallic acid

Conc. (ppm)	Absorbance			Reduction Power (%)			Avg	SD	RP_{50} (ppm)
	1	2	3	1	2	3			
1000	2.500	2.530	2.557	92.233	93.343	94.354	93.31	1.06	92.4
250	1.964	1.979	1.972	72.465	73.022	72.753	72.75	0.28	
125	1.594	1.615	1.617	58.830	59.598	59.672	59.37	0.47	
62.5	1.059	1.112	1.106	39.078	41.033	40.793	40.30	1.07	
31.25	0.647	0.663	0.666	23.889	24.450	24.583	24.31	0.37	
15.63	0.412	0.430	0.436	15.214	15.875	16.089	15.73	0.46	
7.81	0.268	0.299	0.308	9.886	11.022	11.358	10.76	0.77	

Note: SD = Standard Deviation, Avg = Average

The results of this study indicate that EcoE produced from kecapi fruit and matoa fruit peels has weak to moderate antioxidant activity, as indicated by the IC_{50} values from the DPPH test, reduction power, and the total phenolics and flavonoid observed.

3.3 Component Identification in EcoE using LC-HRMS

To determine the chemical components contained in EcoE, an LC-HRMS test was conducted. The Total Ion Chromatogram (TIC) results of EcoE-2 are shown in Figure 4. Both of EcoE used the same fermentation component/ingredients so EcoE-2 was chosen with better antioxidant and reduction power for the LC-HRMS analysis. There were 19 putative chemical components that were identified based on the molecular weight of the compounds (Δ mass $< \pm 3$ ppm) as compared to the database using the Compound Discoverer 3.2. But based on criteria for annotation of putative compounds on the mzCloud best match value is $> 60\%$ as recommended by ThermoFisher Product Communication, only 10 compounds that satisfied this requirement as shown in Table 9. Unfortunately, 10 putative identified compounds are too few to describe characteristics of the ecoenzyme. Because of this we decided to used Sirius 6.2.2 application program to identify more putative compounds based on the raw data from LC-HRMS instrument. The result was shown in Table 10.

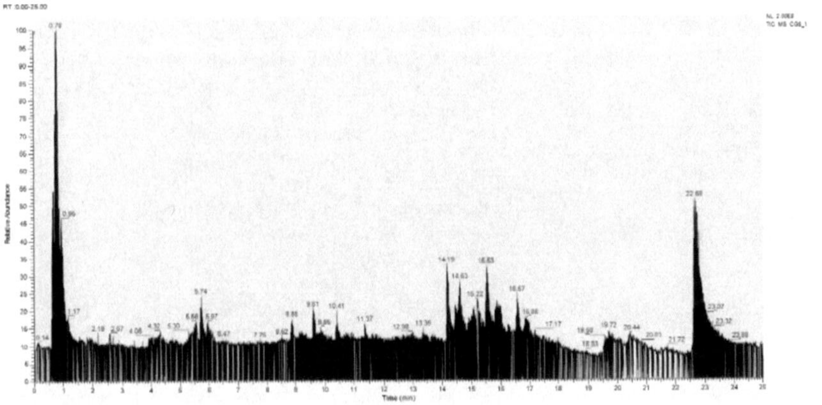

Fig. 4. Total Ion Chromatogram (TIC) of LC-HRMS EcoE-2.

Some setting in Sirius 6.2.2 compute system are as follow: instrument is orbitrap, MS2 mass accuracy is 5 ppm, fallback adduct is $[M+H]^+$, $[M+Na]^+$, $[M+K]^+$, search databases are all listed including PubChem, except for PubMed. Other settings are left as default. From the Sirius results, only decent and good overall Feature Quality, Sirius Score Normalized value > 0.75 are accepted. The duplication was filtered, and we got 108 putative identified compounds.

Table 9. List of putative compounds identified using LC-HRMS and analysed in Compound Discoverer 3.2.

No	Name	Formula	Annot. Δ Mass [ppm]	Calc. MW	RT [min]	Area (Max.)	mzCloud Best Match
1	2-Amino-1,3,4-octadecanetriol	$C_{18}H_{39}NO_3$	-0.99	317.2927	9.743	6683271.01	75.6
2	Tiaprofenic acid	$C_{14}H_{12}O_3S$	0.79	260.0509	0.979	10840735	60.6
3	N,N-Diethyldodecanamide	$C_{16}H_{33}NO$	-0.97	255.256	15.854	27818322.1	88.9
4	N,N-Diisopropylethylamine (DIPEA)	$C_8H_{19}N$	0.36	129.1518	1.944	33043655.8	78.2
5	Bis(4-ethylbenzylidene)sorbitol	$C_{24}H_{30}O_6$	-1.65	414.2036	11.372	40861047.1	99.8
6	Dibenzylamine	$C_{14}H_{15}N$	-0.9	197.1203	5.808	52339669.3	99.9
7	Methyl isonicotinate	$C_7H_7NO_2$	-0.39	137.0476	0.799	57649756.1	98.1
8	2,2,6,6-Tetramethyl-1-piperidinol (TEMPO)	$C_9H_{19}NO$	-0.55	157.1466	9.208	70344617.2	92.7
9	1-Stearoylglycerol	$C_{21}H_{42}O_4$	-1.85	358.3077	15.533	90032358.1	96.3
10	Stearamide	$C_{18}H_{37}NO$	-2.02	283.2869	15.865	182718975	97.9

Table 10. List of putative compounds identified using LC-HRMS and analysed in Sirius 6.2.2 in EcoE2

No	Molecular Formula	adduct	Name	Ion Mass	RT (Min)
1	$C_{24}H_{30}O_6$	$[M+H]^+$	(1R)-1-[(2R,4S,4aR,6R,8aR)-2,6-bis(3,4-dimethylphenyl)-4,4a,8,8a-tetrahydro-	415.211	11.629

			[1,3]dioxino[5,4-d][1,3]dioxin-4-yl]ethane-1,2-diol		
2	$C_9H_{18}O_8$	$[M + K]^+$	(2S,3S,4S,5R,6R)-3-(2,3-dihydroxypropoxy)-6-(hydroxymethyl)oxane-2,4,5-triol	293.063	0.751
3	$C_{19}H_{26}$	$[M + H]^+$	(3E,5E,7E,9E,12Z)-3,5,7,9-tetramethyl-11-methylidenetetradeca-1,3,5,7,9,12-hexaene	255.211	13.484
4	$C_{29}H_{48}$	$[M + H]^+$	(3S,3aR,5aR,5bR,7aS,11aS,11bR,13bS)-5a,5b,8,8,11a-pentamethyl-3-prop-1-en-2-yl-2,3,3a,4,5,6,7,7a,9,10,11,11b,12,13,13a,13b-hexadecahydro-1H-cyclopenta[a]chrysene	397.382	18.518
5	$C_{23}H_{39}NO_2$	$[M + H]^+$	(5Z,8Z,12Z,15Z)-N-(2-hydroxyethyl)henicosa-5,8,12,15-tetraenamide	362.305	15.526
6	$C_8H_{14}O_8$	$[M + Na]^+$	[(3R,4S,5R,6R)-3,4,5-trihydroxy-6-(hydroxymethyl)oxan-2-yl] ethaneperoxoate	261.058	0.984
7	$C_{10}H_8O_3$	$[M + H]^+$	[2-(3-Oxoprop-1-enyl)phenyl] formate	177.055	5.468
8	$C_5H_9N_4O_4P$	$[M + H]^+$	[3-hydroxyimino-3-(1H-1,2,4-triazol-5-yl)propyl]phosphonic acid	221.042	0.742
9	$C_9H_{12}N_4O_2$	$[M + Na]^+$	[4-[[1-(Hydroxymethyl)pyrazol-4-yl]methyl]pyrazol-1-yl]methanol	231.084	0.975
10	$C_{21}H_{33}N_3O_3$	$[M + H]^+$	1,3,5-Tris(2,2-dimethylpropionylamino)benzene	376.259	10.893
11	$C_{15}H_{14}O_{10}$	$[M + H]^+$	1-{[3-(3,4-dihydroxyphenyl)prop-2-enoyl]oxy}propane-1,2,3-tricarboxylic acid	355.066	4.405
12	$C_{20}H_{38}O_3$	$[M + H]^+$	10-(6-Hydroxydec-8-enoxy)dec-2-en-5-ol	327.289	15.535
13	$C_{13}H_{16}$	$[M + H]^+$	1-But-1-ynyl-4-propan-2-ylbenzene	173.132	13.484
14	$C_8H_{19}N$	$[M + H]^+$	1-Butanamine, N-(2-methylpropyl)-	130.159	1.963
15	$C_9H_6O_3$	$[M + H]^+$	1-Hydroxy-3H-2-benzopyran-3-one	163.039	7.658
16	$C_{20}H_{39}NO$	$[M + H]^+$	1-Isocyanatononadecane	310.31	15.973
17	$C_4H_8O_2$	$[M + H]^+$	1-Methoxyethyl(methylidene)oxidanium	89.06	5.747
18	$C_{10}H_{14}N_2$	$[M + H]^+$	1-methyl-2-(pyridin-3-yl)pyrrolidin-1-ium	163.123	0.136
19	$C_{26}H_{31}NO$	$[M + H]^+$	1-Methyl-3,5-bis[(4-propan-2-ylphenyl)methylidene]piperidin-4-one	374.247	16.169
20	$C_{15}H_{20}O_4$	$[M + H]^+$	2-(2,3-Dimethylbut-2-enyl)-5,6-dimethoxy-3-methylcyclohexa-2,5-diene-1,4-dione	265.143	4.722
21	$C_9H_{17}NO_4$	$[M + H]^+$	2-(2-Aminoethyl)malonic acid diethyl ester	204.123	0.788
22	$C_7H_{13}NO_4$	$[M + H]^+$	2-(2-Aminoethyl)pentanedioic acid	176.092	0.872
23	$C_6H_4O_5$	$[M + H]^+$	2-(3-Hydroxy-5-oxofuran-2-ylidene)acetic acid	157.013	0.947
24	$C_{14}H_{31}NO$	$[M + H]^+$	2-(Dodecylamino)ethanol	230.248	9.662
25	$C_7H_9F_2NO$	$[M + H]^+$	2,2-Difluoro-2-(3-methylfuran-2-yl)ethan-1-amine	162.073	13.009
26	$C_{29}H_{37}NO$	$[M + H]^+$	2,4-Di-tert-butyl-6-(dibenzylaminomethyl)phenol	416.294	17.614
27	$C_{15}H_{22}O_2$	$[M + H]^+$	2,5-Di-tert-butyl-3-acetyl-2,4-cyclopentadiene-1-one	235.169	12.655
28	$C_{23}H_{41}NO$	$[M + H]^+$	2,6-DI-Tert-butyl-4-[(dibutylamino)methyl]phenol	348.326	12.962
29	$C_{16}H_{35}NO_3$	$[M + H]^+$	2-[(2,2-Diethoxyethyl)(octyl)amino]ethan-1-ol	290.269	9.784
30	$C_{19}H_{30}$	$[M + H]^+$	2-[(3E,5E)-3,7-dimethylocta-3,5,7-trienyl]-1,3,3-trimethylcyclohexene	259.242	14.761
31	$C_8H_{13}NO_6$	$[M + H]^+$	2-[(3-Hydroxy-2,2-dimethylpropanoyl)amino]propanedioate	220.082	0.872

32	$C_7H_8N_2O_4S_2$	$[M + K]^+$	2-[(5-Methyl-1,3,4-thiadiazol-2-yl)sulfanyl]butanedioic acid	286.957	0.742
33	$C_{12}H_{21}NO_6$	$[M + H]^+$	2-[(R)-1-(Nitromethyl)-2-methylpropyl]malonic acid diethyl ester	276.144	0.993
34	$C_{11}H_{18}N_2O_5$	$[M + H]^+$	2-[[(2S)-piperidine-2-carbonyl]amino]pentanedioic acid	259.129	0.937
35	$C_{10}H_{18}O_8$	$[M + Na]^+$	2-[2-Hydroxy-3-(hydroxymethyl)cyclopropyl]oxy-6-(hydroxymethyl)oxane-3,4,5-triol	289.09	1.003
36	C_8H_8O	$[M + H]^+$	2-ethenylphenol	121.065	15.544
37	$C_{19}H_{32}O_3$	$[M + H]^+$	2-Methyl-3-prop-2-enoylpentadec-2-enoic acid	309.242	14.64
38	$C_8H_9O_3P$	$[M + Na]^+$	2-Phenylethenyl dihydrogen phosphite	207.017	22.666
39	$C_{20}H_{35}NO$	$[M + H]^+$	2-tert-Butyl-6-(dibutylaminomethyl)-4-methylphenol	306.279	11.266
40	$C_8H_{13}NO_6$	$[M + H]^+$	3-(2-Carboxyethylamino)pentanedioic acid	220.082	0.76
41	$C_7H_{13}NO_4$	$[M + H]^+$	3-(Aminomethyl)-3-methylpentanedioic acid	176.092	0.77
42	$C_{24}H_{22}O_{14}$	$[M + H]^+$	3,4-bis({[3-(3,4-dihydroxyphenyl)prop-2-enoyl]oxy})-2,5-dihydroxyhexanedioic acid	535.108	4.414
43	$C_{15}H_{22}O$	$[M + H]^+$	3,5-di-tert-butylbenzaldehyde	219.174	12.99
44	$C_9H_{18}N_2O$	$[M + H]^+$	3-Cyclohexyl-1,1-dimethylurea	171.149	22.983
45	$C_{12}H_{23}NO_7$	$[M + H]^+$	3-methyl-2-({[2,3,4-trihydroxy-5-(hydroxymethyl)oxolan-2-yl]methyl}amino)pentanoic acid	294.155	0.984
46	$C_8H_{17}NO_2$	$[M + H]^+$	4,4-Diethoxybutan-1-imine	160.133	0.788
47	$C_6H_8O_3$	$[M + H]^+$	4-hydroxy-3,5-dimethyl-3H-furan-2-one	129.055	1.17
48	$C_7H_6N_2O_3S_2$	$[M + H]^+$	5-Acetyl-4-methylsulfonyl-1,3-thiazole-2-carbonitrile	230.99	0.956
49	$C_{24}H_{43}NO_2$	$[M + H]^+$	6,6-bis[(Z)-non-2-enoxy]hexanenitrile	378.335	15.898
50	$C_9H_6O_3$	$[M + H]^+$	6-Hydroxycoumarin	163.039	5.3
51	$C_{10}H_8O_2$	$[M + H]^+$	6-methyl-2H-chromen-2-one	161.06	14.752
52	$C_7H_7NO_2$	$[M + H]^+$	6-Methylnicotinic acid	138.055	0.788
53	$C_9H1_7NO_4$	$[M + H]^+$	Acetylcarnitine	204.123	0.937
54	$C_9H_6O_3$	$[M + H]^+$	Benzofuran-4-carboxylic acid	163.039	1.702
55	$C_{20}H_{43}NO_2$	$[M + H]^+$	Bis(2-hydroxyethyl)hexadecylammonium	330.336	11.722
56	C10H8O4	$[M + H]^+$	Buxuletin	193.05	5.607
57	C15H14O6	$[M + H]^+$	Catechin	291.086	3.342
58	$C24H_{20}O_9$	$[M + H]^+$	Catechin-[5,6-E]-4Beta-(3,4-Dihydroxyphenyl)Dihydro-2(3H)-Pyranone	453.118	6.437
59	$C_{27}H_{44}$	$[M + H]^+$	Cholestyne	369.351	17.772
60	$C_{24}H_{38}O_4$	$[M + H]^+$	Dehp	391.284	17.502
61	$C_{16}H_{22}O_4$	$[M + H]^+$	Dibutyl Phthalate	279.159	13.363
62	$C_8H_{19}N$	$[M + H]^+$	dibutylamine	130.159	22.526
63	$C_{22}H_{43}NO$	$[M + H]^+$	docos-13-enamide	338.341	16.98
64	$C_{30}H5_8O_4S$	$[M + H]^+$	dodecyl 3-{[3-(dodecyloxy)-3-oxopropyl]sulfanyl}propanoate	515.412	20.028
65	$C_{21}H_{41}NO$	$[M + H]^+$	Eicosylisocyanate	324.326	16.532

66	$C_{17}H_{35}NO$	$[M + H]^+$	heptadecanamide	270.279	14.64
67	C16H33NO	$[M + H]^+$	Hexadecyl(oxo)azanium	256.263	15.553
68	$C_{18}H_{34}O$	$[M + H]^+$	Hexadecylketene	267.268	15.218
69	$C_{11}H_{19}NO_6$	$[M + H]^+$	Isoleucinopine	262.129	0.937
70	$C_6H_5NO_2$	$[M + H]^+$	Isonicotinic Acid	124.039	0.965
71	$C_{18}H_{32}O_{16}$	$[M + Na]^+$	Lactosucrose	527.158	0.751
72	$C_6H_{10}O_8$	$[M + Na]^+$	Mucic Acid	233.027	0.751
73	$C_{10}H_{23}NO$	$[M + H]^+$	N-(2-butoxyethyl)butan-1-amine	174.185	2.569
74	$C_{20}H_{37}NO$	$[M + H]^+$	N-(2-methylpropyl)hexadeca-2,4-dienamide	308.294	15.414
75	$C_{24}H_{41}NO$	$[M + H]^+$	N-(2-phenylethyl)hexadecanamide	360.325	16.197
76	$C_9H_{12}N_4O_2$	$[M + Na]^+$	N'-(4-methoxypyrimidin-2-yl)cyclopropanecarbohydrazide	231.084	0.798
77	$C_{11}H_{21}NO$	$[M + H]^+$	N-(cyclopropylmethyl)-N-pentylacetamide	184.17	9.131
78	$C_{21}H_{37}NO_2$	$[M + H]^+$	N,N-bis(prop-1-ynoxy)pentadecan-1-amine	336.289	15.414
79	$C_{20}H_{35}NO_2$	$[M + H]^+$	N,N-bis(prop-1-ynoxymethyl)dodecan-1-amine	322.273	14.248
80	$C_{16}H_{17}NO$	$[M + H]^+$	N,N-Dibenzylacetamide	240.138	10.501
81	$C_{16}H_{19}NO$	$[M + H]^+$	N,N-Dibenzylethanolamine	242.154	5.738
82	$C_{30}H_{45}NO$	$[M + H]^+$	N,N-dibenzylhexadecanamide	436.357	17.949
83	$C_{24}H_{49}NO$	$[M + H]^+$	N,N-dibutylhexadecanamide	368.388	17.902
84	$C_{18}H_{37}NO$	$[M + H]^+$	N,N-Dimethylhexadecanamide	284.294	15.535
85	$C_{13}H_{17}NO$	$[M + H]^+$	N-[(1Z,4Z,6Z)-1-[(1E)-buta-1,3-dienoxy]octa-1,4,6-trien-2-yl]methanimine	204.138	12.403
86	$C_{19}H_{33}NO_2$	$[M + H]^+$	N-[bis(prop-1-ynoxy)methyl]dodecan-1-amine	308.258	13.838
87	$C_{15}H_{15}NO$	$[M + H]^+$	N-Benzyl-2-phenylacetamide	226.122	10.408
88	$C_{23}H_{39}NO$	$[M + H]^+$	N-Benzylhexadecanamide	346.31	16.299
89	$C_{25}H_{43}NO$	$[M + H]^+$	N-benzyloctadecanamide	374.341	17.147
90	$C_{10}H_{21}NO$	$[M + H]^+$	N-butyl-N-(3-methylbutyl)formamide	172.17	9.131
91	$C_{12}H_{25}NO$	$[M + H]^+$	N-butyl-N-heptylformamide	200.201	11.704
92	$C_9H_{19}NO$	$[M + Na]^+$	N-cyclopentyloxy-2-methylpropan-1-amine	180.136	8.87
93	$C_{19}H_{39}NO$	$[M + H]^+$	Nonadecyl(oxo)azanium	298.31	15.712
94	$C_{21}H_{43}NO$	$[M + H]^+$	N-pentylhexadecanamide	326.341	16.57
95	$C_{23}H_{47}NO$	$[M + H]^+$	N-pentyloctadecanamide	354.372	17.138
96	$C_9H_{15}N_3O_2$	$[M + H]^+$	N-Propyl-L-histidine	198.124	0.826
97	$C_3H_4N_4O_4$	$[M + H]^+$	O-(1-nitroimidazol-4-yl)oxyhydroxylamine	161.03	0.732
98	$C_{18}H_{32}O$	$[M + H]^+$	Octadec-17-yn-2-one	265.252	15.712
99	$C_{18}H_{34}O$	$[M + H]^+$	Octadec-5-yn-1-ol	267.268	16.001
100	$C_{18}H_{37}NO$	$[M + H]^+$	Octadecyl(oxo)azanium	284.294	16.532
101	$C_{18}H_{35}NO$	$[M + H]^+$	Oleamide	282.279	14.612
102	$C_9H_{17}NO_5$	$[M + H]^+$	Pantothenate	220.118	1.431
103	$C_{18}H_{39}NO_3$	$[M + H]^+$	PEG-3 lauramine	318.3	9.97

104	$C_5H_8O_2$	$[M + Na]^+$	Pent-2-yne-1,5-diol	123.04	22.712
105	$C_8H_4O_3$	$[M + H]^+$	Phthalic Anhydride	149.023	23.132
106	$C_{12}H_{22}O_{11}$	$[M + Na]^+$	Sucrose	365.105	0.751
107	C2H6N2O4	$[M + H]^+$	Trihydroxymethylurea	123.04	0.648
108	$C_{36}H_{75}N$	$[M + H]^+$	Tris-decyl(hexyl)azanium	522.597	19.161

From Table 10, we can observe 7 flavonoids and phenolics compounds were putatively identified i.e. 1-Hydroxy-3H-2-benzopyran-3-one, Catechin, Catechin-[5,6-E]-4Beta-(3,4-Dihydroxyphenyl)Dihydro-2(3H)-Pyranone, Buxuletin, 6-Hydroxycoumarin, and 6-methyl-2H-chromen-2-one, and Benzofuran-4-carboxylic acid. The chemical structure of the 6 phenolics and flvonoid are shown in Figure 5. The flavonoid and phenolics are known to have antioxidant activity.

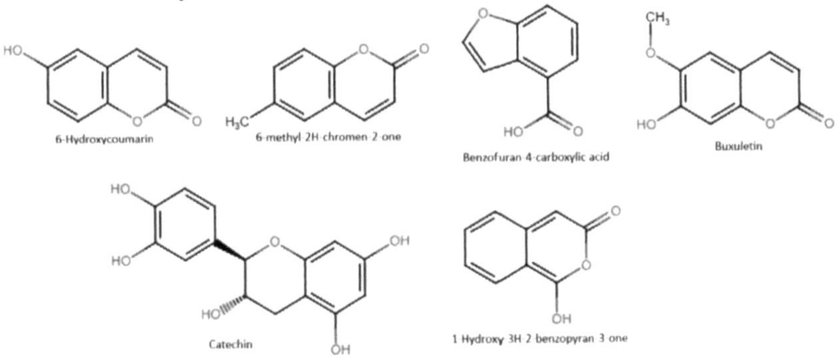

Fig. 5. The six putative identified flavonoid and phenolics compounds from Ecoenzyme (EcoE2) registered in PubChem.

We can observe the Pharmacokinetics, Druglikeness and Medicinal Chemistry of the six flavonoid and phenolics compounds as listed in Table 11 based on SwissADME databased analysis (http://www.swissadme.ch/index.php).

Table 11. The SwissADME results for 6 flavonoid and phenolics identified in EcoE2.

Molecule	#H-bond acceptors	#H-bond donors	GI absorption	BBB permeant	Pgp substrate	CYP1A2 inhibitor	log Kp (cm/s)	Lipinski #violations	Ghose #violations	Veber #violations	Egan #violations	Muegge #violations	Bioavailability Score	PAINS #alerts	Brenk #alerts	Leadlikeness #violations
1-Hydroxy-3H-2-benzopyran-3-on	3	1	High	Yes	No	Yes	-6.56	0	1	0	0	1	0.55	0	0	1
6-Hydroxycoumarin	3	1	High	Yes	No	Yes	-6.56	0	1	0	0	1	0.55	0	1	1
6-methyl-2H-chromen-2-one	2	0	High	Yes	No	Yes	-6.03	0	0	0	0	1	0.55	0	1	1
Benzofuran-4-carboxylic acid	3	1	High	Yes	No	Yes	-5.98	0	1	0	0	1	0.85	0	0	1
Buxuletin	4	1	High	Yes	No	Yes	-6.39	0	0	0	0	1	0.55	0	1	1
Catechin	6	5	High	No	Yes	No	-7.82	0	0	0	0	0	0.55	1	1	0

We can predict the biological activities of the compound by using PASS online Way2Drug Prediction (https://way2drug.com/PassOnline/predict.php) and the result was shown in Table 12.

Table 12. The biological activity of 6 flavonoid and phenolics identified in EcoE2.

Compound Name	Biological activities Prediction	Pa Value
1-Hydroxy-3H-2-benzopyran-3-one	Neuropeptide Y4 antagonist	0,969
	Aspulvinone dimethylallyltransferase inhibitor	0,934
	CYP2C12 substrate	0,900
6-Hydroxycoumarin	CYP2C12 substrate	0,972
	CYP2A11 substrate	0,937
	4-Nitrophenol 2-monooxygenase inhibitor	0,929
	Chlordecone reductase inhibitor	0,927
	Membrane integrity agonist	0,926
	Aspulvinone dimethylallyltransferase inhibitor	0,925
	CYP2A4 substrate	0,921
	Aryl-alcohol dehydrogenase (NADP+) inhibitor	0,909
	CYP2B5 substrate	0,902
6-methyl-2H-chromen-2-one	CYP2C12 substrate	0,952
	CYP2A11 substrate	0,937
	CYP2B5 substrate	0,918
	4-Nitrophenol 2-monooxygenase inhibitor	0,914
Benzofuran-4-carboxylic acid	Methylenetetrahydrofolate reductase (NADPH) inhibitor	0,902
Buxuletin	CYP2C12 substrate	0,958
	Chlordecone reductase inhibitor	0,938
	Aspulvinone dimethylallyltransferase inhibitor	0,931
	CYP2A11 substrate	0,926
	4-Nitrophenol 2-monooxygenase inhibitor	0,921
	UGT1A6 substrate	0,903
Catechin	Membrane integrity agonist	0,983
	Mucomembranous protector	0,962
	TP53 expression enhancer	0,959
	HMOX1 expression enhancer	0,939
	Sulfotransferase substrate	0,927
	CYP1A1 substrate	0,927
	CYP2C12 substrate	0,909

From 108 putative identified compounds, there are 93 compounds that have SMILES in PubChem database. These 93 compounds were analyzed by ADMETLab 2.0 to observe the acceptance of the compounds for 4 drug candicate rules i.e. Lipinski's, Pfizer, GSK, GoldenTriangle rules. There are 17 compounds that accepted in all 4 rules as shown in Table 13.

Table 13. The compounds that accepted based on 4 rules for drugs candidate

No	CompoundName	Lipinski	Pfizer	GSK	GoldenTriangle
1	(5Z,8Z,12Z,15Z)-N-(2-hydroxyethyl)ericosa-5,8,12,15-tetraenamide	Accepted	Accepted	Accepted	Accepted
2	(3R,4S,5R,6R)-3,4,5-trihydroxy-6-(hydroxymethyl)oxan-2-yl)ethanperooxide	Accepted	Accepted	Accepted	Accepted

3	(3-hydroxyimino-3-((1H-1,2,4-triazol-5-yl)propyl)phosphoric acid	Accepted	Accepted	Accepted	Accepted
4	(4-((1-(hydroxymethyl)pyrazol-4-yl)methyl)pyrazol-1-yl)methanol	Accepted	Accepted	Accepted	Accepted
5	2-((3-Hydroxy-2,2-dimethylpropanoyl)amino)propanedioate	Accepted	Accepted	Accepted	Accepted
6	2-((5-Methyl-1,3,4-thiadiazol-2-yl)sulfany)butanedioic acid	Accepted	Accepted	Accepted	Accepted
7	2-(((2S)-piperidine-2-carbonyl)amino)pentanedioic acid	Accepted	Accepted	Accepted	Accepted
8	5-Acetyl-4-methylsulfonyl-1,3-thiazole-2-carbonitrile	Accepted	Accepted	Accepted	Accepted
9	Corvalen	Accepted	Accepted	Accepted	Accepted
10	Isoleucine	Accepted	Accepted	Accepted	Accepted
11	Music Acid	Accepted	Accepted	Accepted	Accepted
12	N'-(4-methoxypyrimidin-2-yl)cyclopropanecarbohydrazide	Accepted	Accepted	Accepted	Accepted
13	N,N-Diterglycetalamide	Accepted	Accepted	Accepted	Accepted
14	N,N-Ditethylanola-mine	Accepted	Accepted	Accepted	Accepted
15	N-((1Z,4Z,6Z)-1-(1-(5-butyl-1,3-dienoxy)octa-1,4,6-trien-2-yl)methanamine	Accepted	Accepted	Accepted	Accepted
16	N-Benzyl-2-phenylacetamide	Accepted	Accepted	Accepted	Accepted
17	Pantothenate	Accepted	Accepted	Accepted	Accepted

We can predict the biological activities of these 17 compounds by using PASS online Way2Drug Prediction. The Table 14 shows the predicted bioactivity of the 17 compounds predicted as the highest Pa value.

Table 14. The biological activities of the 17 compounds

No	CompoundName	Biological activity Prediction Way2drug
1	(5Z,8Z,12Z,15Z)-N-(2-hydroxyethyl)ericosa-5,8,12,15-tetraenamide	Fucosterol-epoxide lyase inhibitor Antieczenatic
2	((3R,4S,5R,6R)-3,4,5-trihydroxy-6-(hydroxymethyl)oxan-2-yl)ethanperooxide	Exoribonuclease II inhibitor Levanase inhibitor
3	(3-hydroxyimino-3-((1H-1,2,4-triazol-5-yl)propyl)phosphoric acid	Glutamate-5-semialdehyde dehydrogenase inhibitor
4	(4-((1-(hydroxymethyl)pyrazol-4-yl)methyl)pyrazol-1-yl)methanol	5-O-(4-coumaroyl)-D-quinate 5'-monooxygenase inhibitor Carboxypeptidase Taq inhibitor

5	2-((3-Hydroxy-2,2-dimethylpropanoyl)amino)propanedioate	Fucosterol-epoxide lyase inhibitor Protein-disulfide reductase (glutathione) inhibitor
6	2-((5-Methyl-1,3,4-thiadiazol-2-yl)sulfany)butanedioic acid	Insulin promoter
7	2-(((2S)-piperidine-2-carbonyl)amino)pentanedioic acid	Proline racemase inhibitor
8	5-Acetyl-4-methylsulfonyl-1,3-thiazole-2-carbonitrile	Mucomembranous protector
9	Corvalen	Membrane integrity agonist Mucomembranous protector
10	Isoleucinphrine	Chymosin inhibitor Acryloylindropepsin inhibitor
11	Music Acid	Aspartate 4-decarboxylase inhibitor Levanase inhibitor
12	N'-(4-methoxypyrimidin-2-yl)cyclopropanecarbohydrazide	HMGCS2 expression enhancer
13	N,N-Dibenzylacetamide	CYP2H substrate Membrane integrity agonist
14	N,N-Dibenzylethanolamine	Glycosylphosphatidylinositol phospholipase D inhibitor Photic disorders treatment
15	N-((1Z,4Z,6Z)-1-(1-(5-butyl-1,3-dienoxy)octa-1,4,6-trien-2-yl)methanamine	Fatty-acyl-CoA synthase inhibitor GGTP substrate
16	N-Benzyl-2-phenylacetamide	Photic disorders treatment Mucomembranous protector
17	Pantothenate	Benzoate-CoA ligase inhibitor Leukopoiesis stimulant

4 Conclusion

The results of this study indicate that EcoE from kecapi and matoa fruit peels has low to moderate antioxidant activity. EcoE-2 has better antioxidant and reducing power than EcoE 1, so it can be seen that the high portion matoa fruit peel content increases the antioxidant and reduction power of ecoenzymes. More than 100 putative organic compounds have been identified by LC-HRMS including sereval flavonoids and phenolics. The total experimental phenolic content of this ecoenzyme is quite high. Although high Folin–Ciocalteu values were observed, these results should be interpreted as overall reducing capacity rather than absolute phenolic content due to the non-specific response of the assay in fermentation-derived matrices. The findings highlight the potential of ecoenzyme fermentation as a sustainable approach to valorize fruit peel waste while underscoring the importance of complementary analytical techniques, such as LC-HRMS, for accurate bioactive compound characterization. Future studies should focus on targeted quantification of key antioxidant compounds and biological validation to better define the functional applications of ecoenzymes.

Acknowledgement

The authors express their heartfelt appreciation to the 2024 Student Creativity Program (PKM), supported by the Directorate General of Higher Education, Research, and Technology through the Directorate of Learning and Student Affairs, for funding this research. Gratitude is also extended to SDAE division of Universitas Pelita Harapan for their additional financial support. We thank We thank Lembaga Penelitian dan Pengabdian Masyarakat (LPPM) Universitas Pelita Harapan for supporting this research by some funding allocated from research proposal grand number of P-062-FIP/XII/2024. Also, LPPM UPH has supported for conference in 2025 and publication fee. Thank you for Happy Chelsy Betaria Napitupulu and Misael Ivan Pangela for helping their junior friends.

References

1. H. C. B. Napitupulu, M. I. Pangela, H. T. Telaumbanua, J. S. P. Lase, S. S. Ringo, and C. Y. Tahya, The Antioxidant Test of Ecoenzymes from Various Fruit Peel Wastes, Indo. J. Chem. Res. **13**, 221–226 (2026)
2. B.P. Samadikun, S. Sudarno, Y.M. Pusparizkita, N. Hardyanti, F.S. Pratama, R.P. Safitri, Organic solid waste management by producing eco-enzymes from fruit skin in Permata Tembalang. J. Presipitasi **20**, 21–30 (2023) https://doi.org/10.14710/vol%25viss%25ipp537-551
3. T. Hayati, M. Mahrudin & R. Irianti, Ethnobotany of Sandoricum koetjape Burm.f. Merr. (Kecapi) among the Bakumpai Dayak in Bantuil, Barito Kuala Regency in the form of a popular scientific book. J. Pengabdi Ilm. Sos. (JUPEIS) **1**, 323 (2022). https://doi.org/10.57218/jupeis.Vol1.Iss4.323
4. Z. D. Nassar, A. A. F. Aisha, and A. M. S. Abdul Majid, The Pharmacological Properties of Terpenoids from Sandoricum koetjape, WebmedCentral Complementary Medicine **1(12)**, WMC001311 (2010)
5. F. N. Hamzah, A. Ahmad, S. Subandi, A. W. Septama, W. Sujarwo, H. Choudhry, M. A. Zamzami, S. Al-Thawadi, T. Mozef, Investigating Sandoricum koetjape leaf extracts for tyrosinase inhibition: flavonoid profiles and mechanisms. E3S Web Conf. **577**, 02008 (2024). https://doi.org/10.1051/e3sconf/202457702008
6. M. N. Tehuayo, A. Ulfa, Identification of the morphological structure of the plant Pometia pinnata in the environment of Muhammadiyah University of Education (UNIMUDA) Sorong campus. Biolearning Journal **10**, 2406–8241 (2023).
7. S. Hajar, W. Rahmah, E. M. Putri, S. Septian Ressandy & H. Hamzah, The potential of matoa fruit extract (Pometia pinnata) as a source of antioxidants. J. Farm. Sains Prakt. **7**, 59–66 (2021). https://doi.org/10.31603/pharmacy.v7i1.4240
8. M. M. Gaspersz & H. Fitrihidajati, Utilization of ecoenzymes made from orange peel and pineapple peel waste as LAS detergent remediation agents. LenteraBio: Berk. Ilm. Biol. **11**, 503–513 (2022).
9. E. Yulistia & R. L. Chimayati, Utilization of organic waste into ecoenzyme. Unb. Env. Eng. J. **2**, 1–6 (2021). https://doi.org/10.54895/ueej.v2i01.1184
10. B. Arifin & S. Ibrahim, Structure, bioactivity, and antioxidant of flavonoids. J. Zarah **6**, 21–29 (2018). https://doi.org/10.31629/zarah.v6i1.313
11. K. Simanjuntak, The role of flavonoid antioxidants in improving health. Maj. Ilm. (UPN) **23**, 135–140 (2012)
12. S. Aryal, M. K. Baniya, K. Danekhu, P. Kunwar, R. Gurung & N. Koirala, Total phenolic content, flavonoid content and antioxidant potential of wild vegetables from Western Nepal. Plants **8**, 96 (2019). https://doi.org/10.3390/plants8040096
13. J. Tang, F. R. Dunshea & H. A. R. Suleria, LC-ESI-QTOF/MS characterization of phenolic compounds from medicinal plants (hops and juniper berries) and their antioxidant activity. Foods **9**, 7 (2020). https://doi.org/10.3390/foods9010007

14. C. Y. Tahya, M. Cornelia, K. Sinaga, F. Purba. Production and application of ecoenzyme from mixture of Persicaria odorata, Bidens pilosa, Macaranga tanarius, Merremia vitifolia, Hibiscus tiliaceus, and Ricinus communis leaves as fertilizer for Brassica rapa L. sprout. AIP Conf Proc. **3206(1)**, 030007 (2025). https://doi.org/10.1063/5.0259021

Determination of carotene content of crude palm oil using Raman spectroscopy

Muhammad Azhar Faqih[1*], *Fauziah Fitri* Agnia[1], *Annida* Amani[1], *Nina Siti* Aminah[1], *Herman*[1], *Sparisoma* Viridi[1], *Anita* Alni[1], and *Mitra* Djamal[1]

[1]Institut Teknologi Bandung, Jawa Barat, Indonesia

Abstract. As one of the world's largest palm oil producers, Indonesia regulates its production through Indonesian National Standards (SNI) to ensure the quality of palm oil available on the market. The main parameters in the SNI standard include color, odor, free fatty acids (FFA), iodine value, water content, and DOBI (Deterioration of Bleachability Index). Beta-carotene, a type of carotenoid, functions as a provitamin A that can be converted into active vitamin A after metabolism in the body. The beta-carotene content in palm oil affects its color and clarity. This research analyzed crude palm oil using two heating stages. Two methods were used to determine the carotene content: the conventional method and Raman spectroscopy combined with Partial Least Squares (PLS) as a modern alternative. Raman spectroscopy uses a light source with a wavelength of 785 nm directed at the sample. The results show that the data obtained from Raman spectroscopy closely matches the data from the conventional method, indicating that Raman spectroscopy can be used as a supporting tool to determine oil quality based on its carotene content.

1 Introduction

Indonesia is one of the world's largest producers of crude palm oil. According to data from the Central Statistics Agency (BPS), Indonesia produces 59% of the world's total palm oil production, or 45.5 million tons per year [1]. In accordance with Indonesian National Standard (SNI) 2901:2021 [2], crude palm oil distributed in Indonesia must meet several requirements aimed at protecting consumers, including Free Fatty Acid (FFA) content, water content, Iodine Value, Deodorization and Bleaching Index (DOBI), odor, and color. Although SNI specifies a reddish-orange color as the quality criterion for CPO, this assessment remains subjective because it can only be evaluated visually. Considering that carotenoid pigments, especially β-carotene, are the main factors responsible for this coloring, determining the carotene content offers a more objective and quantitative alternative for evaluating color quality.

Carotene is a type of natural pigment, specifically a carotenoid that gives certain fruits and vegetables their orange, red, or yellow color [3]. It also acts as an antioxidant and a precursor to vitamin A, meaning your body can convert it into vitamin A, which is essential for eye health, immune system function, and skin health [4]. Crude palm oil (CPO) is rich in

* Corresponding author: muhammadazharfaqih@gmail.com

carotene, especially beta-carotene, which gives the oil its reddish-orange color. This is one of the main indicators of quality in crude palm oil [5].

UV-Vis (Ultraviolet-Visible) spectroscopy is a conventional method commonly used to identify and measure carotene content in palm oil. However, this method has several weaknesses, such as the use of hazardous chemicals (hexane is highly flammable) and the long time required for data collection. Raman spectroscopy has several advantages that make it worth considering for palm oil quality testing. This technique can provide detailed information about the molecular structure and chemical bonds in palm oil samples without requiring complicated sample preparation. The purpose of this research is to test the use of conventional and Raman spectroscopy methods in determining the quality of palm oil circulating in Indonesia based on its carotene content.

Moreover, the data obtained from Raman spectroscopy can be further analyzed using statistical methods to obtain additional information. Partial Least Squares (PLS) is a versatile technique for modeling complex, high-dimensional data. PLS can model the relationship between the chemical composition of palm oil and its quality characteristics, such as carotene content [6]. Since CPO usually undergoes additional processes, such as processing to produce other products, PLS can analyze the impact of these processes on palm oil quality and even suggest optimal conditions for obtaining the best quality. Therefore, PLS is a great method to evaluate palm oil due to its ability to handle multicollinearity, reduce dimensionality, and possess the power to predict [7].

2 Materials and Methods

In this research, we analyzed 15 types of Crude Palm Oil (CPO) from four different companies: Cikasungka, Musim Mas, Kertajaya, and Wilmar, with each sample consisting of 10 Raman spectra due to sample limitations in evaluating quality based on carotene content. Each sample was heated using an oven at a temperature of 200°C, producing three types of data: samples before heating, samples from the first heating stage (200°C for 1 hour), and samples from the second heating stage (200°C for 2 hours). This method is based on research by Sampaio et al., which showed that carotenoids in palm oil undergo significant degradation during heating at temperatures of 170–230 °C with heating times of up to ±140 minutes [8]. The carotene content in each sample was obtained based on ISO 17932:2011 [9]. Test samples were weighed with a precision of 0.1 g according to standard procedures and placed in small tubes. Next, 8 g of isooctane solvent was added, and the mixture was stirred thoroughly to ensure homogeneity. After thorough mixing, the absorbance of the test sample was measured in the range of 400 to 500 nm using an Evolution 220 UV-Visible Spectrophotometer (Thermo Scientific). The carotene content can be calculated using equation 1, where A is the absorbance value at 446 nm and W is the sample weight in grams.

$$\text{Carotene content (ppm)} = \frac{383 \times A \times 25}{W \times 100} \quad (1)$$

MacroRAM Raman Spectroscopy (Horiba) with an excitation wavelength of 785 nm and power beam of 90 mW was used in this research to record the intensity in the Raman shift range from 800 to 3500 cm^{-1}. Then, the spectra were pre-processed by doing baseline, smoothing, and cropping. Next, we used Partial Least Square (PLS) to analyze oil quality based on carotene content. For PLS, we defined variables as predictors and responses, then created a PLS model using Raman spectrum data and a latent variable number of 3. We calculated the intensity in the Raman shift range from 800 to 3000 cm^{-1} for the PLS model of carotene content. Here, we evaluate the model fit using R-squared and RMSE values and visualize the results in graphs.

3 Results and Discussions

Due to the highly sensitive nature of Raman spectroscopy to any changes in molecular symmetry, particularly for symmetric vibrations and non-polar bonds, we observed the Raman spectra of Crude Palm Oil (CPO) as shown in Figure 1. Based on the results, we found that the Raman spectra show peaks such as the CH_2 bending (scissoring) in the fatty acid chains at 1442 cm^{-1} or the bending vibration of C-H (twisting) typically in the saturated fatty acids at around 1310 cm-1. It also presents the existence of the C=C (alkene) bond in unsaturated fatty acids at around 1657 cm^{-1} and the stretching of the carbonyl group (C=O). Furthermore, we noticed the notable peaks at around 1150-1200 cm^{-1} and at 1500 cm^{-1}, associating with the stretching vibrations of the conjugated double bonds in the polyene chain in the carotenoid. Carotenoids are pigments in plants that give vibrant colors thus giving the reddish color of CPO. Compared to other palm oil, CPO has a broader range of fatty acids. Thus, CPO contains abundant amounts of palmitic acid (C16:0), oleic acid (C18:1), and linoleic acid (C18:2).

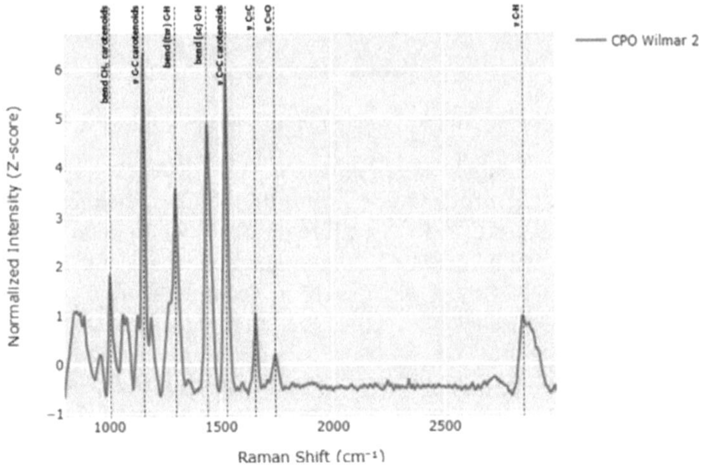

Fig. 1. Raman spectrum of one sample of Crude Palm Oil (CPO).

As part of refining process, CPO is often heated to break down certain compounds such as free fatty acids and make it easier to process. Here, we investigated the effect of heating CPO by comparing the Raman spectra before and after heating as shown in Figure 2. The obvious change is the absence of peaks associated with carotenoids in Raman spectra of heated sample, causing the CPO visually changes to less reddish. The peaks associated with carotenoids appear in the range of 1150–1200 cm^{-1} and around 1500 cm^{-1}, which gradually diminish as the heating process progresses.

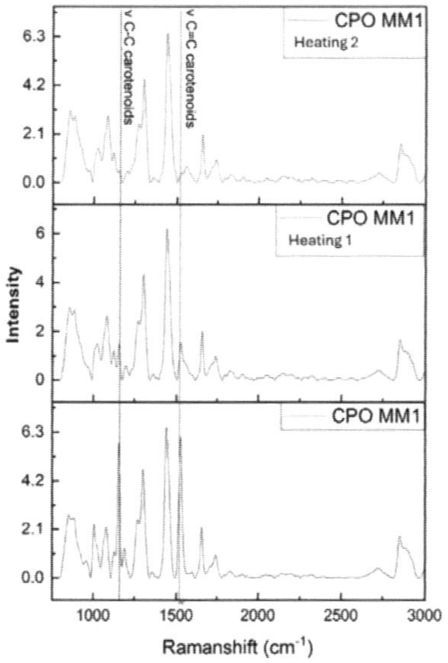

Fig. 2. Raman spectrum of one sample of Crude Palm Oil (CPO) before and after the heating process.

Indeed, Raman is a powerful tool for analyzing oil quality. However, relying only on the Raman spectrum might make it difficult to determine the quality of unknown oils. Here, we propose using Partial Least Square (PLS) to analyze the condition of unknown oils. PLS can model the relationship between oil quality attributes and Raman spectra. By using PLS, we can predict the concentration of these parameters by learning several predictors such as spectra. PLS reduces the dimensions of spectral data but retains the main information from the data.

To predict carotene content, the complete spectral range from 800 to 3000 cm^{-1} and a latent variable number of 3 were used, giving results for samples before heating, samples from heating stage 1, and samples from heating stage 2, as shown in Figure 3-5. The diagonal line (y = x) represents perfect agreement between Raman predictions and conventional methods. The proximity of data points to the y = x line demonstrates the reliability of the PLS model in predicting carotene content within the expected quality range. Figure 3 shows the PLS regression results for evaluating CPO quality before the heating process based on carotene content, with an R^2 value of 0.856 and an RMSE of 0.623, indicating that the Raman spectroscopy model explains 85.6% of the variance and the average prediction error of the model is within the range of 0.623 units compared to the actual value. Subsequently, Figure 4 shows the PLS regression results for CPO after the first heating stage, with an R^2 value of 0.984 and an RMSE of 0.116, indicating that the Raman spectroscopy model explains 98.4% of the variance and the average prediction error of the model is within the range of 0.116 units compared to the actual value. Meanwhile, Figure 5 presents the PLS regression results for CPO after the second heating stage, with an R^2 value of 0.93 and an RMSE of 0.032, indicating that the Raman spectroscopy model explains 93% of the variance and the average prediction error of the model is within the range of 0.032 units compared to the actual value. The R^2 and RMSE values obtained indicate that the PLS model is capable of explaining the variation in carotene content and producing relatively low prediction errors across the entire dataset analyzed.

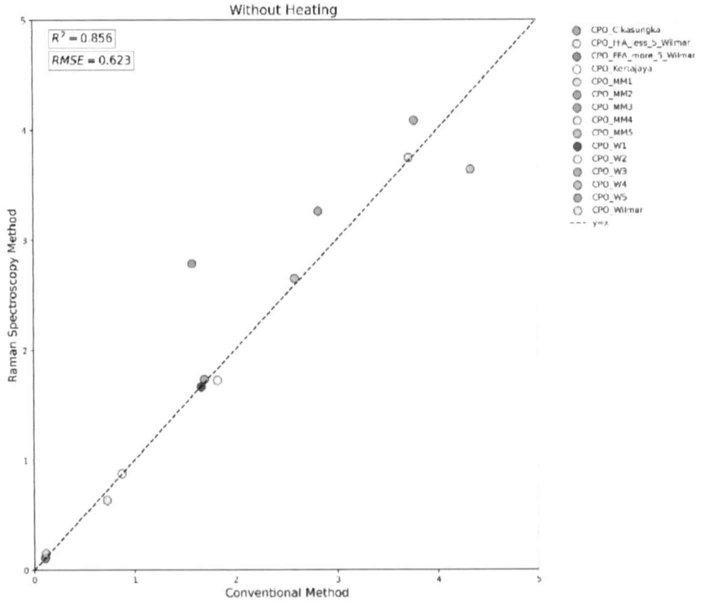

Fig. 3. PLS regression results to evaluate CPO quality before the heating process based on carotene content.

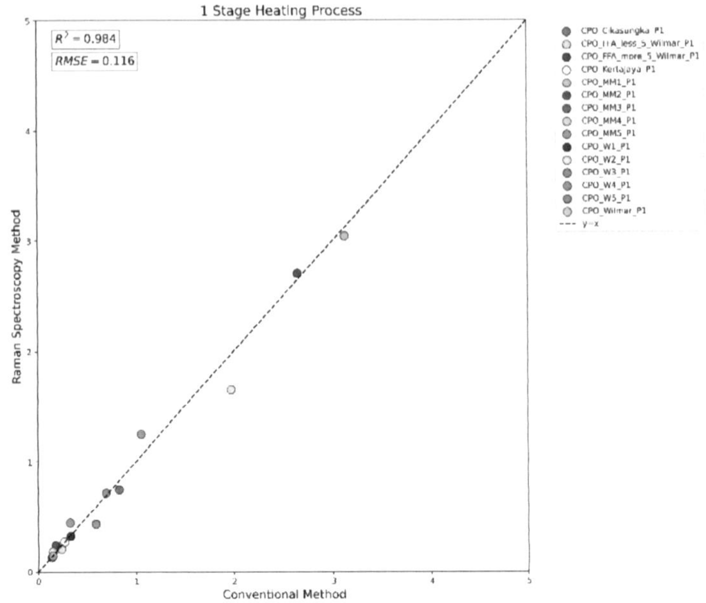

Fig. 4. PLS regression results for evaluate CPO quality from heating process stage 1 based on carotene content.

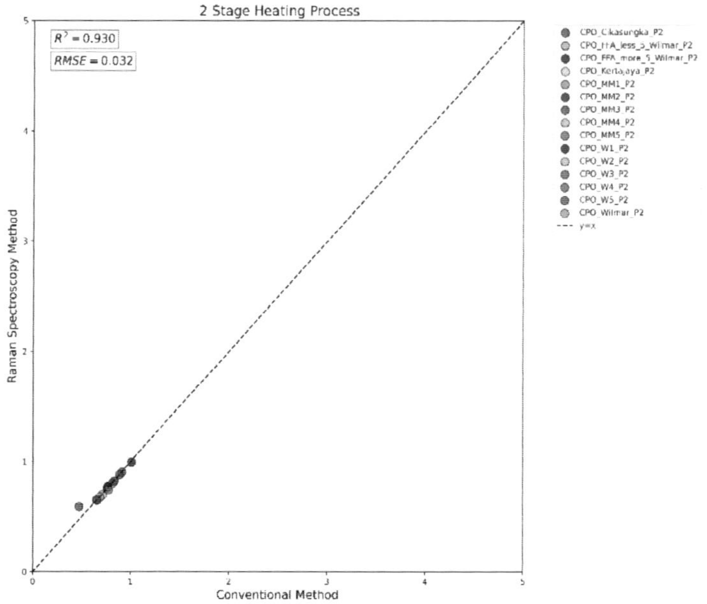

Fig. 5. PLS regression results for evaluate CPO quality from heating process stage 2 based on carotene content.

4 Conclusion

Based on our research, we found that Raman spectroscopy can be used as a supporting or screening tool to determine the quality of palm oil circulating in Indonesia based on its carotene content in accordance with Indonesian National Standards (SNI). In addition, Raman spectroscopy can be used to identify differences between palm oil and its derivatives. Furthermore, the results obtained through the Raman spectroscopy method in carotenoid content can be categorized as fairly close to the results of conventional methods, as indicated by the R-squared and RMSE values obtained.

References

1. Badan Pusat Statistik, Statistik Kelapa Sawit Indonesia 2022 (Badan Pusat Statistik, Jakarta, 2022).
2. Badan Standardisasi Nasional, SNI 2901:2021 – Minyak Sawit Mentah (Crude Palm Oil) (Badan Standardisasi Nasional, Jakarta, 2021).
3. F. Fitra, R. Rasyid, and R. Fadhlany, Pengaruh proses pengolahan terhadap beta karoten pada ubi jalar varietas ungu, in Proc. Semin. Nas. Univ. Negeri Padang (2015), pp. 1–6.
4. P. Rahayu, S. Fathonah, and F. Fajri, Daya terima dan kandungan gizi makanan tambahan berbahan dasar ubi jalar ungu, J. FSCE. **1**, 1 (2012).
5. A. Mortensen, Analysis of a complex mixture of carotenes from oil palm (Elaeis guineensis) fruit extract, Food Res. Int. **38**, 8–9 (2005). https://doi.org/10.1016/j.foodres.2005.01.009
6. K. N. Basri, M. N. Hussain, J. Bakar, Z. Sharif, M. F. A. Khir, and A. S. Zoolfakar, Classification and quantification of palm oil adulteration via portable NIR spectroscopy,

Spectrochim. Acta A Mol. Biomol. Spectrosc. **173**, 335–342 (2017). https://doi.org/10.1016/J.SAA.2016.09.028

7. W. Budiastra, S. Marjan, N. Adiarifia, I. Novianty, and Y. T. Suci, Non-destructive prediction of oil and free fatty acid of oil palm fruitlets using near-infrared spectroscopy and hybrid calibration method, INMATEH – Agric. Eng. **73**, 2 (2024). https://doi.org/10.35633/INMATEH-73-39
8. K. A. Sampaio, J. V. Ayala, S. M. Silva, R. Ceriani, R. Verhé, and A. J. A. Meirelles, Thermal degradation kinetics of carotenoids in palm oil, J. Am. Oil Chem. Soc. **90**, 2 (2012). https://doi.org/10.1007/s11746-012-2156-1
9. International Organization for Standardization, ISO 17932:2011: Animal and vegetable fats and oils — Determination of carotene content by spectrophotometry (ISO, Geneva, 2011).

Immersion time effects on composite propellant burn rate

Dwi Setyaningsih[1]* and *Aprilia* Fitri Yastuti[1]

[1]Research Center for Rocket Technology, National Research and Innovation Agency, Rumpin, 16350, Indonesia

Abstract. Acoustic Emission Strandburner is one of popular methods to measure the solid propellant burn rate. Immersion is a critical factor, as HTPB-based propellants contain ammonium perchlorate, a water-soluble oxidizer. This study aims to evaluate the effect of short-time immersion in water on propellant burn rate. Propellant samples were tested under three conditions: without immersion (normal condition), after 60 seconds of immersion, and after 120 seconds of immersion. The results showed that the burn rate at a chamber pressure of 2–6 MPa ranged from 5.38 – 8.36 mm/s, 5.60–7.93 mm/s, and 5.38 – 8.35 mm/s for normal condition, 60 seconds immersion, and 120 seconds immersion, respectively. Statistical analysis using a one-way ANOVA revealed no significant differences in burn rate among the three conditions. These findings indicate that short-term immersion in water does not significantly affect the combustion performance of the propellant..

1. Introduction

The performance of rocket motors is significantly influenced by the characteristics of the propellants used as fuel. Composite solid propellants are the most widely adopted due to their broad operational temperature range, controlled burn rate, high storage stability, adequate mechanical properties, and relatively low production costs [1]. These propellants typically consist of ammonium perchlorate (AP) as the oxidizer, hydroxyl-terminated polybutadiene (HTPB) as the polymeric binder, aluminum as the metallic fuel, and various additives with a composition of 65–68%, 10–15%, 5–15% and 0-5% by weight, respectively [1, 2].

Burn rate is the most essential internal ballistic parameter in designing and operating of a rocket motor [1]. It is defined as the linear regression rate of the propellant surface in the direction perpendicular to combustion, or the distance traversed by the flame front per unit time under specific pressure and temperature conditions [3]. Burn rate not only governs the rate of gas production but also directly influences the mass flow rate of combustion gases exiting the motor, thereby affecting thrust generation [4] Several factors influence the burn rate, including propellant composition, chamber pressure, and initial grain temperature.

*Corresponding author: dwis011@brin.go.id

Ammonium perchlorate as an oxidizer is particularly significant as it is the dominant component in most composite solid propellants [5, 6].

Burn rate measurements can be performed using various techniques, such as Crawford strand burners, small-scale ballistic motors, and full-scale rocket motors [1]. Other methods include closed bomb testing, accoustic emission, ultrasonic techniques, microwave sensing, X-ray imaging, plasma capacitance, pressure probes, interrupted combustion, and high-speed imaging. Among these, the Acoustic Emission Strand burner (AES) is gaining popularity due to its simplicity, speed, cost-effectiveness, and high accuracy—reportedly up to 98% [3, 7].

AES operates on the principle of detecting acoustic signals generated during the underwater combustion of a propellant at specific pressures. The continous acoustic signals generated by the specimen during combustion are converted into electric signals, amplified by a preamplifier, and sent to the data acquisition system for calculation [7, 8].

However, this immersion condition introduces a new variable: water solubility of AP, which is approximately 24.5 g per 100 g of water at 25 °C [9]. Since AP is water-soluble, immersion may lead to partial leaching of the oxidizer, potentially affecting burn rate measurements. Several investigations have been made to determine the effects of water and humidity on AP or solid propellants. Fournier and Brady determined the AP diffusion coefficients when immersed in different water salinity levels and temperatures for 12 weeks [10]. After 10 days of aging at different relative humidity levels, the burn rate of solid propellants decreased due to the growth of AP crystal size and surface morphology changes, resulting in less accessible surface area. The alterations coincide with the moisture content of the aging environment [11].

The novelty of this study is that short duration water immersion effect on burn rate measurement of composite propellant by an AES has been studied systematically. Even the water solubility of AP is known, but the effect of short time immersion such as underwater AES testing has not been discussed. Therefore, this study was conducted to evaluate the effect of propellant immersion time on the measured burn rate, with specific attention to the influence of short-term water exposure on combustion performance.

2. Methodology

The propellant used in this study was a composite solid propellant composed of an HTPB–TDI binder system, ammonium perchlorate (AP) as the oxidizer, and aluminum (Al) as the metallic fuel, with a total solid content of 82.5% by weight. Test samples were prepared in the form of square rods measuring 5 × 5 mm in cross-section and approximately 100 mm in length, using a dedicated combustion test sampling device.

Acoustic Emission Strandburner (AES) was used to measure propellant burn rate. This device employs an acoustic emission method, with water as the combustion medium. (Figure 1). Each sample was mounted onto a holder using nickelin wire and an inhibitor coating (Figure 2), then immersed in a test chamber filled with approximately 500 mL of water. The sample length and desired test pressure were entered into the AES software before the test began.

Testing was conducted at three pressure levels: 2, 4, and 6 MPa, with a minimum of five replicates per pressure level, depending on the consistency of the data obtained. The immersion conditions applied were 0 seconds (no immersion), 60 seconds, and 120 seconds, where immersion time was recorded from the moment the sample holder entered the chamber, as indicated by a software signal light.

Fig. 1. Burn rate testing apparatus : Acoustic Emission Strand Burner (AES)

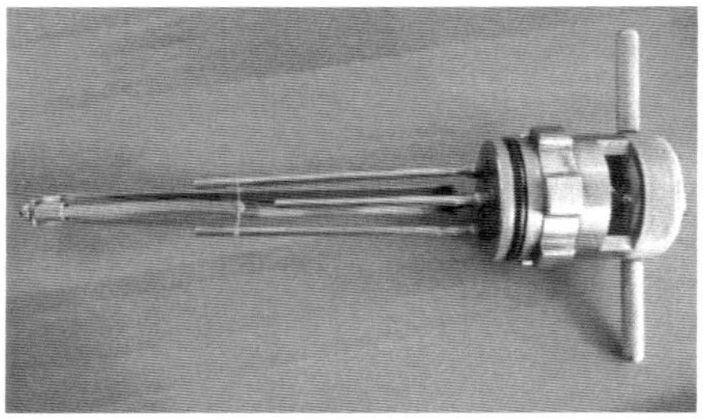

Fig. 2. Sample holder of Accoustic Emission Strandburner

The immersion times of 60 and 120 seconds were chosen to represent realistic short-term conditions during burn rate testing, including specimen insertion, system stabilization, and ignition preparation in water medium. These times were specifically designed to reflect practical handling circumstances rather than protracted moisture aging. Short immersion periods were used to determine if the limited water incorporation in AES procedures could cause burn rate changes through initial AP dissolution or surface reactivity effects, while preventing long-term oxidizer dilation from diffusion.

The output of the testing procedure included combustion time and burn rate (in mm/s). A comparative analysis of burn rate values was then performed across the different pressures and immersion durations. To assess the statistical significance of the results, one-way analysis of variance (ANOVA) was used to determine the influence of each variable on the propellant's burn rate. The primary objective of the statistical analysis was hypothesis testing rather than predictive modelling, specifically determining if short-term water exposure on solid propellant causes noticeable bias in burn rate measurements.

3. Results And Discussion

Burn rate measurements were obtained using the Acoustic Emission Strand burner (AES), yielding data in the form of acoustic emission graphs, combustion time, and burn rate values at various chamber pressures. The relationship between burn rate (r) and combustion pressure (P) was analyzed using the Saint Robert's law, expressed as follows (1):

$$r=aP^n \quad (1)$$

where:

- r = burn rate (mm/s)
- a = pre-exponential constant
- P = combustion chamber pressure (MPa)
- n = pressure exponent

The average burn rate values at pressures ranging from 2 to 6 MPa for the three treatment conditions—normal, 60-second immersion, and 120-second immersion—are shown in Table 1. Based on the Saint Robert equation, the constants (a) and pressure exponents (n) were determined and are presented in Table 2.

Table 1. Average burn rate values for all three test conditions

Normal Condition		Immersion Time			
		60 seconds		120 seconds	
P (MPa)	r (mm/s)	P (MPa)	r (mm/s)	P (MPa)	r (mm/s)
6.308	8,522	6.602	7.991	6.106	8.407
4.562	7.498	4.866	7.682	4.593	7.506
2.855	6.199	2.872	6.202	3.048	6.369

Table 2. Values of the constant a and exponent n for each test condition

Testing Conditions	a	n	R^2
Normal	4.0697	0.4017	0.999
Immerse for 60 seconds	4.5018	0.3157	0.942
Immerse for 120 seconds	4.0806	0.3996	1

The results indicate that the n value, which reflects the sensitivity of the burn rate to pressure, decreased after 60 seconds of immersion. This suggests that combustion became less influenced by pressure, possibly due to localized disturbances on the grain surface resulting from water penetration. In contrast, after 120 seconds of immersion, both a and n returned to values close to those under normal conditions, with a coefficient of determination (R^2) equal to 1—indicating a stable and consistent combustion pattern.

This phenomenon is consistent with the effects of short-term moisture exposure, which, according to the literature, can lead to surface softening and uneven oxidizer distribution on the propellant surface—without significantly degrading overall combustion performance [8]. Meanwhile, longer-term moisture exposure can cause more progressive effects such as oxidizer particle bridging, microcrack formation, and oxidizer agglomeration, which may significantly affect regression and burn behavior [12].

The propellant used in this study contains 75% by weight of ammonium perchlorate (AP) as the primary oxidizer, which plays a dominant role in determining burn rate. A reduction in AP content is theoretically known to lower density, burn rate, and specific impulse [13],[14]. Although AP is water-soluble, its release from the HTPB matrix involves a complex mechanism, including water diffusion into the polymer network, local dissolution of AP, and migration out of the matrix. Among these, diffusion is considered the rate-controlling step [10].

The predicted burn rate values over a pressure range of 2–8 MPa, calculated using the Saint Robert equation, are shown in Table 3 and visualized in Figure 3.

Table 3. Comparison of propellant burn rates at pressures of 2–8 MPa for three immersion conditions

P (MPa)	r (mm/s)		
	normal	60 seconds	120 seconds
2	5.376	5.603	5.383
2.5	5.881	6.012	5.885
3	6.327	6.368	6.330
3.5	6.732	6.686	6.732
4	7.102	6.974	7.101
4.5	7.447	7.238	7.443
5	7.769	7.483	7.763
5.5	8.072	7.711	8.064
6	8.359	7.926	8.350
6.5	8.632	8.129	8.621
7	8.893	8.321	8.880
7.5	9.143	8.504	9.128
8	9.383	8.679	9.367

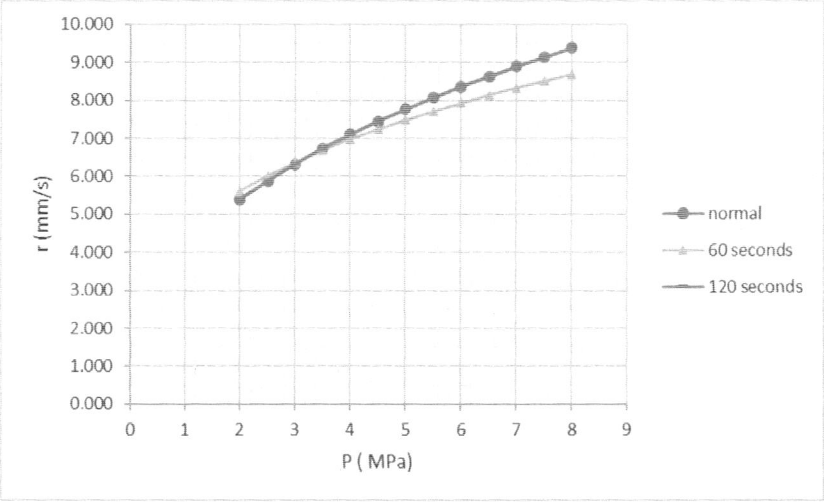

Fig. 3. 1Graph of the relationship between propellant burn rate and pressure for three immersion conditions : normal (0 second), 60 seconds, and 120 seconds

As illustrated in Figure 3, the three burn rate curves—representing normal conditions, 60-second immersion, and 120-second immersion—largely overlap, indicating that the burn rates under the three treatments are relatively similar.

To statistically assess whether the observed differences were significant, a one-way ANOVA test was performed under the null hypothesis (H_0) that there is no significant difference among the three groups. The descriptive statistics and ANOVA results are presented in Tables 4 and 5.

Table 4. Descriptive statistics of the average propellant burn ratein three treatment

Groups	Count	Sum	Average	Variance
normal	13	99.11418	7.624168	1.63834
60 seconds	13	95.63355	7.356427	0.960885
120 seconds	13	99.04716	7.619013	1.619848

Table 5. Results of analysis of variance (ANOVA) of propellant burn rate

Source of Variation	SS	df	MS	F	P-value	F crit (α=0.05)
Between Groups	0.60954	2	0.30477	0.216709	0.806207	3.259446
Within Groups	50.62887	36	1.406358			
Total	51.23841	38				

One way ANOVA is an appropriate and adequate framework for evaluating group means differences given the single factor experimental design and restricted exposure levels. Since no other environmental or composition variables were examined, more complex statistical techniques like regression modeling or multifactor analysis w under realistic ere not required. The results demonstrate that immersion time up to 120 seconds do not cause statistically significant differences in burn rate ($p>0.05$), which supports the trustworthiness of AES mesurements under realistic short-term immersion setting.

The results of ANOVA show that the calculated F-value (F) is less than the critical F-value (Fcrit), supporting the acceptance of H_0. Therefore, it can be concluded that no statistically significant difference exists in the burn rate of the propellant between the normal condition and after immersion for 60 or 120 seconds. This finding aligns with previous studies, which reported that the diffusion coefficient for perchlorate leaching from HTPB-based composite propellants immersed in water ranges between 1.07×10^{-12} and 1.53×10^{-12} m^2/s [10]. Given the short immersion durations of 60 to 120 seconds, the estimated amount of AP released is only around 0.05%, which is insufficient to cause any meaningful alteration in combustion characteristics.

Previous studies examined the amount of AP leached when immersed in a water for long term. After more than 17 hours of immersion, mass of AP reduced about 5%. the longer the immersion time, the higher the reduction in AP mass. Consequently, the surface area of the AP also decreases [10]. Reducing the AP content by 5% by weight in propellant composition, may decrease burn rate by up to 20% [15]. However, in practise, specimen immersion during AES burn rate testing only takes a few minutes, therefore no significant effect on the burn rate as demonstrated in this study.

4. Conclusion

The experimental results indicate that the burn rate of the composite propellant at pressures of 2–6 MPa ranges from 5.38 – 8.36 mm/s under normal conditions, 5.60–7.93 mm/s after

60 seconds of immersion, and 5.38 – 8.35 mm/s after 120 seconds of immersion. Based on one-way ANOVA analysis, no statistically significant differences were found among the three conditions. It can therefore be concluded that immersion in water for up to 120 seconds does not significantly affect the burn rate of the HTPB/AP-based composite propellant.

All authors are main contributors. Dwi Setyaningsih and Aprilia Fitri Yastuti designed and performed the experiments, analyzed the datas, and wrote the manuscript.

References

1. G.P. Sutton, O. Biblarz, Rocket propulsion elements, (John Wiley and Sons, Canada, 2016)
2. J. Ashish, G. Swaroop, and K. Balasubramanian, Effect of ammonium perchlorate particle size on flow, ballistic, and mechanical properties of composite propellant, Nanomaterials in Rocket Propulsion Systems, Elsevier, 299-362, (2018). https://doi.org/10.1016/B978-0-12-813908-0.00008-3
3. G. Gupta, L. Jawale, Mehilal, B. Bhattacharya, Various methods for the determination of the burning rates of solid propellants : an overview, Cent. Eur. J. Energ. Mater. **12(3)**, 593-620 (2015).
4. S. Chaturvedi and P.N. Dave, Solid propellants: AP/HTPB composite propellants, Arabic. J. Chem. **12(8)**, 2061–2068, (2019). https://doi.org/10.1016/j.arabjc.2014.12.033
5. R. Sangtyani, H.S. Saha, A. Kumar, A. Kumar, M. Gupta, and P.V. Chavan, An alternative approach to improve burning rate characteristics and processing parameters of composite propellant, J. Combust. Flame. **209**, 357–362, (2019). https://doi.org/10.1016/j.combustflame.2019.04.054
6. H. Yaman, V. Çelik, and E. Degirmenci, Experimental investigation of the factors affecting the burning rate of solid rocket propellants, J. Fuel. **115**, 794–803, (2014). https://doi.org/10.1016/j.fuel.2013.05.033
7. A. F. Nour Eldin, W. M. Adel, and A. Elsabbagh, Evaluation of solid propellant burning rate by using strand burner and static firing tests, J. Phys. Conf. Ser. **2811** 012027, (2024). https://doi.org/10.1088/1742-6596/2811/1/012027
8. E. Elsaka, S. Elbasuney, H. Mostafa, T. Elhedery, and A. Eldakhakhny, Burning rate measurement of composite propellant using acoustic wave emission in comparison with other techniques, J. Eng. Sci. Mil. Technol. **4(2)**, 226–232, (2020). https://doi.org/10.21608/ejmtc.2021.64610.1167
9. D.R. Lide, CRC handbook of chemistry and physics, (CRC Press, Boca Raton, FL, 2005).
10. E.W. Fournier and B.B. Brady, Perchlorate leaching from solid rocket motor propellant in water, J. Propuls. Power. **21(5)**, 937–941, (2005). https://doi.org/10.2514/1.14246
11. B.A. McDonald, J.R. Rice, and M.W. Kirkham, Humidity induced burning rate degradation of an iron oxide catalyzed ammonium perchlorate/HTPB composite propellant, J. Combust. Flame. **161(1),** 363–369, (2014). https://doi.org/10.1016/j.combustflame.2013.08.014
12. J.C. Thomas, T.E. Sammet, C.A.M. Dillier, A.R. Demko, F.A. Rodriguez, and E.L. Petersen, Aging effects on the burning rates of composite solid propellants with nano-additives, Jt. Propuls. Conf., 1–15, (2018). https://doi.org/10.2514/6.2018-4960
13. A. Manash and P. Kumar, Comparison of burn rate and thermal decomposition of AP as oxidizer and PVC and HTPB as fuel binder based composite solid propellants, J. Def. Technol. **15(2)**, 227–232, (2019). https://doi.org/10.1016/j.dt.2018.08.010
14. S. Park, S. Choi, K. Kim, W. Kim, and J. Park, Effects of ammonium perchlorate particle size, ratio, and total contents on the properties of a composite solid propellant, J.

Propellants Explos. Pyrotech. **45(9)**, 1376–1381, (2020). https://doi.org/10.1002/prep.202000055
15. J. C. Thomas, G. R. Morrow, C. A. M. Dillier, and E. L. Petersen, Comprehensive study of ammonium perchlorate particle size/concentration effects on propellant combustion, J. Propuls. Power **36(1)**, 95–100, (2020). https://doi.org/10.2514/1.B37485

Optimizing ozone solubility for effective surfactant removal and COD reduction in synthetic laundry wastewater

Elfira Diah Ayu Febriani[1], *Maulana Gilar* Nugraha[1], *Avido* Yuliestyan[2], *Rochim Bakti* Cahyono[1], and *Lisendra* Marbelia[1*]

[1]Department of Chemical Engineering, Faculty of Engineering, Universitas Gadjah Mada, Yogyakarta 55281, Indonesia
[2]Department of Chemical Engineering, Faculty of Industrial Technology, Universitas Pembangunan Nasional Veteran Yogyakarta, Yogyakarta 55283, Indonesia

Abstract. Laundry effluents generally contain high concentrations of surfactants and organic substances that can cause serious environmental and health impacts if discharged untreated. This study investigated ozonation as an alternative chemical treatment for synthetic laundry wastewater, focusing on ozone solubility and pollutant removal. Experiments were conducted in a 5 L reactor containing 1,6 mL of detergent/L wastewater with an ozone supply rate of 3 g/h. Operational parameters, including oxidation–reduction potential (ORP), DO, pH, dissolved ozone concentration, surfactant levels (MBAS), and chemical oxygen demand (COD), were monitored for 150 minutes. Based on Henry's law, the theoretical maximum ozone solubility was 7.14 mg/L, but actual values were lower due to limited solubility, instability, and rapid reactions with water constituent. The 5 L reactor at 3 g/h provided the most stable ozone dissolution with lower ozone consumption. Results showed that MBAS decreased by over 90% within 90 minutes, while COD decreased more slowly, reaching about 40% at 150 minutes. The limited COD reduction suggests the presence of complex organics, that are more resistant to ozonation. Overall, ozonation was highly effective for surfactant degradation but less efficient for COD removal, indicating the need for complementary treatments to improve overall wastewater remediation.

1 Introduction

The laundry sector has been rapidly expanding, as reflected in the increasing number of service providers operating in both urban and rural areas. A study by Prochazkova and Masa [1] reported that, on average, 15 liters of freshwater are required to wash 1 kg of linen. Consequently, a facility processing 10 tons of laundry per day could generate up to 150 m^3 of wastewater. Detergent is the main substance used in the washing process. It is a cleaning agent commonly applied in washing activities, with its primary component being surface-active compounds (surfactants) that are capable of reducing the surface tension of water. This

* Corresponding author: lisendra.m@ugm.ac.id

property makes detergents effective in removing dirt from surfaces while keeping it suspended in the solution [2]. In general, detergent formulations consist of three principal components: surfactants, builders such as phosphates, and additives like bleaching agents or fragrances. While surfactants and builders determine cleaning effectiveness, additives are included to improve product performance, stability, and consumer appeal by providing features such as foam stabilizers, optical brighteners, enzymes, fragrances, or dyes [2, 3]. Among the surfactants used, alkylbenzene sulfonates remain the most common. Two structural types of alkylbenzene are employed in their production, namely linear alkylbenzene (LAB) and branched alkylbenzene (BAB) [4].

Water pollution is a persistent issue that affects the entire ecosystem [5].When laundry wastewater is discharged without treatment, it can accumulate as foam on water surfaces, disrupting oxygen transfer and reducing dissolved oxygen levels essential for aquatic organisms such as fish and microorganisms. Beyond ecological concerns, surfactants also threaten human health and can hinder the efficiency of conventional wastewater treatment technologies [6]. Therefore, to mitigate these problems, wastewater treatment must be conducted regularly prior to discharge into the environment. Wastewater treatment technologies generally fall into three categories: physical, chemical, and biological. Biological processes employ microorganisms to degrade organic pollutants, while chemical processes may involve chlorination, ozonation, or ultraviolet (UV) irradiation [7]. Ozonation is particularly attractive due to ozone's high oxidation potential (2.07 V), which enables direct cleavage of chemical bonds in organic molecules [8]. The process also reduces sludge generation compared to chemical coagulants, requires only an ozone generator for operation, and avoids the formation of harmful secondary pollutants since ozone decomposes back into oxygen in the environment [9].

2 Materials and methods

2.1 Materials

The wastewater sample employed in this research was prepared as synthetic laundry wastewater by adding liquid detergent to water. Its formulation was composed of several key components, as shown in Table 1 and was adapted from previous studies on synthetic laundry wastewater composition [10].

Table 1 Composition of the synthetic laundry wastewater

Composition of laundry wastewater	Component composition
Yeast	400 mg/L
Ethanol	103 mg/L
Sodium bicarbonate	480 mg/L
Commercial liquid detergent	1.6 mL/L
Salt solution	5 mL/L

The salt solution itself contained the compounds listed in Table 2.

Table 2 Chemical composition of salt solution

Composition of salt solution	Component composition (g/L)
$NaCl$	50
$MgCl_2.6H_2O$	1.4
$CaCl_2.2H_2O$	1.9

2.2 Experimental procedure

The ozonation experiments were conducted with some variation on the ozone dosage and reactor volume, while temperature and pressure were kept constant. This experiment was done two times, using pure water and synthetic wastewater. Samples were taken at intervals of 0, 30, 60, 90, 120, and 150 minutes during a total reaction time of 2.5 hours. The system was supplied with ozone generated by an ozone generator, as illustrated in Figure 1.

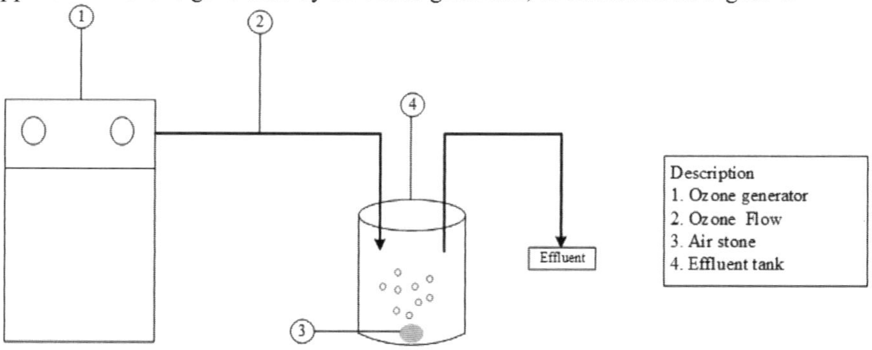

Fig. 1. Experimental setup of the ozonation process

2.3 Analytical Methods

During the ozonation experiments, a number of parameters were recorded, i.e. pH, ORP, DO. ORP and pH measurements were performed during each sampling using a calibrated Lutron pH-208. Additionally, the samples were analysed for its surfactant concentration and COD levels. COD concentration was determined according to the Indonesian National Standard (SNI, 2009), while surfactant concentration was analysed using the Methylene Blue Active Substances (MBAS) method following USEPA 425.1.

The theoretical solubility of ozone in water was estimated with Henry's law, expressed in Equation (1).

$$C = k_H \, x \, p \qquad (1)$$

where C is the dissolved ozone concentration, k_H is Henry's constant, and p is the partial pressure of ozone.

The removal efficiency of MBAS and COD was calculated using Equation (2).

$$Removal\ (\%) = \frac{C_0 - C_t}{C_0} \, x \, 100\% \qquad (2)$$

where C_0 is the initial concentration and C_t is the concentration at time t.

3 Introduction

3.1 Ozone Solubility in Water

The dissolution of ozone in water is primarily governed by gas–liquid mass transfer. Enhancing this transfer process allows higher concentrations of dissolved ozone to be achieved, thereby improving overall ozone utilization efficiency [11]. The extent of ozone solubility can be estimated using Henry's law, which relates the equilibrium concentration of a gas in liquid to its partial pressure in the gas phase. Under ideal conditions, the calculated maximum solubility of ozone was 7.14 mg/L.

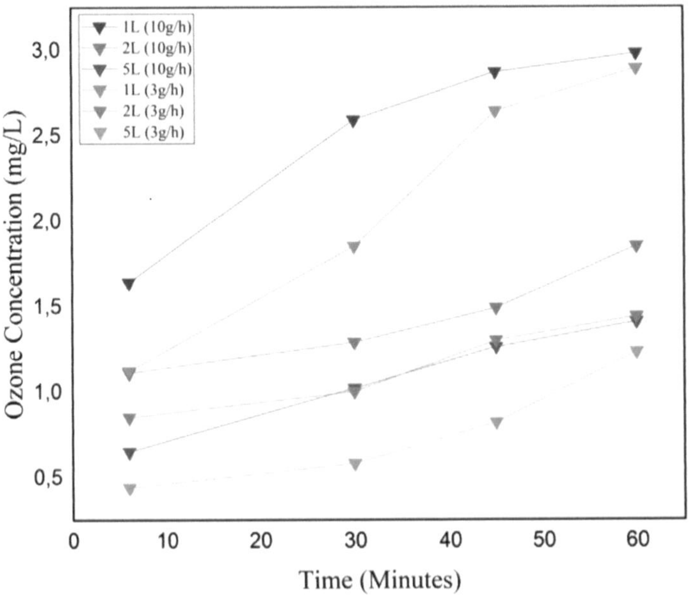

Fig. 2. Ozone concentration at various reactor volume and ozone dosages

As shown in Figure 2, direct measurements indicated that the dissolved ozone concentration was lower than the theoretical value predicted by Henry's law (7.14 mg/L). The lower value suggests that not all ozone present in water remains dissolved. Part of the ozone escapes into the air due to its limited solubility, some decomposes into other compounds, and some directly reacts with substances present in the water. As a result, the measured dissolved ozone concentration is lower than expected. This observation is consistent with Qiqi et al.[12], who reported that ozone exhibits low solubility in water, is highly unstable, readily decomposes, and has a short half-life (20–30 minutes). Figure 2 shows that both reactor volume and ozone dosage influence the dissolved ozone concentration. A smaller volume (1 L) with a high dosage (10 g/h) yielded the highest concentration; however, the difference compared to 5 L at 3 g/h was not significant at the final time point. At 60 minutes, the 5 L volume with a 3 g/h dosage had already reached more than 1 mg/L, approaching the result of 2 L with 10 g/h, despite the much lower dosage. This indicates that at longer contact times, the 5 L–3 g/h condition can produce a stable and efficient increase in dissolved ozone. Therefore, 5 L with a 3 g/h dosage can be considered the optimal condition as it provides adequate results with lower ozone consumption.

3.2 Degradation of Synthetic Laundry Wastewater

Based on the results of ozone solubility in water, the degradation experiments of synthetic laundry wastewater were carried out at a volume of 5 L with a detergent concentration of 1.6 mL/L and an ozone dosage of 3 g/h. Results are shown in Figure 3 and 4.

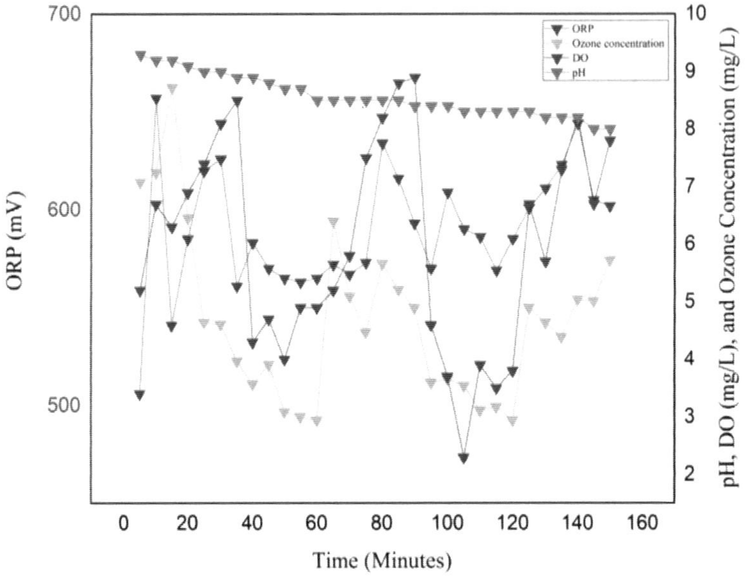

Fig. 3. Process parameters in ozonation

Figure 3 shows the changes in process parameters (ORP, DO, pH, and dissolved ozone concentration) which reflect the reaction conditions during ozonation. ORP values increased as oxidation activity progressed. Dissolved oxygen levels exhibited fluctuations, largely due to the rapid decomposition of ozone into molecular oxygen [13], while the dissolved ozone concentration remained low because it quickly reacted with organic compounds in the water. It is worth noting that the slight decrease in pH indicates that during the ozonation process, acidic intermediates such as carboxylic acids and other oxygenated products were formed, which contributed to the minor reduction in pH [14].

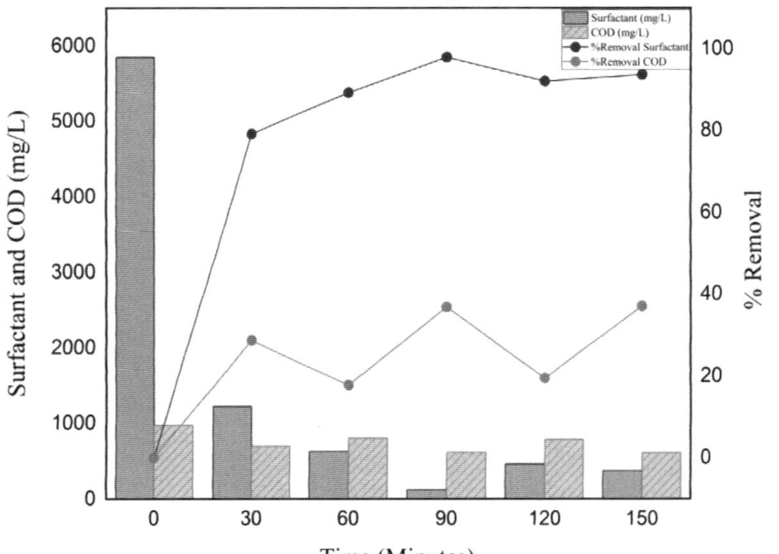

Fig. 4. Degradation of Surfactant and COD During Ozonation

Figure 4 shows the MBAS and COD profile during the ozonation process. A significant decrease in MBAS concentration was observed compared to COD throughout the reaction period. At 0 minutes, MBAS concentration was very high (5800 mg/L), but it dropped sharply to below 100 mg/L at 90 minutes, with a removal efficiency exceeding 90%. This indicates that the surfactants were more easily degraded under the applied ozonation conditions compared to other organic substances measured as COD. In contrast, COD reduction was slower than MBAS.

Despite some fluctuations, COD concentration remained relatively high, with a maximum removal efficiency of around 40% at 150 minutes. This phenomenon suggests that a substantial portion of the COD are more resistant to ozonation and therefore require longer reaction times or additional treatment conditions for optimal degradation. These findings are consistent with previous studies reporting that ozonation is more effective in decomposing surfactants (measured as MBAS) than other organic compounds contributing to COD.

The degradation of MBAS can be attributed to the reaction mechanism of ozone with olefinic and aromatic structures commonly found in surfactant molecules. Ozone selectively attacks carbon–carbon double bonds, forming unstable cyclic ozonide intermediates that subsequently decompose via the Criegee mechanism into smaller compounds such as aldehydes, ketones, and carboxylic acids. The degradation pathway generally involves the formation of primary ozonides followed by zwitterionic intermediates, as described by Joseph et al. [14]. Similar trends have been reported in previous studies, where ozonation was found to be more effective in surfactant removal than in overall COD reduction. Turkay et. al [15] reported a surfactant removal efficiency of 86% by ozonation, while COD reduction after treatment was relatively low, comparable to previous results of 623.53 mg/L.

4 Conclusion

This study demonstrated that ozone solubility in water is strongly influenced by mass transfer and contact conditions, where the theoretical maximum concentration based on Henry's law (7.14 mg/L) was not achieved experimentally due to ozone's limited solubility, instability, and rapid reaction with substances in water. Among the tested conditions, a 5 L reactor volume with an ozone dosage of 3 g/h was found to be optimal, as it produced stable dissolved ozone concentrations comparable to higher dosages while consuming less ozone. Under this condition, ozonation of synthetic laundry wastewater achieved more than 90% removal of surfactants (MBAS) within 90 minutes, whereas COD reduction was slower, reaching around 40% at 150 minutes. This limited COD removal can be attributed to the presence of more complex and recalcitrant organic compounds which are more resistant to ozonation and require longer treatment or additional processes. Overall, these findings confirm that ozonation is highly effective for degrading surfactants but less efficient for complete COD removal, suggesting the need for integrated or complementary treatment methods to enhance overall wastewater treatment performance.

This research was funded by Direktorat Riset, Teknologi, dan Pengabdian kepada Masyarakat, Direktorat Jenderal Pendidikan Tinggi, Riset, dan Teknologi, Kementerian Pendidikan, Kebudayaan, Riset, dan Teknologi Republik Indonesia, under Contract 048/E5/PG.02.00.PL/2024; 2684/UN1/DITLIT/PT.01.03/2024. The authors would like to thank all contributors who supported this research.

References

1. M. Prochazkova and V. M. Masa, Sustainable wastewater management in industrial laundries, Chem. Eng. Trans. **94**, 577–582 (2022) https://doi.org/10.3303/CET2294096

2. O.-W. Achaw and E. Danso-Boateng, Soaps and detergents, in Chemical and Process Industries (Springer, Cham, 2021) https://doi.org/10.1007/978-3-030-79139-1_1
3. H. Susanto, S. Herodian, Y. A. Purwanto, and A. T. Sugiarto, Potential utilization of ultrafine bubbles (UFB) technology in the cleaning process as a solution to replace the use of detergents and environmentally friendly: A review, *IOP Conf. Ser. Earth Environ. Sci.* **1386**, 012005 (2024) https://doi.org/10.1088/1755-1315/1386/1/012005
4. B. V. Vora, G. A. Peterson, S. W. Sohn, and M. G. Riley, Detergent alkylate and detergent olefins production, in Handbook of Detergents, Part F: Production (2009)
5. M. M. Martin, J. R. E. Sumayao, A. N. Soriano, and R. V. C. Rubi, Photocatalytic degradation activity of the biosynthesized *R. rosifolius*-mediated silver nanoparticles in methylene blue dye, ASEAN J. Chem. Eng. **23**, 329–342 (2023) https://doi.org/10.22146/ajche.82506
6. M. Mirwan and R. M. Kusuma, Combination of electrocoagulation and bioethanol to reduce laundry waste surfactant levels, IOP Conf. Ser. Earth Environ. Sci. **1454**, 012027 (2025) https://doi.org/10.1088/1755-1315/1454/1/012027
7. M. Amin, H. Hashemi, A. Bovini, and Y. Hung, A review on wastewater disinfection, Int. J. Environ. Health Eng. **2**, 22 (2013) https://doi.org/10.4103/2277-9183.113209
8. B. S. Panandita, I. Prasetyo, and T. Ariyanto, Removal of metronidazole from simulated wastewater using Fe/C catalyst with a combination of heterogeneous Fenton and ozonation, J. Rekayasa Proses. **17**, 158–165 (2023) https://doi.org/10.22146/jrekpros.75633
9. N. Athikoh, E. Yulianto, A. W. Kinandana, E. Sasmita, A. H. Sanjani, R. W. Mustika, A. P. Pratama, N. F. Amalia, G. Gunawan, and M. Nur, Reduction of methylene blue by using direct continuous ozone, J. Environ. Earth Sci. **10**, 4 (2020) https://doi.org/10.7176/JEES/10-4-07
10. M. F. Carosia, D. Y. Okada, I. K. Sakamoto, E. L. Silva, and M. B. A. Varesche, Microbial characterization and degradation of linear alkylbenzene sulfonate in an anaerobic reactor treating wastewater containing soap powder, Bioresour. Technol. **167**, 316–323 (2014) https://doi.org/10.1016/j.biortech.2014.06.002
11. B. Wang, W. Shi, H. Zhang, H. Ren, and M. Xiong, Promoting the ozone–liquid mass transfer through external physical fields and their applications in wastewater treatment: A review, J. Environ. Chem. Eng. **9**, 106115 (2021) https://doi.org/10.1016/j.jece.2021.106115
12. Q. Zhan, J. Yan, and G. Zhou, Study on dissolution and stability of ozone in acetic acid solution, Guangdong Agric. Sci. **46**, 04 (2019) https://doi.org/10.16768/j.issn.1004-874X.2019.04.018
13. M. C. Galdeano, A. E. Wilhelm, I. B. Goulart, R. V. Tonon, O. Freitas-Silva, R. Germani, and D. W. H. Chávez, Effect of water temperature and pH on the concentration and time of ozone saturation, Braz. J. Food Technol. **21**, e2017156 (2018) https://doi.org/10.1590/1981-6723.15617
14. C. G. Joseph, Y. Y. Farm, Y. H. Taufiq-Yap, C. K. Pang, J. L. Nga, and G. L. Puma, Ozonation treatment processes for the remediation of detergent wastewater: a comprehensive review, J. Environ. Chem. Eng. **9**, 106099 (2021) https://doi.org/10.1016/j.jece.2021.106099
15. O. Turkay, S. Barışçı, and M. Sillanpää, E-peroxone process for the treatment of laundry wastewater: A case study, J. Environ. Chem. Eng. **5**, 4282–4290 (2017) https://doi.org/10.1016/j.jece.2017.08.012

Synthesis of Mg/Al–NO_3 hydrotalcite immobilized by citric acid for $[AuCl_4]^-$ adsorption

Nur Raisah Maddeppungeng[1*], *Sri Juari* Santosa[1], and *Mike* Rahayu[2]

[1]Universitas Gadjah Mada, Department of Chemistry, 55281 Yogyakarta, Special Region of Yogyakarta, Indonesia
[2]IAIN Tulungagung, Department of Chemistry Education, 66221Tulungagung, East Java, Indonesia

Abstract. Synthesis of Mg/Al–NO_3 hydrotalcite immobilized by citric acid (Mg/Al–NO_3 HT–Cit) was synthesized for $[AuCl_4]^-$ adsorption. The Mg/Al–NO_3 HT–Cit was prepared by co-precipitation method. Optimum condition for the synthesis was studied by determining the optimum acidity and concentration of precursors. Application of adsorbent for $[AuCl_4]^-$ adsorption was studied by determining their kinetics model. The results showed that the optimum acidity for Mg/Al–NO_3 HT–Cit preparation was pH 7. The XRD diffractogram showed that synthesis of Mg/Al–NO_3 HT–Cit was successfully immobilized onto the surface of hydrotalcite with basal spacing d(003) was 7.99 Å. Adsorption of $[AuCl_4]^-$ was optimum at pH 3. Isotherm analysis revealed the adsorption of $[AuCl_4]^-$ is best fitted by Langmuir model with q_{max} value of 126.58 mg g^{-1}.

1 Introduction

Gold is a high-value metal widely utilized not only in jewellery but also in electronic components due to its exceptional electrical conductivity and near-zero resistance. The rapid growth of the electronics industry has led to an increase in obsolete equipment, contributing to the accumulation of electronic waste, which contains both hazardous substances and recoverable precious metals such as Au, Ag, and Pd [1]. Given gold's economic and functional significance, its recovery from e-waste has become a critical focus.

Conventional recovery methods including cyanidation, amalgamation, pyrometallurgy, hydrometallurgy, and electrochemical techniques pose environmental risks [2]. Adsorption has emerged as a promising alternative due to its operational simplicity, cost-effectiveness, and recyclability [3]. This process relies on interactions between metal ions and functional groups (–OH, –NH, –SH, –COOH) on the adsorbent surface [4]

Layered Double Hydroxides (LDHs), such as hydrotalcite, are anionic clays with positively charged layers and exchangeable interlayer anions and water molecules, making them effective anion exchangers [5][6]. Prior studies have demonstrated the utility of Mg/Al–NO_3 hydrotalcite for removing humic acid hexacyanoferrate (II) [7][6,8]. LDHs also exhibit a memory effect, enhancing their adsorption performance.

* Corresponding author: raisah.maddeppungeng@gmail.com

Surface modification using organic acids has been shown to improve adsorption capacity [9]. Citric acid, a weak triprotic acid with strong chelating properties, has demonstrated superior performance in metal uptake compared to other organic acids [10]. In this study, Mg/Al–NO_3 hydrotalcite immobilized by citric acid was synthesized via co-precipitation to evaluate its potential for gold recovery from aqueous systems called Mg/Al–NO_3 HT–Cit.

2 Materials and Methods

2.1 Materials and Instrumentation

The materials used are Mg $(NO_3)_2{\bullet}6H_2O$, $Al(NO)_3{\bullet}9H_2O)$, NH_4OH, citric acid monohydrate $(C_6H_8O_7{\bullet}H_2O)$, HCl, NaOH, $FeCl_3{\bullet}6H_2O$, 1000 ppm $HAuCl_4$ solution (destructed from 1 gram of 99.99% ANTAM gold).

The instruments used are X-Ray Diffraction (XRD Rigaku miniflex 600) to determine the the structural phase parameters of Mg/Al–NO_3 HT–Cit, Fourier Transform Infrared (FTIR, Shimadzu Prestige-21) to learn the functional group of Mg/Al–NO_3 HT–Cit, and Ultraviolet-Visible (UV-Vis, Shimadzu UV-1700 pharmaspec) to test stability of citric acid in hydrotalcite matrix.

2.2 Methods

2.2.1 Synthesis of Mg/Al–NO₃ HT

Mg/Al–NO_3 HT was synthesized via a conventional co-precipitation method by dissolving 0.1 mol of Mg $(NO_3)_2{\bullet}6H_2O$ and 0.05 mol $Al(NO)_3{\bullet}9H_2O)$. The pH was adjusted to 10 by the dropwise addition of NH_4OH to initiate the precipitation. The resulting slurry was washed thoroughly with CO_2^- free water to neutral pH and subjected to hydrothermal treatment at 120 °C for 5h. The final product was isolated and dried at mild temperature to yield Mg/Al–NO_3 HT.

2.2.1 Synthesis of Mg/Al–NO_3 HT–Cit

Citric acid was immobilized onto Mg/Al–NO_3 HT via room-temperature adsorption. The process was conducted across a range of pH values (3, 5, 7, 9, and 11), adjusted using HCl or NaOH to determine the optimal pH for synthesis. The mixture was subsequently filtered, and the amount of adsorbed citric acid was quantified using UV-Vis spectrophotometry.

2.2.1 Adsorption of $[AuCl_4]^-$

A solution of $[AuCl_4]^-$ was adjusted to pH 2-11 using HCl or NaOH, then treated with Mg/Al–NO_3 HT–Cit. After 90 minutes of agitation, the mixtures were filtered and residual $[AuCl_4]^-$ concentration were measured via UV-Vis spectrophotometry. Adsorption kinetics were further evaluated by varying contact time under identical conditions.

3 Results and discussion

3.1 Structure and Optimum pH of Immobilisation

Mg/Al–NO_3 HT was synthesized via co-precipitation at pH 10, based on prior findings indicating optimal crystallinity at this condition. The hydrothermal treatment at 120 °C was employed to enhance crystal growth. Structural characterization was conducted using FTIR spectroscopy and XRD and it shows in Figure 1 and Figure 2.

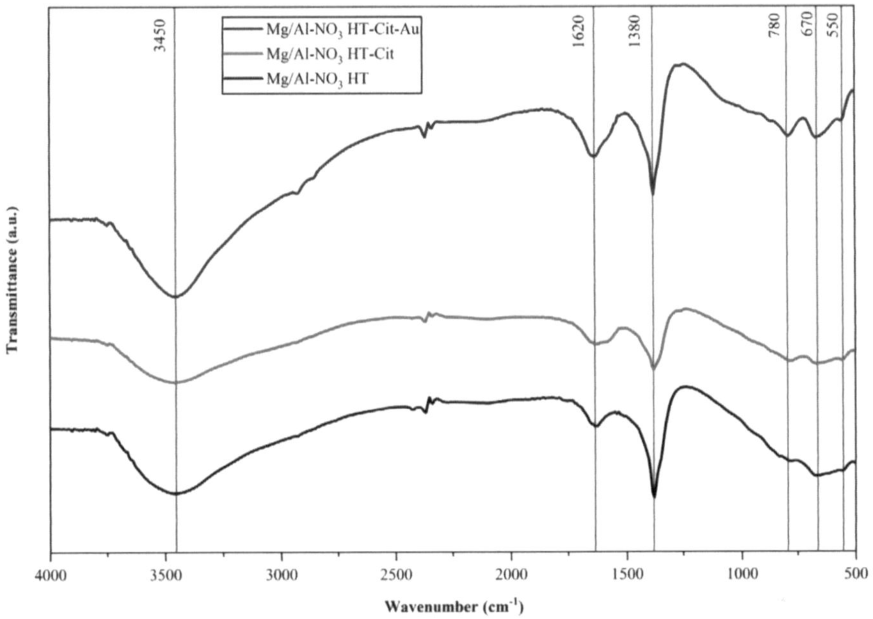

Fig. 1. FTIR Spectra of Mg/Al–NO_3 HT, Mg/Al–NO_3 HT–Cit and Mg/Al–NO_3 HT–Cit–Au

FTIR spectra revealed broad adsorption at 3456 cm^{-1}, corresponding to O–H stretching vibrations, and a band at 1635 cm^{-1} attributed to interlayer water bending. Post-adsorption of citric acid, a new band at 1627 cm^{-1} emerged, indicating asymmetric –COO^- stretching, while $[AuCl_4]^-$ adsorption led to reduced intensity at 3449 cm^{-1} and 1628 cm^1, suggesting displacement of water molecules by the anionic complex. Additional FTIR bands at 1381 cm^{-1} and 671 cm^{-1} confirmed the presence of nitrate ions through N=O stretching and bending vibrations, respectively. The Mg–O–Al framework was evidenced by a peak at 477 cm^{-1}. XRD analysis showed characteristic reflections at 2θ values of 10.72°, 21.87°, 34.29°, and 60.50°, corresponding to d(003), d(006), d(009), and d(110) planes, confirming the formation of a layered hydrotalcite structure. The basal spacing of 8.243 Å matched literature values for nitrate-intercalated LDH. Immobilization of citric acid did not shift the d(003) peak, indicating surface adsorption without structural disruption. Similarly, $[AuCl_4]^-$ adsorption preserved the crystalline framework, as no new peaks were observed.

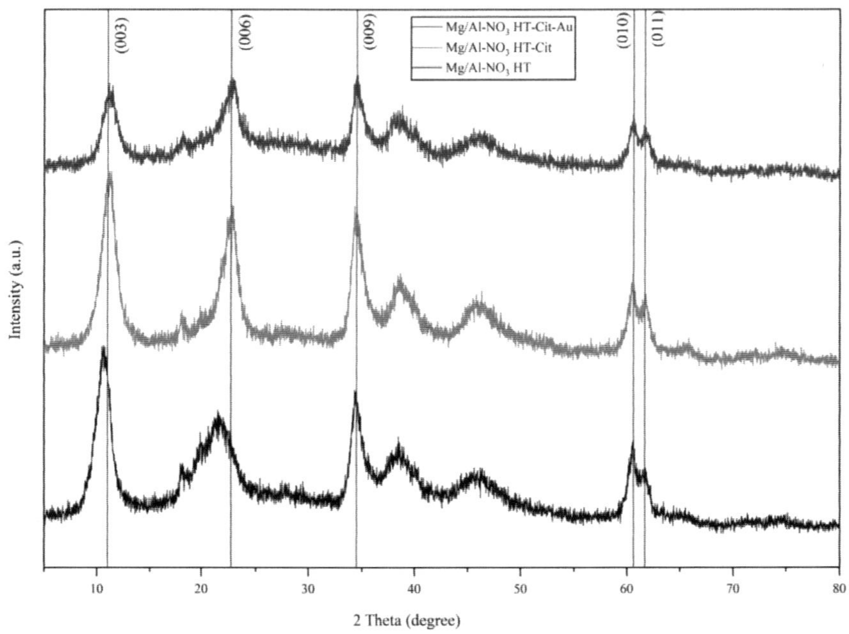

Fig. 2. XRD diffractogram of Mg/Al–NO_3 HT, Mg/Al–NO_3 HT–Cit and Mg/Al–NO_3 HT–Cit–Au

The immobilization efficiency of citric acid on Mg/Al–NO_3 HT was evaluated across pH values ranging from 3 to 11. The pH range was selected to encompass the pKa values of citric acid (3.14, 4.77, and 6.30) and remain below the point of zero charge (pH_{pzc}) of the hydrotalcite ($pH_{pzc} \approx 12$). Figure 3 shows that maximum adsorption (69.51%) was achieved at pH 7, where citric acid predominantly exists as the trivalent anion $C_6H_5O_7^{3-}$, facilitating electrostatic interaction with the positively charged of hydrotalcite surface.

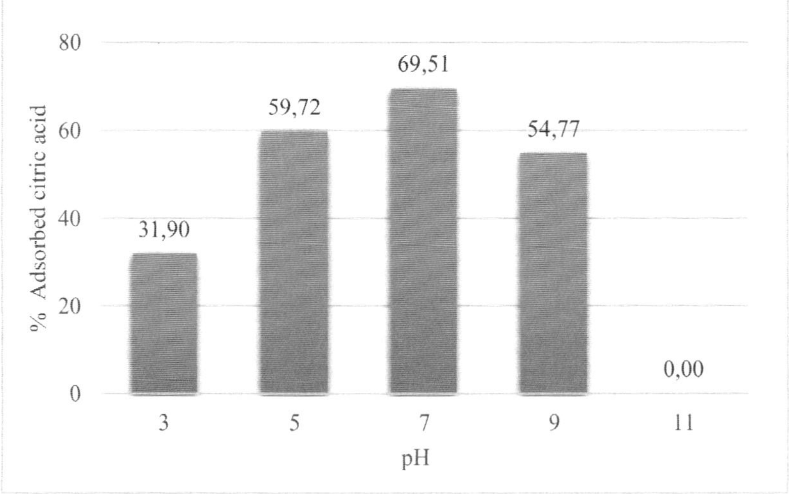

Fig. 3. Adsorption percentage of citric acid onto Mg/Al–NO_3 HT

3.2 Adsorption of $[AuCl_4]^-$

3.2.1 Adsorption Acidity of $[AuCl_4]^-$

The adsorption behaviour of $[AuCl_4]^-$ onto Mg/Al–NO_3 HT–Cit was investigated across a pH range of 2–11 to evaluate both the speciation of gold complexes and the stability of the adsorbent. Figure 4 shows the maximum adsorption of $[AuCl_4]^-$, observing at pH 3, where $[AuCl_4]^-$ is the dominant species. At this pH, protonation of hydroxyl groups on immobilized citrate generates partial positive charges, promoting electrostatic interaction with the negatively charged gold complex. In contrast, adsorption efficiency declined at pH 2 due to poor structural stability of the adsorbent. At higher pH values, competitive interaction with OH^- ions and the formation of bulkier $AuCl_{4-n}(OH)_n^-$ species hindered adsorption due to steric limitations and reduced accessibility to active sites.

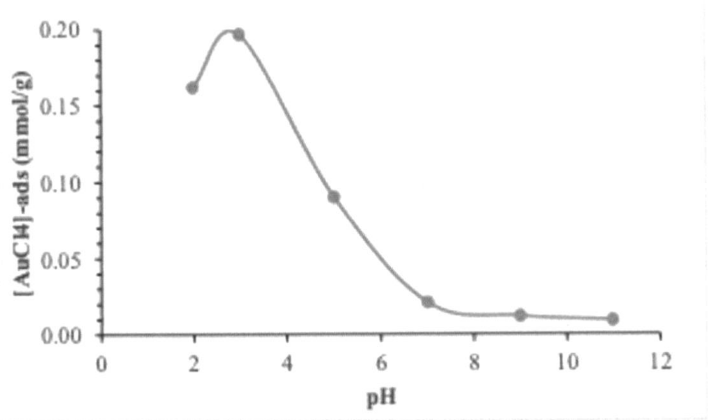

Fig. 4. pH influence on adsorption of $[AuCl_4]^-$ onto Mg/Al–NO_3 HT–Cit

3.2.2 Kinetic and Isoterm Studies of $[AuCl_4]^-$ Adsorption

Kinetic studies were conducted at pH 3 to determine the adsorption rate and equilibrium time for $[AuCl_4]^-$ on Mg/Al-NO_3 HT–Cit (Figure 5). Rapid uptake occurred within the first 180 minutes, followed by a plateau, indicating saturation of active sites.

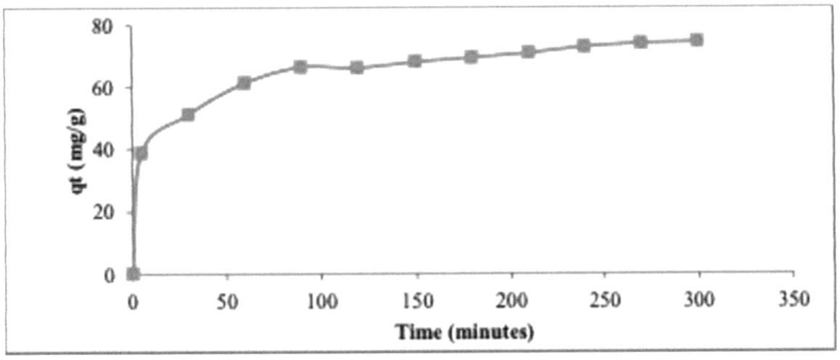

Fig. 5. The influence of interaction time on adsorption of $[AuCl_4]^-$ onto Mg/Al–NO_3 HT–Cit

Adsorption isotherms represent the relationship between the quantity of adsorbate retained on the adsorbent surface and its equilibrium concentration in the liquid phase. The adsorption behaviour of $[AuCl_4]^-$ on Mg/Al–NO_3 HT–Cit was investigated to evaluate its adsorption capacities. The obtained adsorption data were subsequently fitted using the Freundlich and Langmuir isotherm models. The Freundlich equation was developed to describe adsorption on heterogeneous systems and is expressed as follows:

$$\log q_e = \log K_F + \frac{1}{n}\log C_e \quad (1)$$

In this equation, q_e represents the amount of $[AuCl_4]^-$ adsorbed at equilibrium, C_e is the equilibrium concentration of $[AuCl_4]^-$ in solution, and $\frac{1}{n}$ is the Freundlich constant related to adsorption intensity. The amount of $[AuCl_4]^-$ adsorbed by Mg/Al–NO_3 HT–Cit was determined by plotting $\log q_e$ versus $\log C_e$ as the y- and x-axes, respectively as shown in Figure 6. The slope of the resulting linear regression corresponds to $\frac{1}{n}$, while the intercept represents $\log K_F$.
The Langmuir isotherm model quantitatively describes monolayer adsorption on a homogeneous adsorbent surface, if no further adsorption occurs once the active sites are fully occupied. This model is expressed by the following equation:

$$\frac{C_e}{q_e} = \frac{1}{q_{max}K_L} + \frac{C_e}{q_{max}} \quad (2)$$

where q_e is the amount of $[AuCl_4]^-$ adsorbed at equilibrium, C_e is the equilibrium concentration, q_{max} denotes the maximum adsorption capacity, and K_L is the Langmuir constant. The Langmuir isotherm was evaluated by plotting $\frac{C_e}{q_e}$ (g/L) against C_e (mg/L) as the ordinate and abscissa, respectively as shown in Figure 7. The slope of the linear plot corresponds to $\frac{1}{q_{max}}$, while the intercept is equal to $\frac{1}{q_{max}K_L}$. The Adsorption isotherm parameters for $[AuCl_4]^-$ on Mg/Al–NO_3 HT–Cit are summarized in Table 1.

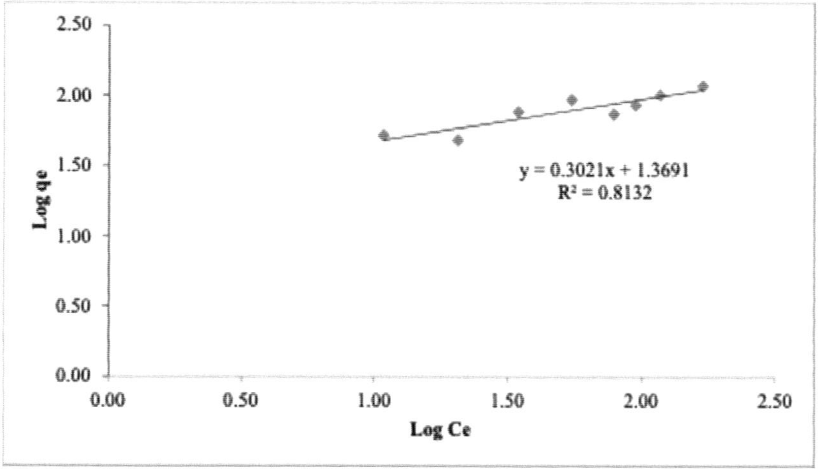

Fig. 6. The plotting of $\log q_e$ versus $\log C_e$ on adsorption of $[AuCl_4]^-$ onto Mg/Al–NO_3 HT–Cit

Table 1. Adsorption isotherm parameters for $[AuCl_4]^-$ on Mg/Al–NO_3 HT–Cit

Parameters					
Langmuir			Freundlich		
R^2	K_L (L mol^{-1})	E (kJ mol^{-1})	R^2	K_F (mg g^{-1})	n
0,9263	11471,71	23,3149	0,8132	23,3938	3,3102

Based on the data presented in **Table 1**, Mg/Al–NO_3 HT–Cit exhibit a better fit to the Langmuir isotherm model, as indicated by correlation coefficients (R^2) approaching unity. The R^2 values of 0.9263 were obtained for Mg/Al–NO_3 HT–Cit. These results suggest that the adsorption process occurs on a homogeneous adsorbent surface and that the adsorbate forms a monolayer coverage on the adsorbent surface and fits with the previous reports [3,6] Furthermore, from the equation known that q_{max} of $[AuCl_4]^-$ adsorbed by Mg/Al–NO_3 HT–Cit was 126.58 mg g^{-1}.

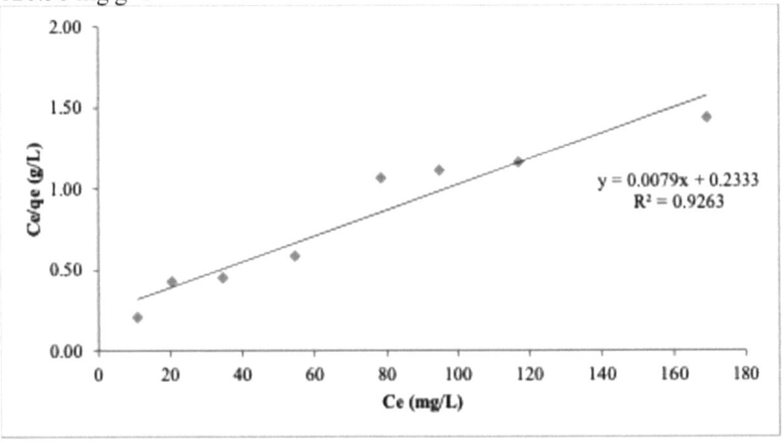

Fig. 7. The plotting of $\frac{C_e}{q_e}$ against C_e on adsorption of $[AuCl_4]^-$ onto Mg/Al–NO_3 HT–Cit

4 Conclusions

Mg/Al–NO_3 HT–Cit was successfully synthesized via co-precipitation method at pH 7. The adsorbent demonstrated chemical stability within the pH range of 3–7. Optimal adsorption of $[AuCl_4]^-$ occurred at pH 3. Isotherm adsorption of $[AuCl_4]^-$ onto Mg/Al–NO_3 HT–Cit fits to the Langmuir isotherm model indicate that adsorption take place on a homogeneous adsorbent surface through monolayer formation with q_{max} was 126.58 mg g^{-1}.

References

1. P. Gramatyka, R. Nowosielski, P. Sakiewicz, Recycling of waste electrical and electronic equipment. J. Achiev. Mater. Manuf. Eng. **20(1-2)**, 535-538 (2007)
2. H. Antrekowitsch, M. Potesser, W. Spruzina, F. Prior, Metallurgical recycling of electronic scrap, In Proc. of 135th TMS Annual Meeting & Exhibition. (2006)
3. M. Keller. Assessment of Gold Recovery Processes in Bangalore, India: And Evaluation of an Alternative Recycling Path for Printed Wiring Boards: a Case Study, Ph.D thesis (2006)
4. Z. Chang, C. Wu, S. Song, Y. Kuang, X. Lei, L. Wang, and X. Sun, Synthesis mechanism study of layered double hydroxides based on nanoseparation, Inorg. Chem. **52,** 8694–8698 (2013) https://doi.org/10.1021/ic4008763
5. I. Pavlovic, C. Barriga, M.C. Hermosín, J. Cornejo, M.A. Ulibarri, Adsorption of acidic pesticides 2,4-D, Clopyralid and Picloram on calcined hydrotalcite, Appl. Clay Sci. **30,** 125–133 (2005) https://doi.org/10.1016/j.clay.2005.04.004
6. S.J. Santosa, E.S. Kunarti, and Karmanto, Synthesis and utilization of Mg/Al hydrotalcite for removing dissolved humic acid, Appl. Surf. Sci. **254,** 7612–7617 (2008) https://doi.org/10.1016/j.apsusc.2008.01.122

7. R. Roto, F. Nindiyasari, and I. Tahir, Removal of hexacyanoferrate (II) using Zn-Al-O-Ac Hydrotalcite as an anion exchanger. J. Phys. Sci. **20**, 73-84 (2009)
8. S. J. Santosa and E. S. Kunarti, Synthesis and utilization of Mg/Al hydrotalcite for removing dissolved humic acid. Appl. Surf. Sci. **254**,7612-7617 (2008)
9. S. Hokkanen, A. Bhatnagar, M. Sillanpää, A review on modification methods to cellulose-based adsorbents to improve adsorption capacity, Water Res. **91** (2016) 156–173. https://doi.org/10.1016/j.watres.2016.01.008
10. J. Perera, M. Weerasekera, N. Kottegoda, Slow release anti-fungal skin formulations based on citric acid intercalated layered double hydroxides nanohybrids, Chem. Cent. J. **9**, 27 (2015). https://doi.org/10.1186/S13065-015-0106-3

Remediation of Contaminated Soil with Crude Oil Using Solvent Extraction Method

Riyanto[1]*, *Indah Hadzar* Febriari[2], *Imam* Sahroni[2], *Muhaimin*[3]

[1]Department of Chemistry, Magister Chemistry Study Program, Universitas Islam Indonesia, Jalan Kaliurang KM 14,5 Sleman, Yogyakarta, 55584, Indonesia
[2]Department of Chemistry, Chemistry Study Program, Universitas Islam Indonesia, Jalan Kaliurang KM 14,5 Sleman, Yogyakarta, 55584, Indonesia
[3]Department of Chemistry, Chemistry Education Study Program, Universitas Islam Indonesia, Jalan Kaliurang KM 14,5 Sleman, Yogyakarta, 55584, Indonesia

Abstract. An innovation for removing crude oil from contaminated soil was studied. A mixture of hexane, acetone, and water was used to investigate the ability of solvent extraction techniques to clean soil under various extraction conditions. The study was conducted with initial processing stages using batch and ultrasonic methods, followed by GC-MS and FTIR analysis. The results showed that the soil was contaminated with saturated hydrocarbons, aromatic naphthene's, polar aromatics, and asphaltenes. The effect of solvent ratio and extraction time was studied, and the results showed that the best solvent ratio and extraction time were 1:6 (w/v) and 60 minutes, with a solvent efficiency of 70% and successfully extracting contaminants (saturated fractions of 97.25% and aromatics of 2.75%). Repeated extraction was better than a single extraction where the solvent efficiency was 63.8% and the contaminants were paraffin compounds of 71.96% and naphtha of 28.04%.

1 Introduction

Soil contaminated with crude oil contains organic compounds such as aromatic, aliphatic, and heterocyclic compounds, which impact the environment and living organisms [1]. Crude oil spills in soil occur through various stages, including processing, storage, and use. The impacts of crude oil spills not only pollute the soil but also water resources and aquatic ecosystems [2,3]. Crude oil spills also occurred at the oil refinery of the Oil and Gas Education and Training Centre in the Cepu area, Blora Regency, Central Java. The refinery has an average capacity of 200 barrels per day (bpd) with an average oil loss of 0.4%. Crude oil released into the environment can cause oil spills on land.

Crude oil is a highly complex mixture of hydrocarbons. It typically contains small amounts of heteroatoms such as nitrogen, oxygen, and sulphur, as well as metals such as vanadium and nickel. Oil pollutants are classified into four types: saturated, naphthene aromatic, polar aromatic, and nC7-asphaltenes based on polarity and solubility.

* Corresponding author: riyanto@uii.ac.id

Soil contaminated by oil spills can pollute the environment. Oil spills contaminate the soil surface and evaporate, potentially even seeping into the ground. As a result, various chemicals will mix with the soil and contaminate groundwater, which can harm human health and the environment. Although they are low or non-toxic, they have the capacity to absorb into the soil surface, changing the surface from water-wet to oil-wet, and pose potential risks to human health and the environment [4].

Several methods for treating soil contaminated by oil spills include physical, chemical, and biological methods. The most common method for treating soil contaminated with crude oil is remediation using bacterial consortia [5] phytoremediation [6] ozonation [7] bio washing and bio pile [8] biochar and biosurfactant [9] bioremediation [10] microbial and phytoremediation [11]. Bioremediation methods have several drawbacks, such as long processing times and the need for special handling. One highly effective chemical method is solvent extraction. Solvent extraction is a cleanup method that uses solvents to extract or remove hazardous chemicals from contaminated material. Therefore, this method is efficient in removing hydrophobic organic contaminants from soil. The solvent extraction method is expected to quickly and simply treat oil-contaminated soil. Solvent extraction can treat soil contaminated with heavy crude oil. The extraction method can also be used to treat sludge contaminated with crude oil [12].

A solvent mixture of hexane and acetone (25% by volume) is very effective in removing crude oil hydrocarbons from contaminated soil containing aromatic naphthene, polar aromatics, and n-asphaltenes. However, further research is needed regarding solvent mixtures, as the hexane-acetone mixture does not produce a stable emulsion phase in the soil, making oil removal less effective. Extraction with water solvents can recycle organic solvents used for the extraction of crude oil-contaminated soil. In addition, water is readily available and safe to use. Based on the above, researchers will conduct research on the treatment of crude oil-contaminated soil using the solvent extraction method using a mixture of n-hexane, acetone, and water. The parameters used are extraction time, solvent-to-soil ratio, solvent efficiency, and extraction method. The novelty of this research is the use of water as a polar solvent to dissolve soil containing many polar compounds. Water can facilitate the extraction process. Furthermore, the novelty of this research is the soil profile after extraction and the comparison of extraction processes using magnetic stirrers and ultrasonic.

2 Experimental Section

2.1 Materials and instrumentation

The research equipment used was Magnetic Stirrer, Rotary Evaporator Buchi, GC-MS (Gas Chromatography Mass Spectrometry) QP 2010, FTIR (Fourier Transform Infrared Spectroscopy) Perkin Elmer and Ney Ultrasonic 28H. The research materials were n-hexane pro analysis and acetone pro analysis from JT Baker. Crude oil contaminated soil samples were taken from the Sumur-Ldk 175 well of PT. Blora Patra Energi, Blora Regency, Central Java. Sampling was carried out using a simple random sampling system. Soil samples were taken using a shovel. Samples were taken from the topmost layer of soil. To avoid evaporation of the contained substances and contamination from other materials, soil samples were put into plastic and stored in an ice box. The sampling location is shown in Figure 1.

Fig. 1. Sampling area of the old oil well Ldk-175 in the Ledok Field, Ledok Village, Sambong District, Blora Regency, Central Java.

2.2 Solvent extraction of contaminated soil samples

A 50 g contaminated soil sample was placed in a 250 mL beaker glass. Then, n-hexane, acetone, and water were added to a ratio of 5:4:1. The beaker was then covered with aluminum foil. The extraction process was stirred using a magnetic stirrer for 20 minutes. The mixture was filtered using filter paper. The filtration was analyzed using FTIR. The volume of the filtration was determined using a measuring cylinder. The filtrate was transferred into a separating funnel to separate the organic compounds and water. The volume of the oil and organic solvent mixture was then determined. The solvent remaining in the filtrate was separated using a rotary evaporator at 69 °C. The volume of the crude oil obtained was determined and analyzed using GC-MS. This study also carried out three repeated soil sample extractions. The remediation process procedure is shown in Figure 2.

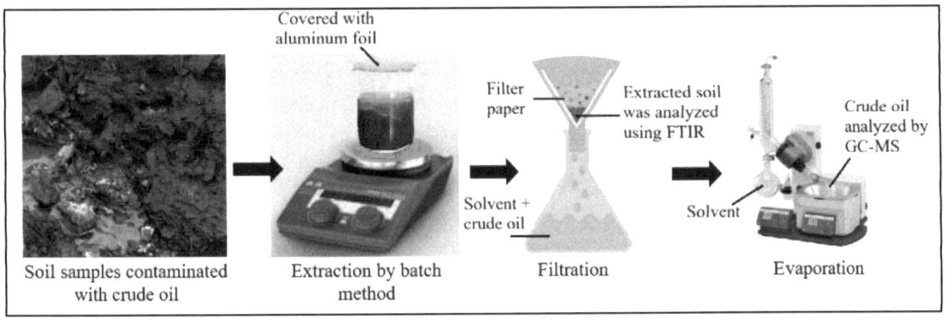

Fig. 2. Remediation process for crude oil-contaminated soil using the batch method

2.3 Ultrasonic oil extraction of soil samples

The extraction process was carried out with a sample-to-solvent ratio of 1:4 (w/v). A 50 g soil sample was placed in a 250 mL beaker. Then, n-hexane, acetone, and water were added, totaling 160 mL of solvent (100 mL n-hexane, 40 mL acetone, and 20 mL water). Water was added to the soil sample to form an emulsion to facilitate the extraction process. Next, a mixture of n-hexane and acetone was added. The beaker containing the soil sample and solvent was covered with aluminum foil. The beaker was placed in a Ney Ultrasonic 28 H reactor for 20 minutes at an operating temperature of 5-40 °C and a frequency of 28 Hz.

3 Result and Discussion

3.1 Extraction results with varying sample-to-solvent ratios

The solvent extraction process on crude oil-contaminated soil samples was carried out in three variations of soil sample-to-solvent ratios (1:2), (1:4) and (1:6) (w/v). Compound content analysis in the extracted results was performed using GC-MS (Gas Chromatography Mass Spectrometry) QP 2010 with the following operating conditions are injection temperature of 285 °C, column oven temperature of 50 °C, and a Flame Thermionic Detector (FTD). The GC-MS chromatogram of the extracted crude oil is shown in Figure 3.

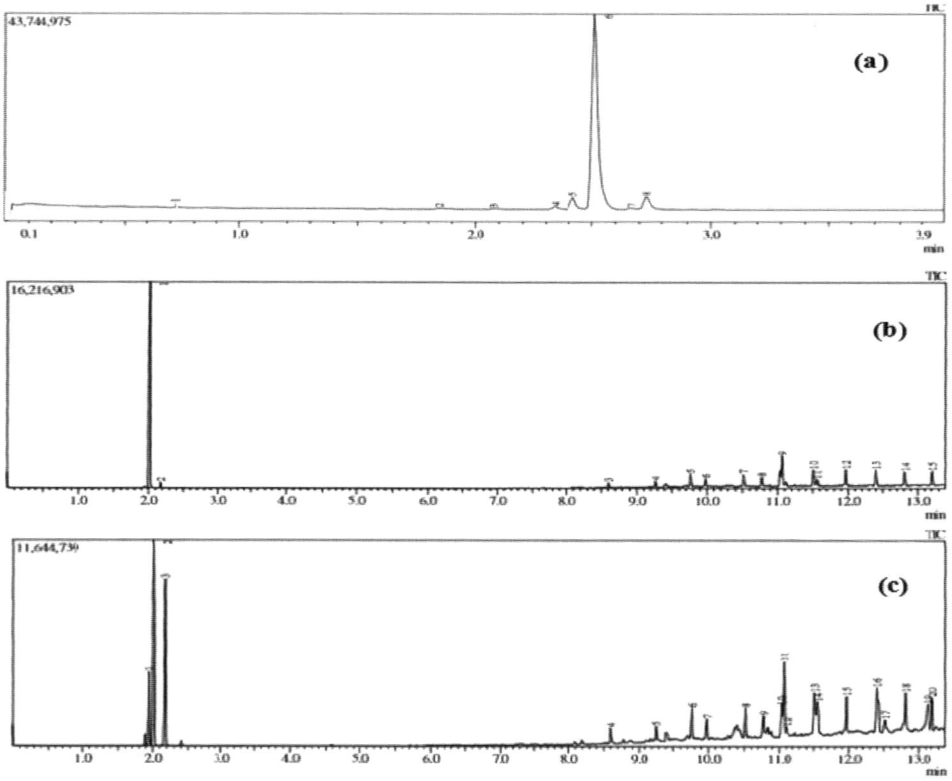

Fig. 3. Chromatograms of soil extract analysis results with soil sample: solvent ratios of (a) 1:2 (b) 1:4 (c) 1:6.

Chromatogram analysis was performed using mass spectra. The results are shown in Table 1. The three ratios (1:2), (1:4), and (1:6) were able to extract organic compounds from the contaminated soil samples. These compounds are saturated hydrocarbon fractions (saturates) and naphtha (aromatics). Based on the chromatograms and MS, it can be concluded that the best ratio was 1:6 (w/v). This conclusion is the extracted organic compounds contain paraffins (78%), naphthene (20.3%), and aromatics (1.2%).

Table 1. Content of the top five chemical compounds extracted with varying soil sample: solvent ratios

No.	Comparison of soil sample weight with solvent (w/v)					
	1:2		1:4		1:6	
	Compound name	(%)	Compound name	(%)	Compound name	(%)
1	Methylcyclopentane	38.58	2,6,10,14-tetramethylpentadecane	26.48	Methylcyclopent anee	20.30
2	3-Methylpentane	31.49	Octadecan	8.40	Eicosane	10.20
3	1-propen-1-ol	14.68	Eicosane	8.27	2,6,10,14-tetramethylpentad e cane	9.60
4	2-Methylpentan	8.46	Nonadecan	7.69	Octadecane	6.60
5	1-[8-(3-octyloxiranyl)-1-oxooctyl]Pyrolidine	3.45	2,6,10,15-Tetramethylheptadecane	6.82	Heptadecane	4.80

The more solvents used, the more organic compounds are extracted. The greater the liquid phase, the greater the mass transfer, increasing the contact area between the phases and the concentration gradient. The oil extraction process in crude oil-contaminated soil samples is significantly influenced by the amount of solvent. Based on variations in the sample-to-solvent ratio, the solvent efficiency is shown in Table 2.

Table 2. The results of extraction of soil contaminated with crude oil with variations in the sample and solvent ratios

No	Comparison	Initial solvent volume (mL)	Final solvent volume (mL)	Solvent recovery (%)	Crude oil volume extracted (mL)
1	1:2	90	47	52	2.1
2	1:4	180	113	62	4.4
3	1:6	270	200	74	5.6

Based on Table 2, the best solvent efficiency is at a ratio of 1:6 (w/v) with a crude oil extract yield of 5.6 mL. The correct ratio of solids to solvents will provide maximum extraction results [13]. The ratio of solvent used to solids must be in accordance with the solubility of the solute. The smaller the solubility of the solid to the solvent, the greater the ratio of solvent to solids. Water is essential in the extraction of organic compounds from soil samples. Extraction of soil contaminated with crude oil using organic solvents requires water. Extraction of soil contaminated with crude oil without water will cause clumping. Water is also easily separated because it is insoluble in n-hexane and acetone.

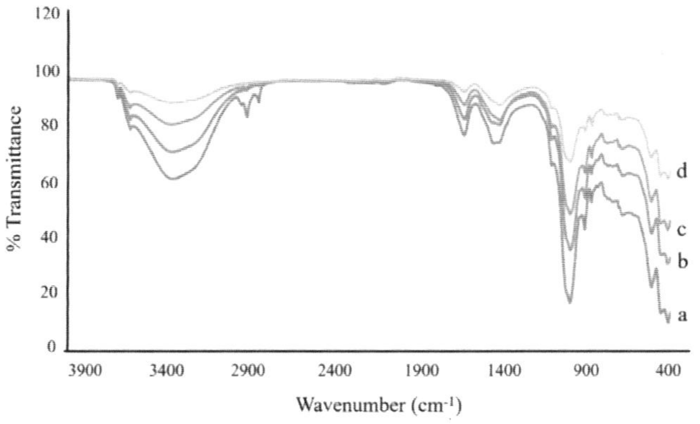

Fig. 4. FTIR spectra of (a) soil sample before extraction and soil sample after extraction with sample:solvent ratios of (b) 1:2 (c) 1:4 and (d) 1:6.

Figure 4 shows that the soil before extraction had an absorption peak around 2850.26-2955.40 cm^{-1}, indicating C-H stretching vibrations (alkanes). After extraction, this absorption peak was not visible, indicating that the extraction process had occurred. Absorption at wavenumbers around 1455.27 cm^{-1} indicates C-H bending vibrations (cycloalkanes), which are classified as naphthene saturated hydrocarbons. However, in the spectra of the three sample ratios, the absorption peak shifted successively to 1420.33, 1420.28, and 1420.25 cm^{-1}. This indicates a reduction in the surface area of hydrocarbon compounds in the soil sample after extraction.

3.2 Extraction results with variations in extraction time

Extraction of crude oil-contaminated soil samples was carried out using varying extraction times of 20, 30, and 60 minutes. The sample-to-solvent ratio selected for the extraction time variation was 1:6 (w/v). Figure 5 shows that extraction times of 20, 30, and 60 minutes were effective in extracting organic compounds from crude oil. These compounds include the saturated hydrocarbon fraction (paraffin and naphthene) and the aromatic fraction. Longer extraction times increase solvent penetration into the material, increasing the number of extracted compounds [14]. This is demonstrated in the research results, where 60 minutes was the optimal extraction time. Based on GC-MS data, the extracted organic compounds included 82.53% of the paraffin fraction, 14.72% of the naphthene fraction, and 2.75% of the aromatic hydrocarbon fraction. Solvent efficiency is shown in Table 3. Table 3 shows that as extraction time increases, the volume of crude oil extracted decreases. The longer the extraction time, the more degradation and evaporation occur in crude oil. Components in the oil can degrade due to the influence of heat, light, and oxygen.

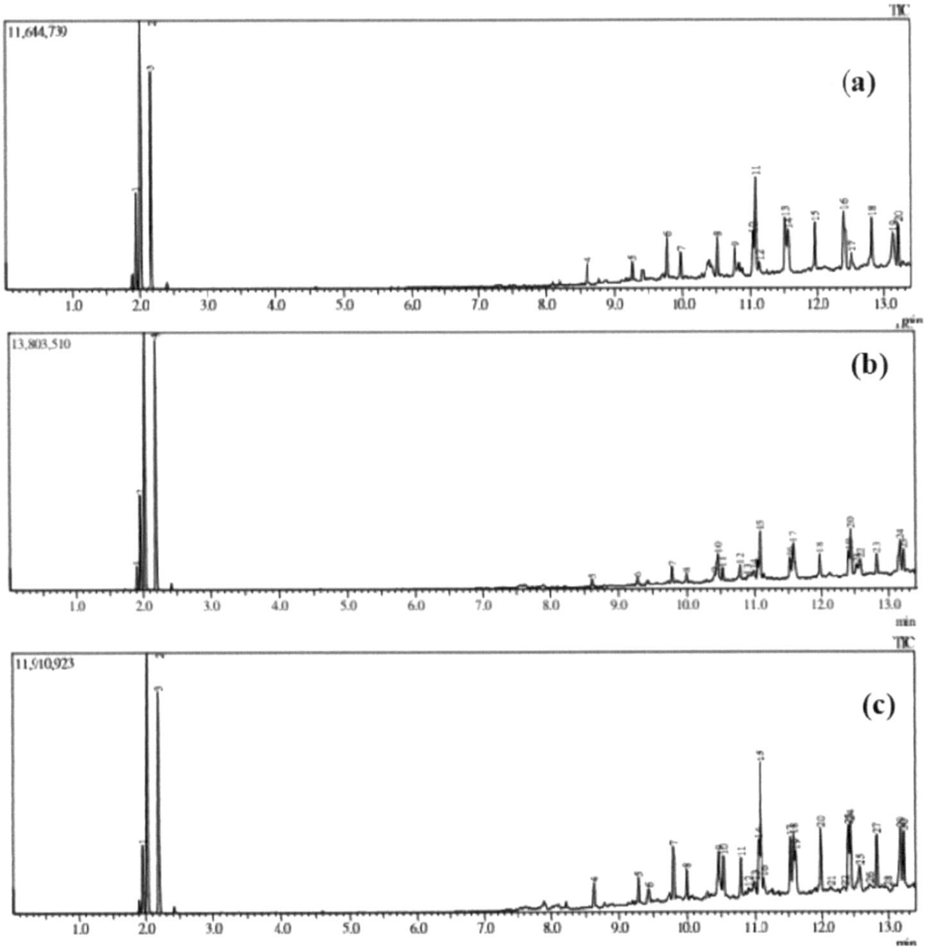

Fig. 5. Chromatogram of extract analysis results with varying extraction times (a) 20 (b) 30 (c) 60 minutes

Table 3. Data on the results of extraction of soil contaminated with crude oil with variations in extraction time

No	Extraction time (minutes)	Initial solvent volume (mL)	Final solvent volume (mL)	Solvent recovery (%)	Crude oil volume extracted (mL)
1	20	270	200	74.0	5.6
2	30	270	178	65.9	5.0
3	60	270	190	70.0	4.2

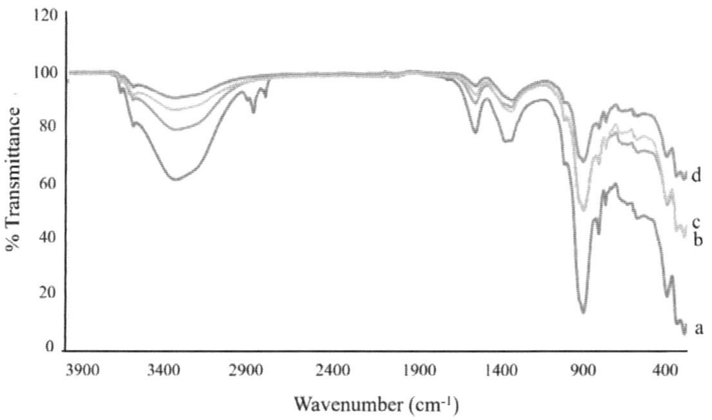

Fig. 6. Spectra of (a) soil samples before and after extraction with varying extraction times of (b) 20, (c) 30, and (d) 60 minutes.

Extraction of soil samples with a mixture of n-hexane, acetone, and water at extraction times of 20, 30, and 60 minutes reveals changes in absorption peaks indicating hydrocarbon compounds (Figure 6). Before extraction, the soil had an absorption peak around 2850.26-2955.40 cm^{-1}, indicating C-H stretching vibrations (alkanes). After extraction, absorption at this wavenumber disappeared. Absorption at a wavenumber around 1455.27 cm^{-1} indicates C-H bending vibrations (cycloalkanes) of naphthenes.

3.3 Extraction of soil samples using batch (stirring) and ultrasonic methods

Table 4 shows the differences in extraction results using the batch and ultrasound methods. The ultrasonic method has a faster extraction process than conventional methods. The mechanical effect caused by ultrasonic waves can increase the solvent penetration ability into the material cells, thereby increasing the number of components that diffuse into the solvent. The results of GC-MS analysis showed that paraffin compounds were 74.97%, naphthenes were 23.55%. The ultrasonic method produced a lower yield than the batch method (stirring). Sonication can significantly increase pollutant removal, and the increase depends on various factors such as sonication strength, flow rate, and soil type. Extraction with low power, ultrasonic frequency, and short time is not enough to produce high yields [15]. This study used a frequency of 28 kHz, while the ideal frequency for ultrasonic extraction is 37 kHz.

Table 4. Data on the results of extraction of soil contaminated with crude oil using the batch and ultrasonic methods

No	Method	Initial solvent volume (mL)	Final solvent volume (mL)	Solvent recovery (%)	Crude oil volume extracted (mL)
1	Batch (Stirring)	180	102	62.7	4.4
2	Ultrasound	180	112	62.2	2.0

Based on the FTIR results shown in Figure 7, it can be seen that the soil before extraction showed absorption in several peaks. The absorption of C–H (alkane) is shown by the appearance of three strong bands below 3000 cm^{-1}. The band with the strongest absorption intensity, namely at a wave number of 2923.9 cm^{-1}, is caused by absorption from the C–H group (alkane). This peak indicates a group of saturated hydrocarbon compounds. There is also a peak at a wave number around 1455.27 cm^{-1} indicating the bending vibration of C–H (cycloalkane) which includes saturated hydrocarbon compounds of the naphthene fraction (naphthene aromatics).

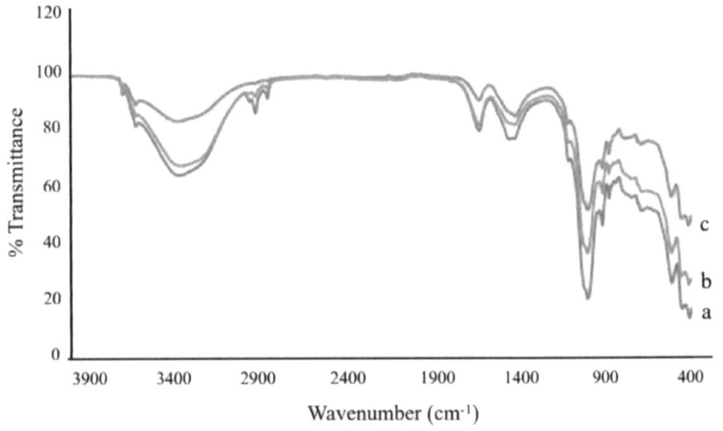

Fig. 7. FTIR spectra of soil samples before extraction (a), after extraction using the ultrasonic method (b), and in batch (c).

Extraction is more efficient when repeated with smaller amounts of solvent than when using a large amount of solvent but only extracted once. The extraction results of soil samples contaminated with crude oil with different extraction rates are shown in Table 5. Three extractions resulted in a solvent efficiency of 63.8% and a crude oil yield of 6.9 mL.

Table 5. Data on the results of extraction of soil contaminated with crude oil with the number of extractions one and three times

No	Number of extractions (times)	Initial solvent volume (mL)	Final solvent volume (mL)	Solvent recovery (%)	Crude oil volume extracted (mL)
1	1	180	113	62.7	4.4
2	3	360	230	63.8	6.9

4 Conclusions

Remediation of crude oil-contaminated soil was carried out by solvent extraction using a mixture of n-hexane, acetone, and water. This solvents mixture is very effective for extracting hydrocarbon compounds. The ratio of soil samples to solvent in extracting soil samples was 1:6 (w/v) with a solvent efficiency of 74% and the results of GC-MS analysis showed a paraffin saturated hydrocarbon content of 30.84% and a naphthene saturated hydrocarbon fraction of 69.19%. The best time to extract soil samples was 60 minutes. The extraction process using the ultrasonic method was better than the batch method. The repeated method had better results than a single extraction.

References

1. L. Feng, X. Jiang, Y. Huang, D. Wen, T. Fu, R. Fu, Petroleum hydrocarbon-contaminated soil bioremediation assisted by isolated bacterial consortium and sophorolipid. Environ. Pollut. **273**, 116476 (2021)

2. U. Michael-Igolima, S.J. Abbey, A.O. Ifelebuegu, A systematic review on the effectiveness of remediation methods for oil contaminated soils. Environ. Adv. **9**, 100319 (2022)
3. K.M. McFarlin, R.C. Prince, Petrodiesel Fuels: Science, Technology, Health, and Environment Vol. 3 (ed Ozcan Konur) Ch. 44, 911–935 (CRC Press: Taylor & Francis Group, 2021)
4. C. Dumitran, I. Onutu, F. Dinu, Extraction of hydrophobic organic compounds from soils contaminated with crude oil. Rev. Chim. **60**, 1224-1227 (2009)
5. Varjani, J. Sunita, N.U. Vivek, Comparative Studies on Bacterial Consortia for Hydrocarbon Degradation, Int. J. Innov. Res. Sci. Eng. Technol. **2(10)**, 5377-5380 (2015)
6. M. Aakula, M.M. Kedare, S. Patra, T. Gangar, S. Singha, S. Meher, L. Sahoo, S. Rao, P. Radhakrishnanand, Restoration of crude oil-contaminated soil using microbes with degradative and plant-growth-promoting abilities, J. Microbiol. Methods. **241**, 107377 (2026)
7. K. Christodoulis, M. Theodoropoulou, S. Kavouri, M. Dourou, C.D. Tsakiroglou, In situ remediation of oil-contaminated soils by ozonation: Experimental study and numerical modeling. Chemosphere. **372**, 144133 (2025)
8. T. Kim, J.K Hong, E.H. Jho, G.K. Kang, D.J. Yang, S.J. Lee, Sequential biowashing-biopile processes for remediation of crude oil contaminated soil in Kuwait. J. Hazard. Mater. **378**, 120710 (2019)
9. Z. Wei, J.J. Wang, L.A. Gaston, J. Li, L.M. Fultz, R.D. DeLaune, S.K. Dodla, Remediation ofcrude oil-contaminated coastal marsh soil: Integrated effect of biochar, rhamnolipid biosurfactant and nitrogen application. J. Hazard. Mater. **396**, 122595 (2020)
10. S. Hosseini, R. Sharifi, A. Habibi, Efficient bioremediation of crude oil contaminated soil by a consortium of in-situ biosurfactant producing hydrocarbon-degraders. Sci. Rep.. **15**, 19852 (2025)
11. D. Xuezhi, A.A. Ahmad, I. Muhammad, T. Shah, Q. Kalwar, Remediation Methods of Crude Oil Contaminated Soil. World J. Agri. & Soil. Sci. **4(4)** (2020)
12. L.C. Okorie, Resource Recovery from Crude Oil Sludge Using Solvent Extraction and Delayed Cooking Method. Int. J. Petroleum Gas Eng. Res. **6(1)**, 10-20 (2023)
13. P. Rai, B. Mishra, U. Jayakrishnan, A. Mukherjee, S. Moulik, S. Roy, A. Kumari, Ultrasonic cavitation assisted deep eutectic solvent extraction of pectin from waste sweet lime peel: Statistical optimization and energetic analysis. Ind. Crops Prod. **228,** 120895 (2025)
14. I.T. Moreira, O. Oliveira, J.A. Triguis, A.F. Queiroz, S.L. Ferreira, C. Martins, A. Silva, B.A. Falcao, Phytoremediation in mangrove sediments impacted by persistent total petroleum hydrocarbons (TPH's) using *Avicennia schaueriana.* Mar. Pollut. Bull. **67**, 130-136 (2013)
15 R.C. Gholam, M. Dariush, Theoretical and experimental investigation of desalting and dehydration of crude oil by assistance of ultrasonic irradiation. Sonochem. **20**, 378-385 (2013)

Application of polyaluminium chloride (PAC) as a Heterogeneous catalyst for glycerol formal synthesis

Tatang Shabur Julianto[1*], *Riskiyatul* Mauludiyah[1], and *Cutwan Iynayah* Putri[1]

[1]Department of Chemistry. Faculty of Mathematics and Natural Sciences, Universitas Islam Indonesia, Indonesia

Abstract. Acetalization of glycerol with formaldehyde is one approach to produce glycerol formal compounds that have the potential to be used as fuel additives and environmentally friendly solvents. This study explores the use of polyaluminium chloride (PAC) as a heterogeneous catalyst in the acetalization reaction with a focus on its effectiveness in accelerating the reaction and product yield. Variations in reaction conditions studied include the molar ratio of glycerol : formaldehyde, PAC catalyst concentration, temperature, and reaction time. Characterization was carried out using GC-MS to identify the main products and evaluate the conversion of reactants. In addition, SEM-EDX was used to evaluate the surface conditions of PAC before and after the reaction. The results showed that PAC was able to activate the carbonyl group of formaldehyde through a Lewis acid mechanism, facilitating the formation of hemi-acetals and cyclic products. The optimum conditions were obtained at a ratio of 1:5, 1% PAC concentration, a temperature of 70°C, and a reaction time of 3 hours, with the main product being 1,3-dioxane-5-ol or glycerol formal. The use of PAC has been proven to be efficient and selective, and has the potential to be developed as an environmentally friendly heterogeneous catalyst in the synthesis of glycerol formal.

1 Introduction

Crude glycerol is the major byproduct of biodiesel production, accounting for about 10% of the total mass in the transesterification of triglycerides with methanol. The continuous growth of biodiesel production has led to a significant surplus of crude glycerol. Due to its functional properties, glycerol is utilized in pharmaceuticals, food and beverages, alkyd resins, animal feed, and as a precursor for value-added chemicals such as epichlorohydrin, propylene glycol, bio-methanol, and biohydrogen. However, its utilization strongly depends on purity, making purification a crucial step to enhance its industrial value [1].

One promising valorization route is the acetalization of glycerol with formaldehyde under acidic conditions, yielding glycerol formal, a mixture of two cyclic isomers: 1,3-dioxan-5-ol and 1,3-dioxolan-4-methanol. This derivative functions as an oxygenated fuel additive,

* Corresponding author: tatang_shabur@uii.ac.id

where hydroxyl and acetal groups improve fuel properties, combustion efficiency, and reduce emissions of CO_2, NO_x, particulates, and unburned hydrocarbons [2]. Its water solubility and stability also make it applicable in polymer synthesis, surfactants, and solvents [3]. From a circular economy perspective, converting waste glycerol into glycerol formal adds value while minimizing waste, with advantages over ethanol or MTBE due to its renewable origin and higher thermal and oxidative stability [4,5].

Traditionally, acetalization has employed homogeneous catalysts such as sulfuric acid, which, although effective, present drawbacks including difficult separation, non-reusability, and harmful waste generation [6,7]. Heterogeneous catalysts provide a more sustainable alternative as they are reusable and environmentally benign. Common examples include Amberlyst resins, zeolites, and mesoporous silicates. Polyaluminium chloride (PAC) has also emerged as a promising heterogeneous catalyst owing to its low cost, ease of preparation, and high efficiency. Its catalytic activity in organic synthesis, highlighting PAC's potential in green chemistry approaches that emphasize recyclable catalysts, non-toxic reagents, and sustainable processes [8]. Recently, PAC known as inorganic material that commonly used as coogulant in water treatment. Base on the evaluation on the previous researches, PAC have not been yet used as a catalyst for the synthesis of glycerol formal from glycerol and formaldehyde. This study therefore proposes the use of PAC as a heterogeneous acid catalyst of glycerol formal to substitute conventional homogeneous catalysts. Acting as a polyprotic Lewis acid, PAC is anticipated to efficiently promote the acetalization reaction while offering advantages of simple separation and potential reusability.

2 Methods

2.1 Materials and instruments

The materials employed in this work consisted of glycerol (99% purity), formaldehyde solution (37%), and polyaluminum chloride (PAC) containing approximately 28–30% Al_2O_3. The experimental isntruments included standard Pyrex glassware such as a three-neck flask, thermometer, magnetic stirrer, Erlenmeyer flask, boiling stones, condenser, water bath, watch glass, glass rod, and dropper pipette. Analytical characterization was performed using, GC–MS (Shimadzu GCMS-QP2010 SE), and SEM–EDX (Phenom Pro-X).

2.2 Methods

2.2.1 Synthesis of Glycerol Formal via Acetalization Reaction

The synthesis/acetalization process begins with mixing glycerol and formaldehyde in a three-necked flask, followed by the addition of PAC as a heterogeneous catalyst after both reactants have become homogeneous. Reflux is then carried out at various temperatures of 50, 70, and 90°C. The evaporated solvent is collected back due to condensation through a condenser connected to the same flask. The reaction is carried out at time intervals of 3, 5, and 7 hours with glycerol/formaldehyde ratios of 1:1, 1:3, and 1:5, respectively, and PAC at 1% of the glycerol weight. Finally, the glycerol formal is separated from the solvent using a rotary vacuum evaporator.

After acetalization, the solution is cooled and separated using a rotary vacuum evaporator at a specific time and temperature. The solution is transferred to a round-bottom flask and installed in a rotary vacuum evaporator. The device is set at 90°C, 130-170 mbar, and 50 rpm. The working principle of a rotary vacuum evaporator is the separation of compounds and their solvents based on the difference in boiling points by heating and vacuum [9].

2.2.2 Characterization of the Chemical Structure of Glycerol Formal

Glycerol formal samples are then analyzed, resulting in an Fourier Transform Infra Red (FTIR) spectrum by measuring the absorption of infrared radiation over a range of wavelengths (usually in the mid-infrared region, around 4000 to 400 cm-1). FTIR spectrum analysis is both qualitative and quantitative. Qualitative analysis identifies absorption peaks that characterize functional groups and chemical bonds in glycerol formal. For example, formaldehyde glycerol generally exhibits peaks associated with hydroxyl groups (OH) and ether bonds (C-O-C). Quantitative analysis measures the concentration or abundance of glycerol formal in a sample based on the intensity of a particular peak. This involves comparing the height or area of the peak with a calibration standard or a known concentration of glycerol formal. The results of the analysis determine the composition, purity, and structural features of glycerol formal. Experimental spectra are compared with reference spectra or databases to confirm the identity of the glycerol formal and detect impurities or contaminants.

2.2.3 Characterization of the Chemical Composition of Glycerol Formal

Gas Chromatography Mass Spectrometry (GC-MS) analysis confirmed that glycerol formal was formed from the reaction between glycerol and formaldehyde in the presence of a PAC catalyst. In this technique, gas chromatography separates the distillate components according to their boiling points and retention times, after which each fraction is subjected to mass spectrometry to determine its molecular structure and mass-to-charge ratio. The resulting chromatogram is correlated with mass spectra to identify peaks corresponding to glycerol formal and other products. Each spectrum was matched against a reference database to verify compound identities, enabling the determination of glycerol formal conversion and concentration in the reaction mixture. Retention times were utilized to distinguish compounds, while mass spectra provided structural confirmation. The GC–MS analysis was carried out on a Shimadzu GCMS-QP2010 SE, operating within a 50–300°C temperature range, equipped with a quadrupole mass analyzer and a scan speed of 10,000 u/sec [6].

2.2.4 PAC Surface Characterization

The surface characteristics of the PAC catalyst were analyzed using Scanning Electron Microscope-Energy Dispersive X-Ray (SEM-EDX) to determine the changes that occurred before and after synthesis. The PAC sample used for synthesis was dried at 90°C in a vacuum for 24 hours. After that, it was cooled to room temperature and reweighed. The sample was then analyzed using SEM-EDX, producing images of the sample at different magnifications to observe its morphology.

3 Results and Discussion

3.1 Synthesis of Glycerol Formal via Acetalization Reaction

With the rest in lower case. You should leave 22 mm of space above the title and 6 mm after the title. The acetalization product of glycerol and formaldehyde was characterized by FTIR and GC–MS to confirm the formation of glycerol formal. This condensation reaction involves the nucleophilic attack of glycerol hydroxyl groups on the carbonyl carbon of formaldehyde, producing cyclic acetals, namely 1,3-dioxolan-4-methanol (five-membered ring) and 1,3-dioxan-5-ol/glycerol formal (six-membered ring) [10, 11]. FTIR spectra indicate ring formation through the disappearance of free –OH absorption and the emergence of C–O–C bands typical of cyclic ethers. As shown in Figure 1, the PAC catalyst, acting as a Lewis acid, protonates the carbonyl oxygen, enhancing electrophilicity and accelerating hemiacetal formation, which subsequently undergoes dehydration and intramolecular cyclization into stable acetals [12, 13]. Unlike homogeneous acids, PAC is less prone to deactivation by water, thereby maintaining catalytic efficiency. The resulting products are mainly the five- and six-membered cyclic isomers collectively known as glycerol formal [14].

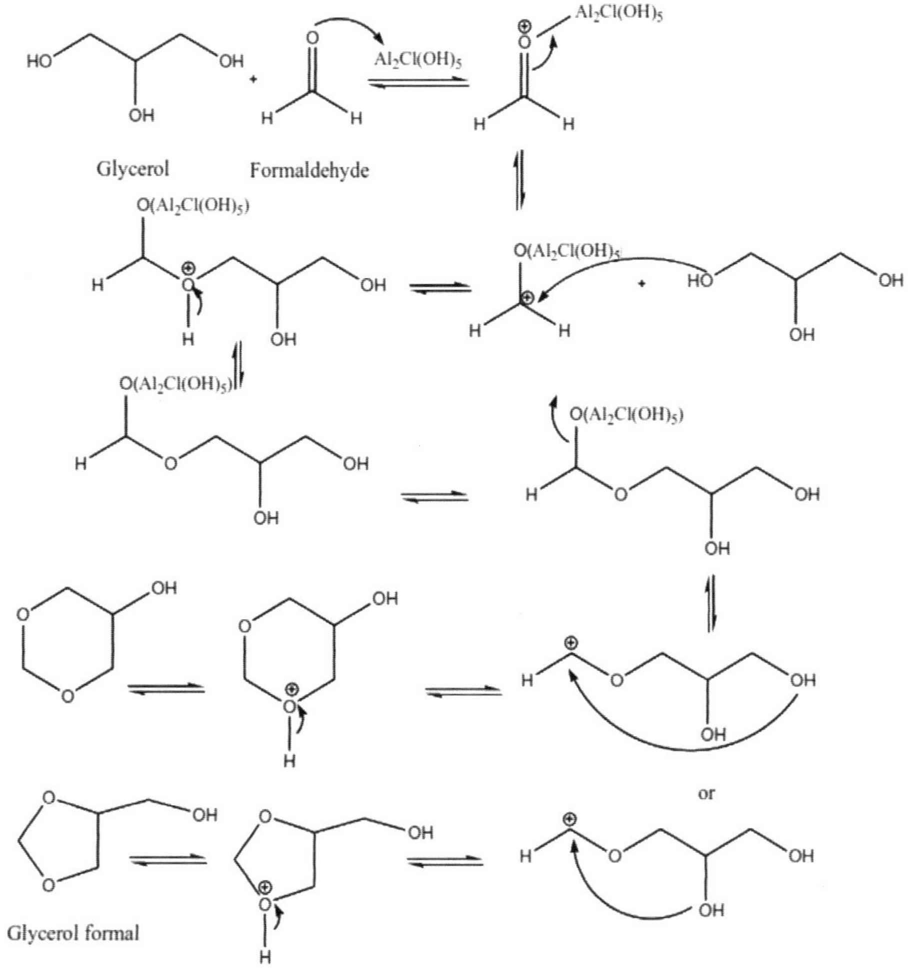

Fig. 1. The mechanism of glycerol formal synthesis using glycerol and formaldehyde

The acetalization of glycerol with formaldehyde is an equilibrium process, where conversion is significantly affected by factors such as reactant molar ratio, temperature, catalyst selection, and water removal strategy. Since water formed as a byproduct drives the equilibrium backward, the reaction is commonly performed under vacuum or in the presence of dehydrating agents like zeolites to enhance glycerol formal yield [15]. Thus, optimizing these parameters is essential to maximize yield and valorize glycerol as a biodiesel byproduct in line with green chemistry principles.

3.2 Characterization of the Chemical Structure of Glycerol Formal

The effect of the formaldehyde-to-glycerol molar ratio on product formation was evaluated at three ratios: 1:1 (GF1), 3:1 (GF2), and 5:1 (GF3), using 1% PAC catalyst and a reaction time of 3 h. After distillation, the products were characterized by FTIR. All spectra (FD1, FD2, FD3) showed characteristic C–O–C stretching bands at 1020–1150 cm^{-1} and ring C–H vibrations at 920–950 cm^{-1}, confirming the formation of cyclic acetal structures (dioxolane and dioxane). The C–O–C absorption in this region is a key indicator of five-membered ring formation in polyol–formaldehyde acetalization, supporting the presence of glycerol formal at all tested ratios [11]. A broad -OH absorption band (3200–3600 cm^{-1}) was also observed in all samples, indicating residual unreacted glycerol. In contrast, the absence of aldehyde carbonyl absorption near 1700 cm^{-1} suggests nearly complete consumption of formaldehyde, reducing the likelihood of aldehyde contamination. These results align with the reversible nature of acetalization, where excess reactant or water removal drives the equilibrium toward acetal formation.

3.3 Characterization of the Chemical Composition of Glycerol Formal

The variation in the molar ratio of formaldehyde to glycerol is intended to identify the optimal proportion for producing cyclic acetal compounds. This ratio determines the availability of formaldehyde molecules to react with the hydroxyl groups of glycerol, leading to the formation of 1,3-dioxolane or 1,3-dioxane structures. The reaction relies on the presence of the carbonyl group in formaldehyde, which can interact with two hydroxyl groups of glycerol to generate cyclic acetal frameworks such as 1,3-dioxane or 1,3-dioxolane. An insufficient formaldehyde ratio may result in incomplete conversion, whereas an excessive amount can shift the equilibrium toward greater product formation [11]. In this study, the molar ratios of formaldehyde to glycerol were varied at 1:1, 3:1, and 5:1, under conditions of 3 hours reaction time, 50 °C, and 1% PAC catalyst.

Table 1. Concentration of main compounds based on variations in molar ratio

Compound Name	Retention Time (min)	% Area Molar Ratio Variation (glycerol:formaldehyde)		
		1:1	1:3	1:5
1,3-dioxane-5-ol	**2.622**	**5.11**	**18.42**	**18.70**
2,3-propandiol, 1-metoksi	4.325	0.33	-	-
Diethyl acetal	4.810	-	-	4.81
Glycerol	**5.908**	**85.52**	**58.30**	**73.47**
Propyl glycol	12.650	-	12.65	-

GC-MS analysis as shown in Table 1, revealed that at a 1:1 molar ratio, glycerol dominated (85.52%) with only 5.11% 1,3-dioxan-5-ol (glycerol formal), showing limited acetal formation due to insufficient formaldehyde. At 3:1, glycerol formal rose to 18.42% as excess formaldehyde enhanced hemiacetal and cyclic acetal formation. At 5:1, the product slightly increased to 18.70% with reduced glycerol (73.47%), indicating the reaction equilibrium had shifted maximally toward the product. Thus, a 5:1 ratio is considered near-optimal for glycerol formal production, though excessive formaldehyde may trigger side reactions.

Reaction temperature plays a crucial role in determining the activation energy and rate of the acetalization reaction. Since the process is mildly endothermic, an increase in temperature enhances molecular motion, accelerates effective collisions, and facilitates the formation of the C–O–C bond. However, excessively high temperatures may cause side reactions such as formaldehyde degradation or polymerization. As shown in Table 2, the reaction was conducted at three temperature variations: 50°C, 70°C, and 90°C, with a fixed molar ratio of formaldehyde to glycerol at 5:1, a reaction time of 3 hours, and 1% PAC catalyst. At 50°C, GC-MS analysis revealed that the main product, 1,3-dioxan-5-ol, was only formed at 5.11%, while unreacted glycerol remained as high as 85.52%. This low conversion indicates that the temperature was insufficient to effectively activate the carbonyl group of formaldehyde and the hydroxyl groups of glycerol, resulting in slow formation of the hemiacetal intermediate. Consequently, the cyclization step toward glycerol formal was not efficient, leading to low product yield. In contrast, the reaction at 70°C displayed the formation of additional products, suggesting that the increased kinetic energy improved collision frequency and facilitated further transformation. This result highlights the importance of selecting an optimal reaction temperature that balances sufficient activation energy for acetalization without promoting unwanted side reactions.

Table 2. Concentration of main compounds based on reaction temperature variations

Compound Name	**Retention Time (min)**	**% Area Reaction Temperature Variation**		
		50°C	**70°C**	**90°C**
1,3-dioxan-5-ol	**3.175**	**18.70**	**74.15**	**35.88**
Glycerol	**3.392**	**73.47**	**13.12**	**25.85**
Diethyl acetal	4.244	4.81	-	-

At 70 °C, the chromatogram revealed a higher formation of 1,3-dioxan-5-ol along with new compounds, suggesting that this intermediate temperature is close to the optimum for acetalization. At this point, the supplied energy is sufficient to overcome activation energy without causing major degradation or side reactions. However, raising the temperature to 90 °C significantly accelerates the reaction but also risks polymerization of formaldehyde into paraformaldehyde and partial decomposition of the acetal due to thermal instability. Previous studies confirm that excessive temperatures lower yield by favoring linear products and aldehyde oligomers. Thus, temperature not only influences product yield but also its composition, governed by activation energy, hemiacetal stability, and competing side reactions.

Reaction time plays a crucial role in the acetalization of glycerol with formaldehyde, as it influences the reaction pathway, the stability of intermediate species, and the distribution of final products. It also determines how long the reaction proceeds within a closed system. Under mild acidic conditions, this transformation typically occurs quickly; however, extending the duration excessively may promote side reactions or lead to volatilization of light compounds. As shown in Table 3, reaction times of 3, 5, and 7 hours were examined at a formaldehyde-to-glycerol molar ratio of 5:1, a reaction temperature of 70°C, and with 1% PAC as the catalyst. The resulting concentrations of the main products as a function of reaction time are presented in the following table.

Table 3. Concentration of main compounds based on the effect of reaction time

Compound Name	**Retention Time (min)**	**% Area Reaction Time Variation**		
		3 hours	**5 hours**	**7 hours**
2,2-Dimethyl-1,3-dioxolane-4-methanol	2,125	2,75	2,26	0,65
1,3-dioxan-5-ol	**3,175**	**74,15**	**14,95**	**49,15**
Glycerol	**3,392**	**13,12**	**47,06**	**21,52**
2-methyl-4-methoxy-1,3-dioxane	9,30	5,80	16,56	-

GC-MS results revealed that at 3 hours, 1,3-dioxan-5-ol reached its highest yield (74.15%), with only 13.12% glycerol remaining. This indicates that 3 hours provides optimum conditions for hemiacetal formation and cyclization, consistent with reports that equilibrium is typically achieved within 120–180 minutes. At 5 hours, the yield of 1,3-dioxan-5-ol dropped to 14.95%, while other derivatives such as 2-methyl-4-methoxy-1,3-dioxane increased, suggesting secondary reactions like trans-acetalyzation or substitution [11]. Extending the reaction to 7 hours produced 49.15% 1,3-dioxan-5-ol, 21.52% glycerol, and a significant amount of 1-acetoxy-2-propanol (17.49%), likely formed via glycerol degradation and esterification. Overall, a 3-hour reaction was the most selective, yielding the stable six-membered cyclic acetal as the dominant product.

The PAC catalyst provides Al^{3+} and Cl^- ions that introduce Lewis and Brønsted acidity, promoting activation of the formaldehyde carbonyl group and electron withdrawal from the –OH group of glycerol. Catalyst concentration strongly influences reaction direction and potential side reactions. As shown in Table 4, PAC concentrations of 1%, 3%, and 5% were tested at a formaldehyde:glycerol ratio of 5:1, 70°C, and 3 h reaction time to identify the optimal conditions for maximizing cyclic acetal yield while minimizing by-products. At 1% PAC, the major products included 1,3-dioxan-5-ol (74.15%), glycerol (13.12%), 2-methyl-4-methoxy-1,3-dioxane (5.80%), and 2,2-dimethyl-1,3-dioxolane-4-methanol (2.75%). The results for 3% and 5% PAC concentrations are summarized in the following table.

Table 4. Concentration of main compounds based on the influence of PAC catalyst concentration

Compound Name	Retention Time (min)	% Area Variation of PAC Catalyst Concentration		
		1%	3%	5%
1,3-dioxan-5-ol	**3.227**	**74.15**	**-**	**6.86**
Glycerol	**3.392**	**13.12**	**47.06**	**21.52**
2-methyl-4-methoxy-1,3-dioxane	9.347	-	-	54.11
2,2-Dimethyl-1,3-dioxolane-4-methanol	18.320	-	10.05	-
Di-(9-octadecenoyl)-glycerol	21.935	-	19.84	-

The best results for 1,3-dioxane-5-ol synthesis were obtained with 1% PAC, giving the highest yield, purity, and yield. Higher catalyst concentrations reduced efficiency and promoted side reactions such as esterification or derivative formation. GC and GC-MS analyses showed that the optimal conditions were a formaldehyde:glycerol ratio of 5:1, temperature of 70°C, reaction time of 3 h, and 1% PAC, producing 1,3-dioxane-5-ol at 74.15%. Longer reaction times (5–7 h) or PAC concentrations above 1% decreased yield and generated by-products like 2-methyl-4-methoxy-1,3-dioxane and 2,2-dimethyl-1,3-dioxolane-4-methanol. At 3% PAC, the target product was absent, likely due to reduced active Lewis acid sites from structural changes in the aluminum species. At 5% PAC, glycerol formal reappeared in small amounts, suggesting partial reactivation of catalytic sites after structural rearrangement.

3.4 Characterization of PAC Catalyst Surface

The surface characterization of PAC was conducted to examine its stability, activity, and structural changes during the catalytic process. SEM-EDX was used to analyze PAC as a heterogeneous catalyst in the acetalization of glycerol and formaldehyde, focusing on surface morphology and elemental composition before and after the reaction as shown in Figure 2. The results show that PAC initially contained high levels of Al and Cl, reflecting the presence of polycationic $AlO_4Cl_{12}(OH)_{24}{}^{7+}$ as the main structure. Al remained stable, suggesting that Lewis active sites were preserved [8, 11]. However, Cl decreased significantly (9.89%), indicating partial hydrolysis of Al–Cl bonds in aqueous glycerol medium, leading to Cl replacement by hydroxyl or oxide groups. Meanwhile, the increase in C after the reaction suggests adsorption of organic residues, including intermediates and cyclic acetal products, on the catalyst surface. This accumulation likely caused site blocking, reducing reactant access to active centers.

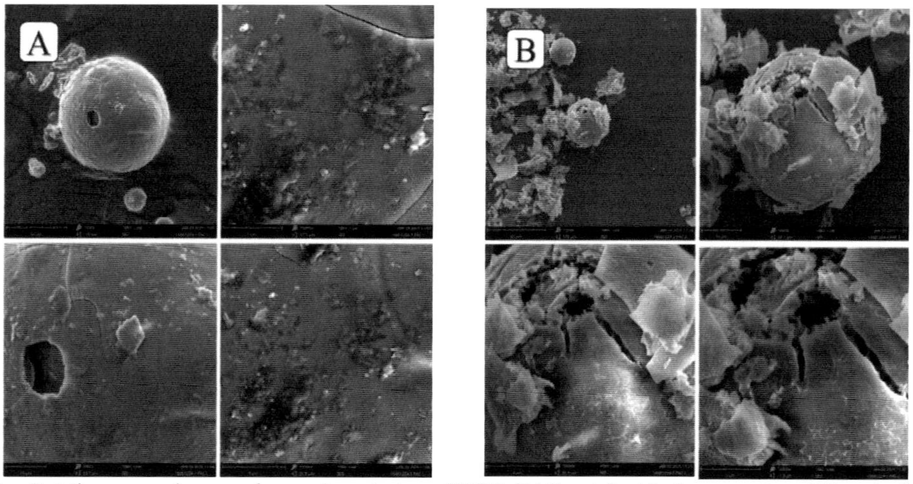

Fig. 2. Microscope image of scanning electron (SEM) PAC catalyst before reaction (A) and after reaction (B)

Other elements such as oxygen and carbon increased. Oxygen increased by 1.78%. The high O indicates the presence of hydroxyl and oxide groups on the PAC surface, which are important in the formation of Al-O-Al bridges and the adsorption capacity of the reaction water. Carbon increased by 4.3%, which is most likely due to the adsorption of residual organic products such as glycerol formal or hemi-acetal intermediates.

4 Conclusion

Overall, optimum conditions were obtained at a formaldehyde:glycerol molar ratio of 5:1, a temperature of 70°C, a reaction time of 3 hours, and a PAC catalyst concentration of 1%, which resulted in the most selective conversion of glycerol to glycerol formal (1,3-dioxan-5-ol).

References

1. T. Attarbachi, M. D. Kingsley, and V. Spallina, New trends on crude glycerol purification: A review, *Fuel* **340**, 127485 (2023). https://doi.org/10.1016/j.fuel.2023.127485
2. C. R. Chilakamarry, A. M. M. Sakinah, A. W. Zularisam, A. Pandey, and D.-V. N. Vo, Technological perspectives for utilisation of waste glycerol for the production of biofuels: A review, *Environ. Technol. Innov.* **24**, 101902 (2021). https://doi.org/10.1016/j.eti.2021.101902
3. A. L. Olson, M. Tunér, and S. Verhelst, A concise review of glycerol derivatives for use as fuel additives, *Heliyon* **9**, e12987 (2023).
4. I. Fatimah, I. Sahroni, G. Fadillah, M. M. Musawwa, T. M. I. Mahlia, and O. Muraza, Glycerol to solketal for fuel additive: Recent progress in heterogeneous catalysts, *Energies* **12**, 2872 (2019). https://doi.org/10.3390/en12152872

5. R. Estevez, L. Aguado-Deblas, D. Luna, and F. M. Bautista, An overview of the production of oxygenated fuel additives by glycerol etherification over heterogeneous catalysts, *Energies* **12**, 2364 (2019).
6. Mahreni, L. S. Aji, S. Gunawan, and M. A. Ikhsana, Solketal from glycerol, acetone, and ethanol solution using wolfram acid phosphotungstic catalyst, *MATEC Web Conf.* **372**, 03003 (2022). https://doi.org/10.1051/matecconf/202237203003
7. M. J. da Silva, M. G. Teixeira, D. M. Chaves, and L. Siqueira, An efficient process to synthesize solketal from glycerol over tin(II) silicotungstate catalyst, *Fuel* **281**, 118724 (2020). https://doi.org/10.1016/j.fuel.2020.118724
8. Y. Liang, G. Wang, Z. Wu, W. Liu, M. Song, Y. Sun, X. Chen, H. Zhan, and S. Bi, Polyaluminum chloride as highly efficient and green catalyst for Friedel–Crafts alkylation of bis(indolyl)methanes, *Catal. Commun.* **147**, 106136 (2020). https://doi.org/10.1016/j.catcom.2020.106136
9. F. S. Rikantara, M. R. Utami, and A. Kasasiah, Aktivitas antioksidan kombinasi ekstrak daun sirsak dan daun pepaya dengan metode DPPH, *Lumbung Farm. J. Ilmu Kefarmasian* **3**, 124 (2022).
10. O. Ilgen, S. Yerlikaya, and F. O. Akyurek, Synthesis of solketal from glycerol and acetone over Amberlyst-46, *Period. Polytech. Chem. Eng.* **61**, 144 (2017). https://doi.org/10.3311/PPch.8895
11. K. Yamamoto , A. M. Kiyan , J. C. Bagio, K. A. B. Rossi , F. D. Berezuk, and M. E. Berezuk, Green cyclic acetals production by glycerol etherification reaction with benzaldehyde using cationic acidic resin, *Green Process Synth*, 8: 183–190 (2018). https://doi.org/10.1515/gps-2018-0059
12. M. Checa, S. N. Delgado, V. Montes, and J. M. Encinar, Recent Advances in Glycerol Catalytic Valorization: A Review, Catalysts 10, 1279 (2020). http://doi.org/10.3390/catal10111279
13. M. Bayat, and D. Gheidari, Green Lewis Acid Catalysis in Organic Reactions, *ChemistrySelect*, 7, 28 (2022). https://doi.org/10.1002/slct.202200774
14. A. R. Trifoi, P. Ş. Agachi, and T. Pap, Glycerol acetals and ketals as possible diesel additives. A review of their synthesis protocols, *Renew. Sustain. Energy Rev.* **62** (2016). https://doi.org/10.1016/j.rser.2016.05.013
15. L. Chen, B. Nohair, and S. Kaliaguine, Glycerol acetalization with formaldehyde using water-tolerant solid acids, *Applied Catalysis A: General*, 509 (2016). https://doi.org/10.1016/j.apcata.2015.08.014

Formulation and hedonic evaluation of fennel leaf (foeniculum vulgare) essential oil aromatherapy candles using soy and paraffin wax

Gani Purwiandono[1*], *Revando Ahmad* Afiifi[1], *Hafizh* Royhan[1], *Nurcahyo Iman* Prakoso[1], and *Maisari* Utami[1]

[1]Department of Chemistry, Universitas Islam Indonesia, Yogyakarta 55584, Indonesia

Abstract. This study explores the formulation of aromatherapy candles using fennel leaf (Foeniculum vulgare) essential oil with soy and paraffin wax bases. The essential oil, extracted by steam-water distillation and analyzed by GC-MS, was found to contain α-phellandrene (62.27%) as the main stress-relieving compound. Candles were made using three different wax-to-citric acid ratios and tested for aroma preference and stress-relief effectiveness. Physical properties of the oil were also evaluated based on Food Chemical Codex standards. A 1:1 ratio (15 g wax : 15 g citric acid) showed the best results, with soy wax candles receiving the highest approval for scent and long-lasting calming effects. Although paraffin candles had a stronger initial aroma, their effect faded faster. Overall, soy wax proved to be a cleaner-burning and more effective base for stress-relief aromatherapy candles. Further research should optimize the formulation and assess commercial potential for sustainable products.

1 Introduction

Stress is a major issue in contemporary culture, frequently resulting in adverse impacts on both physical and mental well-being. Chronic stress is associated with several health difficulties, including anxiety disorders, depression, and physiological conditions such as hypertension and cardiovascular illnesses [1]. Given these extensive consequences, identifying strategies to reduce stress has become necessary. Aromatherapy, an alternative and increasingly favoured approach, employs the natural ingredients in essential oils to facilitate relaxation and reduce stress [2]. Aromatherapy functions by inhaling volatile molecules that engage odour receptors and trigger the brain's limbic system, resulting in change in mood and emotional well-being [3].

Aromatherapy candles provide a practical and visually appealing method to provide the therapeutic advantages of essential oils [4]. Upon combustion, these candles emit essential oils into the atmosphere, facilitating prolonged exposure to the volatile molecules that aid in

*Corresponding author: gani_purwiandono@uii.ac.id

stress reduction [5]. Fennel leaf (Foeniculum vulgare) essential oil was known for its unique fragrance and comfort characteristics among the array of essential oils [6]. The extraction of fennel leaf essential oil is generally conducted via steam-water distillation, a technique that effectively preserves the oil's volatile components [7].

The selection of wax was crucial, alongside the appropriate essential oil, in the creation of excellent aromatherapy candles [8]. Soy wax and paraffin wax were among the most often utilised materials [9]. Soy wax, obtained from hydrogenated soybean oil, was regarded as an eco-friendlier alternative because to its lower pollutant emissions in comparison to paraffin wax [10]. It additionally combusts at a reduced rate, facilitating a prolonged release of the essential oils. Conversely, paraffin wax was more frequently utilised owing to its cost-effectiveness and ease of manipulation. Nonetheless, it generates increased pollutants and combusts at elevated temperatures, potentially influencing the emission of essential oils [11].

The research aimed to evaluate if the essential oil properties conformed to the criteria established by the Food Chemicals Codex (FCC) for density, refractive index, and optical rotation. The essential oil analysis using GC−MS to determine the principal components contributing to its fragrance and medicinal attributes. A hedonic test was performed to assess panellist preferences about the aroma and perceived stress-relief benefits of the candles. This study's findings will enhance the existing information regarding the production of aromatherapy products and offer insights into the appropriate therapeutic applications of fennel leaf essential oil.

2 Research Methods

2.1 Extraction and Characterization of Essential Oils

This study extracted essential oil from fennel leaves (Foeniculum vulgare) using steam-water distillation and made aromatherapy candles from the oil. Steam-water distillation used a stainless-steel kettle, condenser, and collection flask. In the kettle, 2.2 kg of fennel leaves and 10 litres of distilled water were distilled for 6 hours. Steam carried volatile compounds through plant material during distillation to the condenser for cooling and separation. The essential oil was then removed from the separator and stored in a dark glass bottle to prevent light and air damage. Oil yield was calculated as a proportion of plant material weight and the extracted essential oil was analysed three times.

Standard laboratory equipment including a refractometer and polarimeter analysed the oil's density, refractive index, and optical rotation. A pycnometer measured density, while a refractometer recorded refractive index at 28.83°C. A polarimeter at the same temperature measured optical rotation. GC-MS (Shimadzu QP 2010 SE, with Rtx-5MS column) was used to analyse the essential oil's chemical makeup in addition to these physical attribute measurements. GC-MS analysis identified the oil's key volatile constituents, revealing its scent and medicinal effects.

2.2 Hedonic Assessment

Aromatherapy candles were made using the essential oil after characterisation. Candles were made with soy and paraffin wax. Each wax was mixed with citric acid and fennel leaf essential oil in different amounts. Three soy and paraffin wax compositions were made. Formulation 1 had 15 g wax, 15 g citric acid, and 5 mL fennel leaf essential oil. Formulations 2 and 3 mixed 20 g of wax, 10 g of citric acid, and 5 mL of essential oil. After melting the wax and citric acid in a double boiler, the essential oil was added. After pouring the fluid into moulds with wicks in the centre, it was let to cool and solidify at room temperature. A hedonic test assessed aromatherapy candles' sensory and stress-relieving qualities. The survey included 20-25 years old panellists. Each panellist received all three soy and paraffin candle compositions. Panellists assessed candles based on scent and stress-relief effectiveness. Google Forms surveys were used to collect and analyse assessments to identify the best formulation. Soy wax and paraffin wax candles were compared for fennel leaf essential oil therapeutic benefits using the hedonic test.

3 Results and Discussion

3.1 Extraction and Characterization of Essential Oils

Fennel leaf essential oil was extracted using steam-water distillation, yielding 0.7% of oil. Although this yield is low, it is typical for fennel leaves, as they contain small amounts of volatile compounds. The extracted oil was assessed for its physical properties, which were compared to the standards set by the FCC [12],[13].

Table 1. Characterization of Fennel Leaf oil

% Yield	Density	Refractive index	Optical rotation
This research	0.88 g/mL	1.4834	+22.1 degrees
FCC [12],[13]	0.987−0.998 g/mL	1.5230−15430	+12 − +24 degrees

The density of the fennel leaf essential oil was found to be 0.88 g/mL (Table 1), slightly below the FCC standard of 0.987-0.998 g/mL, possibly due to environmental factors or differences in the distillation process. The refractive index of 1.5321 was within the FCC range of 1.5230-1.5430, though slight deviations may indicate differences in composition or impurities. The optical rotation (+22.1°) also fell within the FCC standard range, confirming the oil's optical activity.

Figure 1 showed the chromatograms of fennel leaf essential oil. The GC-MS analysis, in Table 2, showed that α-phellandrene (62.27%) was the dominant component of the oil, which is consistent with other studies, highlighting its aromatic and medicinal properties. Other notable components included Methyl Thymylether (5.39%), Benzenemethanol (4.29%), and Germacrene D (13.53%), all known for their calming effects [14].

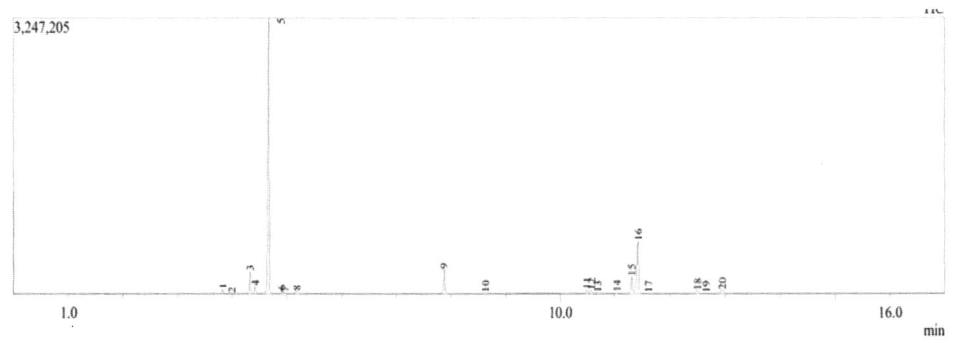

Fig. 1. Chromatograms of Fennel Leaf Essential Oil

Table 2. Chemical compounds in the Fennel Leaf Essential Oil

Peak	Retention time	Compounds	% Area	Molecular weight	Molecular formula
1	3.825	*Cis-ocimene*	0.92	136	$C_{10} H_{16}$
2	4.007	*Camphene*	0.18	136	$C_{10} H_{16}$
3	4.339	*2-β-PINENE*	3.91	136	$C_{10} H_{16}$
4	4.432	*β-Myrcene*	1.53	136	$C_{10} H_{16}$
5	4.681	*α-Phellandrene*	62.27	136	$C_{10} H_{16}$
6	4.921	*Benzene, 1-methyl-4-1-methylethyl*	0.67	134	$C_{10}H_{14}$
7	4.979	*l-Limonene*	0.31	134	$C_{10}H_{14}$
8	5.184	*1,3,6-Octatriene*	0.68	134	$C_{10}H_{14}$
9	7.870	*Methyl thymylether*	5.39	134	$C_{10}H_{14}$
10	8.635	*Endobornyl acetate*	0.24	196	$C_{12}H_{20}O_2$
11	10.497	*Thymohydroquinone dimethyl ethe*	1.04	194	$C_{12}H_{18}O_2$
12	10.577	*Trans-caryophyllene*	0.94	204	$C_{15}H_{24}$
13	10.70	*Trans-α-bergamotene*	0.29	204	$C_{15}H_{24}$
14	11.043	*α-Humulene*	0.42	204	$C_{15}H_{24}$
15	11.304	*Benzenemethanol, 4-(1-methylethyl)*	4.29	150	$C_{10}H_{14}O$
16	11.414	*Germacrene D*	13.53	204	$C_{15}H_{24}$
17	11.605	*Germacrene B*	0.28	204	$C_{15}H_{24}$
18	12.482	*Propanoic acid,*	1.15	234	$C_{15}H_{22}O_2$
19	12.642	*4,5-Methano-4,7-dimethyl-6-octen-2-1*	0.21	166	$C_{11}H_{18}O$
20	12.93	*Cyclohexane, bromo*	1.72	162	$C_6H_{11}Br$

3.2 Aromatherapy Candle Formulation

The fennel leaf essential oil was incorporated into two types of wax: soy wax and paraffin wax. Three formulations with different ratios of wax to citric acid were prepared to test the effect of these ingredients on the candle's fragrance and stress-relief properties. Table 3 presents the different formulations of soy wax, paraffin wax, citric acid, and fennel leaf essential oil. These variations aimed to evaluate how the wax's melting properties and the acid's effects on preservation and scent enhancement influenced the final product. After solidification, the candles were tested for their sensory properties, including aroma and texture. Soy wax candles had a matte finish and a softer texture, which allowed for a slower, more even burn. In contrast, paraffin wax candles had a glossy, harder surface, which led to a faster burn and quicker scent release.

Table 3. Aromatherapy candle formulation of fennel leaves using soy wax and paraffin wax

Treatment	Soy wax mass (g)	Paraffin wax mass (g)	Citric acid mass (g)	Fennel leaf essential oil mass (g)
Formulation 1	15	15	15	4.4
Formulation 2	20	20	10	4.4
Formulation 3	10	10	20	4.4

The highest performance of the 1 : 1 wax to citric acid ratio is attributed to the balanced thermal and structural properties of the candle matrix, which enabled a controlled and sustained release of essential oil during combustion. Although α-phellandrene constituted 62.27% of the fennel leaf essential oil, the optimal formulation ratio is not directly determined by its concentration. Instead, the 1 : 1 ratio promoted effective volatilization of α-phellandrene and other aromatic components, resulting in enhanced aroma perception and prolonged stress-relief effects as observed in the hedonic evaluation. The representative appearances of the soy and paraffin wax candle formulations are shown in Figure 2 and 3.

Fig. 2. Representative images of the aromatherapy soy wax candle of (a) formulation 1, (b) formulation 2, and (c) formulation 3.

Fig. 3. Representative images of the aromatherapy paraffin wax candle of (a) formulation 1, (b) formulation 2, and (c) formulation 3.

3.3 Hedonic Assessment

The hedonic test provided clear insights into panellists' preferences regarding the scent and stress-relief effects of the candles. Eighteen panellists aged 20 to 25 evaluated each candle using a five-point Likert scale for both fragrance and stress-relief potential [15]. In this research, the hedonic assessment was conducted using 20 semi-trained panelists, consisting of 7 male and 13 female participants aged between 20 and 25 years. The number of panelists was selected based on standard sensory evaluation guidelines, which recommend the use of 15–25 semi-trained panelists for hedonic testing to obtain reliable and consistent data. Prior to the evaluation, the panelists were given a brief explanation to familiarize them with the characteristics of fennel leaf aromatherapy candles, including the aroma profile of fennel leaves, the chemical composition of the essential oil and the intended function of the aromatherapy candles for stress relief.

During the test, panelists were asked to inhale the aroma of fennel leaf aromatherapy candles prepared using soy wax and paraffin wax. To minimize bias, the panelists were alternately exposed to both candle types, allowing each participant to evaluate both formulations. After completing the aroma exposure, the panelists filled out a structured questionnaire to assess their level of preference and perceived stress-relief effect. The hedonic test in this study was designed to evaluate both aroma acceptance and the effectiveness of the candles as stress-relief aromatherapy products.

All panellists rated Soy Wax Formulation 1 as having the most pleasant scent (100% approval), as shown in Figure 4 (a). This may be due to the balanced ratio of wax and citric acid, which helped release the aroma of fennel essential oil more effectively. Citric acid is known to lower the melting point of wax, allowing essential oils to evaporate more steadily. Among the paraffin candles, Formulation 3 was preferred, with 70% approval for fragrance, though it still ranked below the soy wax candle in overall scent preference.

Similar results were observed for the stress-relief effect. Soy Wax Formulation 1 again received full approval (100%) for its calming effect. Panellists described the scent as strong enough to be relaxing without being overwhelming. In contrast, Paraffin Formulation 3 received 65% approval, as shown in Figure 4 (b). While its fragrance was initially strong, it faded quickly, reducing its effectiveness over time. The higher preference for soy wax candles appears related to their slower burning and more controlled release of essential oils, offering a longer-lasting aromatherapy experience. Soy wax also burned cleaner, producing less smoke and soot than paraffin wax, which contributed to a more pleasant environment. Paraffin wax, on the other hand, burned faster and released a stronger initial aroma that quickly diminished. These results suggest that fennel leaf essential oil, when used with soy wax, creates an effective aromatherapy candle for stress relief. The main active compound, α-phellandrene, is known for its calming effects, supporting its use in relaxation treatments. Although the oil's physical properties slightly differed from FCC standards in density and refractive index, its quality was still acceptable for therapeutic use.

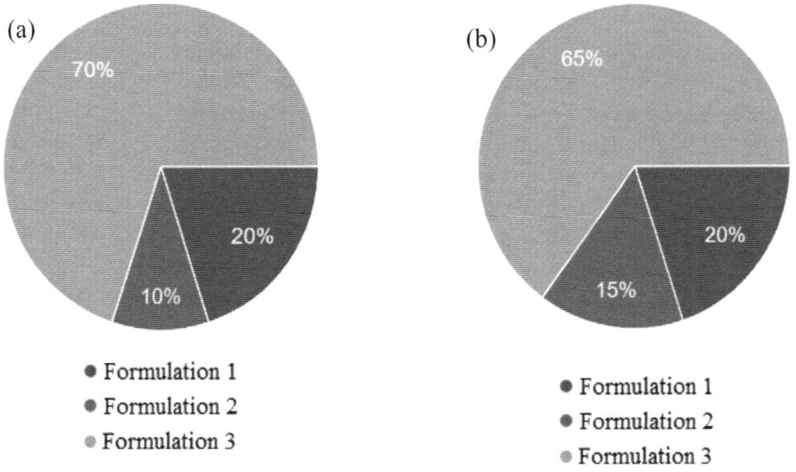

Fig. 4. The aromatherapy candle scent with fennel leaf made from: (a) soy wax and (b) paraffin wax.

The distinctions between soy wax and paraffin wax formulations were notable for scent emission and perceived stress-relief advantages. Soy wax surpassed paraffin wax in both categories, presumably owing to its better combustion properties and more regulated emission of essential oils [15]. The pliable nature and gradual combustion of soy wax facilitated a prolonged and more uniform fragrance, which was favoured by the panellists [15]. Paraffin wax, although initially more potent in aroma release, failed to maintain its smell over time, rendering it less appropriate for extended stress-relief purposes. This study effectively developed an aromatherapy candle utilising fennel leaf essential oil, which exhibited considerable stress-relief advantages, with soy wax identified as the most efficient medium for conveying these therapeutic effects.

4 Conclusion

This study showed that aromatherapy candles help relieve stress with steam-distilled fennel leaf (Foeniculum vulgare) essential oil. The essential oil extraction recovered 0.7%, and GC-MS analysis validated its chemical makeup, dominated by α-phellandrene (62.27%). While the essential oil's density and refractive index slightly from Food Chemical Codex (FCC) norms, its optical rotation (+22.1 degrees) was within the allowed range, validating its therapeutic usage. Soy wax candles were more efficient in delivering a pleasant scent and relieving tension than paraffin wax candles. The formulation with 15 g of soy wax, 15 g of citric acid, and 5 mL of fennel leaf essential oil was favourited by all panellists for scent and stress alleviation. Soy wax burns slowly, releasing essential oils gradually and prolonging aromatherapy. While paraffin candles had a stronger initial scent, they did not last as long as soy candles. The findings aid aromatherapy product formulation for efficacy and sustainability. Further investigations could optimise the wax-to-oil ratio, long-term stability, and commercial scalability of these compositions.

References

1. M. Swathi, S. Manjusha, I.J. Vadakkiniath, and A. Gururaj, Prevalence and correlates of stress, anxiety, and depression in patients with chronic diseases: a cross-sectional study. Middle East Curr. Psychiatry. **31**, 66 (2023).

https://doi.org/10.1186/s43045-023-00340-2

2. K. Caballero-Gallardo, P. Quintero-Rincón, and J. Olivero-Verbel, Aromatherapy and essential oils: holistic strategies in complementary and alternative medicine for integral wellbeing. Plants. **14**, 400 (2025).
https://doi.org/10.3390/plants14030400
3. S. Ahmad and P.D. Pratap, Unravelling the environment, aroma, and molecular effects on mood fluctuations: a review. Neuroscience. **14**, 14 (2024).
https://doi.org/10.37591/rrjons
4. N. Halligudi and M. Al Ojaili, The science and art of aromatherapy: a brief review. J. Biomed. Pharm. Res. **2**, 6 (2013)
5. S.C. Sikka and A.R. Bartolome, Perfumery, essential oils, and household chemicals affecting reproductive and sexual health, (Bioenvironmental issues affecting men's reproductive sexual health, Elsevier, 2018), pp. 557–569.
https://doi.org/10.1016/b978-0-12-801299-4.00036-0
6. S. Musaddiq, F. Imama, and M. Kiran, Therapeutic and pharmacological applications of ethnobotany (IGI Global, pp. 155–188, 2024)
7. F. Ghasemy-Piranloo, F. Kavousi, and M. Kazemi-Abharian, Comparison for the production of essential oil by conventional, novel and biotechnology methods. J. Essent. Oil Res. **34**, 455 (2022).
https://doi.org/10.1080/10412905.2022.2120557
8. S.N.Z. Ahmmad, M.T. Mokhtar, F. Muchtar, and P.K. Singh, Implementation of automated aroma therapy candle process planting using IoT and WSN, (Cham: Springer International Publishing, 2020)
9. S. Romero, R.J. Minari, and S.E. Collins, Bio-paraffin from soybean oil as eco-friendly alternative to mineral waxes. Ind. Eng. Chem. Res. **60**, 5364 (2021).
https://doi.org/10.1021/acs.iecr.1c00322
10. K. Rezaei, T. Wang, and L.A. Johnson, Hydrogenated vegetable oils as candle wax. J. Am. Oil Chem. Soc. **79**, 1241 (2002).
https://doi.org/10.1007/s11746-002-0634-z
11. R.-M. Chivu, J. Martins, F. Popescu, I. V Ion, And M. Fratita, Mixtures of essential oils and alcohols with diesel oil use and impact on engine performance and pollutant emissions. a review. Innov. Technol. Transp. Eng. **1**, 71 (2024).
https://doi.org/10.36910/conf_avto.v1i1.1395
12. B.S. Ahmad, T. Talou, Z. Saad, A. Hijazi, M. Cerny, H. Kanaan, A. Chokr, and O. Merah, Fennel oil and by-products seed characterization and their potential applications. Ind. Crop. Prod. **111**, pp.92-98 (2018).
https://doi.org/10.1016/j.indcrop.2017.10.008
13. M. Pencheva, Z. Petkova, I. Dincheva, I. Kostova, S. Damyanova, A. Stoyanova, and L. Gaceu, Phytochemical and biological profiles of fennel fruits (Foeniculum vulgare Mill. var. dulce Mill.). Carpath. J. Food Sci. Technol. **14**, 4 (2022).
https://doi.org/10.34302/crpjfst/2022.14.4.3
14. A. Sivaphongthongchai, Effects of selected aroma compounds on physiological activities and emotions, Ph.D. thesis, Chulalongkorn University (2021)
15. E.J. Krendlinger and U.H. Wolfmeier, Natural and Synthetic Waxes: Origin, Production, Technology, and Applications (John Wiley and Sons, New York, 2022)

Effect of tofu wastewater concentration on lipid content of *chlorella vulgaris*

Fatharani Rahmani Putri[1], *Tria* Amalia Sari[1], *Herawati* Budiastuti[1], *Rintis* Manfaati[1] and *Alfiana* Adhitasari[1*]

[1]Department of Chemical Engineering, Politeknik Negeri Bandung, Jalan Gegerkalong Hilir, Bandung, 40559, Indonesia

Abstract. *Chlorella vulgaris* has emerged as a promising microalgal species for alternative energy development due to its ability to accumulate lipids ranging from 5% to 40% of its dry biomass. In this study, tofu wastewater was employed as a cultivation medium, given its rich content of essential nutrients including nitrogen, phosphate (P_2O_5), potassium (K_2O), and organic carbon, which are critical for supporting microalgal growth. Beyond nutrient composition, light wavelength is also recognized as a key factor influencing photosynthetic energy absorption, thereby affecting both growth performance and lipid productivity. The objective of this research was to identify the optimal concentration of tofu wastewater and light wavelength for maximizing the growth and lipid content of *Chlorella vulgaris*. Cultivation experiments were conducted over a 10-day period using tofu wastewater concentrations of 0%, 10%, 15%, and 20%. Biomass growth was evaluated via gravimetric dry weight analysis, and lipid accumulation was quantified using the Bligh–Dyer method. Results indicated that a 15% tofu wastewater concentration yielded the highest biomass (2.054 g/L), with a specific growth rate of 0.354 day^{-1} and a lipid content of 10.23%, suggesting that moderate nutrient enrichment supports optimal microalgal productivity.

1 Introduction

Chlorella vulgaris is a photosynthetic microorganism that utilizes water, light, carbon dioxide, and mineral nutrients to support cell growth and biomass production. Under optimal cultivation conditions, it is capable of synthesizing lipids comprising 5–40% of its dry biomass, primarily in the form of hydrocarbons, glycolipids, and minor quantities of free fatty acids [1]. The microalga *Chlorella vulgaris* exhibits strong adaptability to a wide range of environmental conditions due to its ability to absorb nutrients, particularly nitrogen and phosphorus, from wastewater, as well as its tolerance to heavy metals such as Mn, Cr, Zn, and Cd. A study by Darwin [2] demonstrated that *Chlorella vulgaris* can remove up to 91.96% of total nitrogen and reduce Chemical Oxygen Demand (COD) by 81.58% in wastewater media, indicating its effective utilization of organic nutrients for cell growth. One type of wastewater rich in various organic compounds is tofu wastewater.

[*]Corresponding author: alfiana.adhitasari@polban.ac.id

The wastewater generated from tofu production and soybean washing processes has the potential to pollute the environment, especially when it is not properly managed and directly discharged into water bodies. This wastewater can alter the physical, chemical, and biological properties of water, thereby affecting the surrounding aquatic life. Therefore, it is crucial to treat or utilize tofu wastewater before disposal to prevent environmental contamination. The nutrients found in tofu wastewater—such as organic carbon (5.803%), phosphate (P_2O_5, 5.54%), nitrogen (1.24%), and potassium oxide (K_2O, 1.34%)—are essential elements required by plants for photosynthesis [3]. These characteristics make tofu wastewater a potential medium for the cultivation of the microalga *Chlorella vulgaris*.

In the process of photosynthesis, the nutrient ratio of organic carbon, nitrogen, and phosphorus plays a crucial role in supporting microalgal growth and lipid production. The ideal C:N:P ratio for microalgal growth, commonly referred to as the Redfield ratio, is 106:16:1 in molar form [4]. An imbalance in this ratio can induce stress responses in microalgae, which in turn promotes lipid accumulation. Tofu wastewater typically exhibits an unbalanced nutrient composition. This was demonstrated in a study by Syaichurrozi [5], where tofu wastewater used as a cultivation medium for *Spirulina platensis* contained 7666.21 mg/L COD, 2874.83 mg/L total organic carbon, 179.68 mg/L total nitrogen, and 33.43 mg/L P–PO_4. Such imbalances can inhibit optimal microalgal growth. However, after the addition of synthetic nutrients—specifically 1 g/L $NaHCO_3$ and 0.05 g/L urea—to tofu wastewater media at various concentrations, the 6% (v/v) concentration with added nutrients resulted in the highest specific growth rate of 0.1107/day and the highest OD_{680} value of 0.91. This condition yielded a C:N:P ratio of 161:17:1, which closely approaches the ideal ratio for microalgal growth. Similarly, a study by Cahyani [6] demonstrated the potential of tofu wastewater as a nutrient source for microalgal cultivation. In this study, *Chlorella pyrenoidosa* was cultivated for 7 days using tofu wastewater at concentrations of 0%, 10%, 15%, and 20%. The best microalgal growth was observed at 20% tofu wastewater concentration, reaching a cell density of 3.235×10^4 cells/mL and producing a lipid content of 1.86%.

Previous studies on the use of tofu wastewater as a cultivation medium for the microalga *Chlorella vulgaris* remain limited. In this study, tofu wastewater was utilized as the cultivation medium to determine the optimal lipid content.

2 Materials and Methods

The main materials used included tofu wastewater sourced from a tofu industry in Cibogo, *Chlorella vulgaris* culture obtained from Ugo Plankton Shop (Central Java), and F/2 Guillard medium used as a control. For lipid extraction, analytical-grade chloroform ($CHCl_3$) and methanol (CH_3OH) were used. The equipment used consisted of 1000 mL Erlenmeyer flasks, white LED lamps (T5, 3 Watt, 500 lumens), CO_2 gas cylinders equipped with AFR 2000 + PM 20 regulators, aerator pumps (Amara AA–350), oven, centrifuge (2000 rpm), and autoclave.

Tofu wastewater was analyzed for its nitrogen, phosphorus, and organic carbon content prior to use. It was then diluted with distilled water to obtain concentrations of 10%, 15%, and 20% (v/v). Each concentration was prepared in 1000 mL Erlenmeyer flasks containing 900 mL of the respective tofu wastewater dilution. The media were sterilized using an autoclave at 121°C and 1 atm for 15–20 minutes. A 100 mL inoculum of *Chlorella vulgaris* (10% v/v) was then added to each flask. At the end of the cultivation period, microalgal biomass was harvested by centrifugation at 2000 rpm for 15 minutes and dried in an oven at 105°C until a constant weight was achieved. Lipid extraction was conducted using a modified Bligh and Dyer method. A mixture of methanol and chloroform (2:1 v/v) was added to the dried biomass, followed by incubation for 24 hours. The mixture was then stirred, centrifuged, and the lower lipid-rich layer was separated, dried, and weighed to determine the total lipid content. Data on optical density and dry biomass were used to calculate the specific growth rate. The biomass and lipid content from each tofu wastewater concentration were compared to determine the most effective concentration for cultivation.

3 Results and Discussions

Moreover, this study is divided into three main sections: the characterization of tofu wastewater and F/2 Guillard medium, the effect of tofu wastewater concentration on the growth of *Chlorella vulgaris*, and its effect on lipid content.

3.1 Characteristics of Tofu Wastewater and F2 Guillard Medium

Tofu wastewater used in this study was collected from the tofu production process, where a portion of the liquid is sometimes reused in the next production batch. This wastewater contains essential nutrients for microalgae growth, including 5.803% organic carbon, 1.24% nitrogen, 1.34% K_2O, and 5.54% P_2O_5 [3]. Nitrogen supports protein, lipid, and carbohydrate synthesis, while phosphorus is critical for DNA, RNA, ATP, and phospholipid formation [7] [8]. Organic carbon is a key energy source for biomass production. Based on laboratory analysis, the tofu wastewater had 0.183 mol organic carbon (reducing sugars), 0.3 mol nitrogen, and undetectable phosphorus content, likely due to sampling limitations or sedimentation. Additionally, a 10% CO_2 gas supply at 50 mL/min (equivalent to 29.4 mol over 10 days) was provided to support photosynthesis. This resulted in a C:N:P ratio of 29.583:0.3:0, which is significantly imbalanced compared to the ideal Redfield ratio of 106:16:1 for optimal microalgae growth. Despite this, growth still occurred, possibly due to trace phosphorus and the resilience of *Chlorella vulgaris* under nutrient-limited conditions. In contrast, the synthetic F/2 Guillard medium, also supplied with CO_2, had much lower nitrogen and phosphorus levels (C:N:P ratio of 100:0.03:0.0001), making tofu wastewater more favorable as a nutrient source. Previous studies have also confirmed tofu wastewater's potential for supporting biomass and lipid production in *Chlorella* cultures [9] [10]. Thus, tofu wastewater shows strong potential as an alternative nutrient source for microalgae cultivation.

3.2 Effect of Tofu Wastewater Concentration on the Growth of Chlorella vulgaris

Tofu wastewater is an organic effluent rich in nutrients, making it a potential medium for microalgae cultivation. However, microalgal growth is significantly influenced by nutrient availability and environmental factors such as light intensity and pH stability. In this study, *Chlorella vulgaris* was cultivated for ten days in tofu wastewater media at concentrations of 10%, 15%, and 20%. The pH was maintained between 7 and 8 using NaOH to counter CO_2-induced acidification, and soybean oil was used as a natural antifoam during aeration. The exponential growth phase varied across concentrations, with the 15% tofu wastewater medium showing the highest biomass concentration (2.054 g/L) and specific growth rate (0.354/day), suggesting optimal conditions for cell proliferation despite a non-ideal C:N:P ratio. Conversely, the 10% concentration resulted in limited growth, likely due to nutrient deficiency, especially nitrogen and phosphorus, leading to a high C:N ratio (98.61:1), which is known to inhibit chlorophyll and protein synthesis [11] [12]. Excessive nutrient levels in the 20% concentration were also detrimental, potentially causing osmotic stress and light penetration inhibition [1] [7]. Compared to the control medium F/2 Guillard, which provided a stable nutrient composition and supported sustained growth (OD: 2.110 g/L, μ: 0.177/day), tofu wastewater media exhibited a higher growth rate at 15% concentration, highlighting its potential as a low-cost cultivation medium. These results align with prior findings by Munir [13] and Arianto [14], which reported optimal microalgal growth and biomass production at similar tofu wastewater concentrations.

3.3 Effect of Tofu Wastewater Concentration on Lipid Content

The cultivation medium plays a critical role in determining the growth rate and metabolite production of microalgae, including lipid accumulation. Lipid content analysis is essential to assess the potential of microalgae for bioresource applications. Figure 1 presents the lipid content of *Chlorella vulgaris* cultivated in F/2 Guillard medium and tofu wastewater at various concentrations.

Among the tofu wastewater treatments, the highest lipid content was observed at 15% concentration (10.23%), followed by 10% (6.01%) and 20% (4.92%). These results suggest that at 15%, the nutrient levels were sufficient to support growth while inducing moderate nutrient stress that promotes lipid accumulation. Similar findings were reported by Harahap [10], who observed a lipid yield of 0.05160 g/L from *Chlorella sp.* cultivated in 15% tofu wastewater. In contrast, the 10% concentration provided insufficient nitrogen and phosphorus, resulting in lower biomass and growth rates, which limited lipid synthesis. On the other hand, the 20% concentration, despite containing more organic matter, may have led to cell stress or inhibition due to excessive nutrients or antifoam toxicity, thus reducing lipid productivity. Interestingly, while 20% tofu wastewater supported higher biomass than the 10% concentration, its lipid content was lower. This inverse relationship aligns with the understanding that mild nitrogen limitation can trigger lipid accumulation, whereas abundant nitrogen favors continuous growth over lipid storage [15] [11]. The 15% concentration thus represents an optimal balance, allowing moderate stress conditions under which cells divert carbon flux from growth to lipid storage. Elystia [15] explain that under nitrogen limitation, cells activate alternative biosynthetic pathways that enhance the production of triacylglycerol (TAG), including glycolysis-derived G3P and acetyl-CoA, along with membrane lipid degradation and enzyme activation (e.g., acyl hydrolase, DGAT). Furthermore, the highest lipid content overall was recorded in the F/2 Guillard medium (30.2%), which contains a complete nutrient profile including macro- and micronutrients and trace elements essential for lipid biosynthesis, thus supporting higher lipid accumulation than tofu wastewater media.

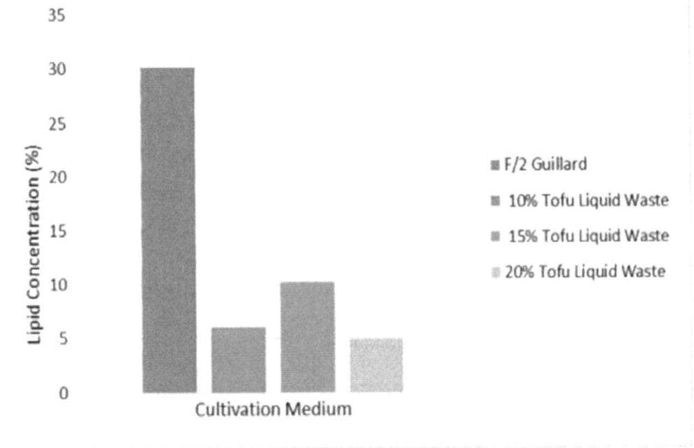

Fig. 1. Effect of Concentration on Lipid Content.

4 Conclusions

Based on the experimental results, the tofu wastewater concentration of 15% was found to be the most effective for supporting the growth and lipid accumulation of *Chlorella vulgaris*. At this concentration, the microalgae achieved the highest biomass concentration of 2.054 g/L, with a specific growth rate of 0.354/day. In addition to promoting rapid cell division, this concentration also provided sufficient nutrients to induce moderate nutrient stress conditions, which are favorable for lipid biosynthesis. As a result, *Chlorella vulgaris* cultivated in the 15% tofu wastewater medium accumulated a lipid content of 10.2%, significantly higher than the values obtained at other concentrations. These findings indicate that tofu wastewater at 15% concentration offers a balanced nutrient environment that supports both optimal growth and enhanced lipid production, making it a promising alternative cultivation medium for sustainable microalgae-based bioresource applications.

References

1 L.G.C. Wijaya, S. Prabaningtyas, Effect of phosphate source on growth of *Chlorella vulgaris* cultured with indole-3-acetic acid-producing bacteria. *J. Ilm. Sains* **24**, 148–159 (2024). https://doi.org/10.35799/jis.v24i2.57375

2 A.S. Afifah, G. Prajati, D. Darwin, Kultivasi mikroalga *Chlorella* dengan media air limbah (studi literatur untuk produksi biomassa dan pengolahan air limbah). *Serambi Eng.* **8**, 4474–4482 (2023).

3 R.N. Amalia, S.D. Devy, A.S. Kurniawan, N. Hasanah, E.D. Salsabila, D.A.A. Ratnawati, F.M. Fadil, N.A. Syarif, G.A. Aturdin, Potensi limbah cair tahu sebagai pupuk organik cair di RT. 31 Kelurahan Lempake Kota Samarinda. *ABDIKU: J. Pengabdian Masy.* **1**, 36–41 (2022). https://doi.org/10.32522/abdiku.v1i1.38

4 L. Setoaji, J. Hermana, Pengaruh aerasi dan sumber nutrien terhadap kemampuan alga filum *Chlorophyta* dalam menyerap karbon (carbon sink) untuk mengurangi emisi CO_2 di kawasan perkotaan. *J. Tek. Pomits* **2**, D69–D73 (2013).

5 I. Syaichurrozi, J. Jayanudin, Potensi limbah cair tahu sebagai media tumbuh *Spirulina platensis*, J. Ilm. Pangan, **6**, (2016).

6 D. Cahyani, Pengaruh pemberian konsentrasi limbah tahu yang berbeda terhadap kandungan lipid mikroalga *Chlorella pyrenoidosa*, Sarjana Thesis, Universitas Brawijaya, Indonesia, 2020.

7 M. J. Zarrinmehr, O. Farhadian, F. P. Heyrati, J. Keramat, E. Koutra, M. Kornaros, E. Daneshvar, Effect of nitrogen concentration on the growth rate and biochemical composition of the microalga, *Isochrysis galbana*. *Egypt. J. Aquat. Res.* **46**, 153–158 (2020). https://doi.org/10.1016/j.ejar.2019.11.003.

8 H. Atiku, R. M. S. R. Mohamed, A. A. Al-Gheethi, A. A. Wurochekke & A. H. M. Kassim, Harvesting of microalgae biomass from the phycoremediation process of greywater. *Environ. Sci. Pollut. Res.* **23**, 24624–24641 (2016). https://doi.org/10.1007/s11356-016-7456-9

9 S. R. Muria, F. M. Shiddiq, I. Damayanti, I. Purnama, Kultivasi mikroalga *Chlorella* sp. secara fed-batch menggunakan limbah cair tahu untuk produksi lipid. *J. Bioprocess Chem. Environ. Eng. Sci.* **4**, 37–56 (2023). https://doi.org/10.31258/jbchees.4.1.37-56

10 P. S. Harahap, A. B. Susanto, D. Susilaningsih, D. Y. Rahma, Pengaruh substitusi limbah cair tahu untuk menstimulasi pembentukan lipida pada *Chlorella* sp. *J. Mar. Res.* **2**, 80–86 (2013).

11 M. J. Griffiths, S. T. L. Harrison, Lipid productivity as a key characteristic for choosing algal species for biodiesel production. *J. Appl. Phycol.* **21**, 493–507 (2009). https://doi.org/10.1007/s10811-008-9392-7

12 F. Gao, H.-L. Yang, C. Li, Y.-Y. Peng, M.-M. Lu, W.-H. Jin, J.-J. Bao & Y.-M. Guo, Effect of organic carbon to nitrogen ratio in wastewater on growth, nutrient uptake and lipid accumulation of a mixotrophic microalgae *Chlorella* sp. *Bioresour. Technol.* **282**, 118–124 (2019). https://doi.org/10.1016/j.biortech.2019.03.011

13 F. Munir, R. Hariyati, E. Wiryani, Pengaruh limbah cair tahu terhadap pertumbuhan populasi *Chlorella pyrenoidosa* H. Chick dalam skala laboratorium. *J. Akad. Biol.* **6**, 84–92 (2017).

14 D. J. Arinto, H. P. Paramastri, D. Soetrisnanto, Potensi air dadih (whey) tahu sebagai nutrien dalam kultivasi *Chlorella* sp. untuk bahan baku pembuatan biodisel. *J. Teknol. Kim. Ind.* **2**, 233–242 (2013)

15 S. Elystia, M. D. Amelia, S. R. Muria, Pengaruh variasi laju alir gas CO_2 terhadap penyisihan COD dan penyerapan CO_2 oleh *Chlorella* sp. menggunakan flat-photobioreactor. *J. Teknol. Pertan. Andalas* **24**, 43–53 (2020). https://doi.org/10.25077/jtpa.24.1.43-53.2020

A Reproducible sol-gel synthesis of magnetic-coated silica nanoparticles for reliable SARS-CoV-2 RNA extraction

Grace Tjungirai Sulungbudi[1*], *Muflikhah*[1], *Ahadi Damar* Prasetya[1], *Mujamilah*[1], *Andon* Insani[1], *Wildan Zakiah* Lubis[2], and *Uus* Saepulloh[3]

[1]Research Center for Nuclear Beam Analysis Technology, National Research and Innovation Agency (BRIN), South Tangerang, 15314, Indonesia
[2]Directorate for Nuclear Facility Management, National Research and Innovation Agency (BRIN), South Tangerang, 15314, Indonesia
[3]Primate Animal Research Center, IPB University, Bogor 16151, Indonesia

Abstract. The development of magnetic-coated silica (MAGSi) nanoparticles has shown significant promise for the efficient extraction of viral RNA in reverse transcription-polymerase chain reaction (RT-PCR) tests, a key method for COVID-19 detection. To facilitate large-scale production, it is crucial to establish a synthesis method that is both high-quality and reproducible. This study investigates the reproducibility of a sol-gel synthesis method, assisted by an ultrasonic bath, for preparing MAGSi nanoparticles and assesses their performance in extracting RNA from the SARS-CoV-2 virus. The synthesis protocol was repeated across five distinct batches, and the resulting nanoparticles were characterized for yield, morphology, crystallite phase, and magnetic moment. The results demonstrated no significant differences in the physical properties or quantity among the replicated batches, confirming the method's good reproducibility. Furthermore, duplicate RNA extraction tests performed on samples from each batch revealed consistent adsorption performance, highlighting the method's efficacy. This work underscores the importance of validating a synthesis method's reproducibility before scaling up production for biomedical applications, such as a cost-effective viral RNA extraction kit.

1 Introduction

Over the past decade, research on the synthesis of magnetic nanoparticles (MNPs) has gained significant momentum, accompanied by their application in various fields such as catalysis, adsorbents, biomedical applications, and medicine [1]. Among all magnetic nanoparticles, magnetite (Fe_3O_4) stands out due to its remarkable properties, particularly its ferromagnetic nature at temperatures below 858 K [2]. Recently, magnetite nanoparticles have been intensively studied in biomedical fields, especially for hyperthermia treatment, diagnostics, and therapeutic delivery platforms due to their favorable biocompatibility and

*Corresponding author: grac001@brin.go.id

biodegradability [3]. Current reports highlight their potential in binding, extraction, and purification of biomolecules, including RNA, DNA, proteins, enzymes, and organic small molecules [4,5]. Moreover, a recent and novel application of MNPs has been in COVID-19 detection by extracting RNA viruses in the reverse transcription-polymerase chain reaction (RT-PCR) tests [4,5]. The advantage of the magnetic property of MNPs allows the movement of particles to be controlled by an external magnetic field. Upon applying an external magnetic field, allowing the separation of biomolecules bound to the surface of MNPs, like enzymes, RNA, DNA, and proteins, from the solution. This method provides rapid separation, high specificity, and good reproducibility, making it a popular technique in the purification and separation of nucleic acids as well as in various biotechnological and medical applications [4,6]. However, magnetite has a notable drawback: it is prone to oxidation, which transforms it into hematite (Fe_2O_3). This transformation alters its magnetic properties from superparamagnetic to ferromagnetic and leads to a decline in its saturation magnetization. Several methods have been carried out to protect the MNPs from oxidation, one of them is introducing other functional groups onto the MNPs' surfaces using organic materials such as polymers, surfactants, and inorganic substances such as silica [7].

Silica has been widely used as a coating material for magnetic surfaces in various applications due to its biocompatibility, chemical and physical stability, and cost-effectiveness abundant hydroxyl groups in silica facilitate further functionalization, making it an excellent adsorbent. Recently, magnetic-silica beads have been developed for RNA and DNA extraction, gaining increased attention since the SARS-CoV-2 pandemic [5]. Reverse transcription-polymerase chain reaction (RT-PCR) has been widely used and has become the gold standard for confirming SARS-CoV-2 infections due to its specificity and simplicity as a qualitative assay. As a nucleic acid-based assay, it includes RNA separation via extraction. Initially, silica-based spin columns have become the RNA extraction method that is widely used, requiring multiple centrifugation and column-transfer steps. Despite being effective, this process is labor-intensive, time-consuming, and requires relatively expensive equipment, toxic chemicals, and disposable laboratory supplies. These limitations hinder the potential for rapid diagnosis and mass analysis [4]. The magnetic-based separation method offers an attractive option for isolating viral RNA using magnetic-silica hybrid particles, which have the advantage of enabling automation of the process. Furthermore, the preparation of magnetic-silica hybrid particles in nano- or micron-size forms is relatively simple.

Several works have been carried out in the development of magnetic-silica hybrid, ranging from bead, micrometer-sized particles to nano-sized particles [4]. The binding between nucleic acids and silica surfaces relies on the intermolecular electrostatic interaction and hydrogen bonding with silanol groups on the silica surface. Therefore, a larger surface area of magnetic-silica particles can more effectively isolate DNA and RNA from samples [8]. The synthesis of silica-coated MNPs is generally conducted using the sol-gel method, which is widely employed for modifying silica into magnetic surfaces [4]. The most common approach involves a two-step synthesis where MNPs are first prepared and then coated with silica [1]. This method is preferred because it is relatively simple and effective to do on a large-scale production compared to other techniques such as microemulsion, aerosol pyrolysis, or electrochemical [2, 9]. We successfully obtained magnetic-silica (MagSi) for RNA SARS-CoV-2 extraction using the sol-gel method combined with mechanical stirring, achieving particle sizes ranging from 50 nm to 1.5 μm [4]. A similar method has also been used by Capriotti et al. to prepare MNPs-silica with a size of 2 microns for RNA SARS-CoV-2 extraction.

In our previously developed method, approximately 1 gram of magnetic-silica (MAGSi) material with a concentration of 50 mg/ml could be prepared in a single batch. This amount is sufficient for 100 RNA extractions [4]. The extraction result indicated that, at the same concentration of magnetic beads, MAGSi performed comparably to commercial magnetic beads. Furthermore, commercial magnetic beads not only have a smaller surface area but also require additional surface functionalization with other molecules to enhance adsorption capability. This functionalization complicates the sample preparation process and increases costs. Based on our findings, we see significant potential for developing MAGSi nanoparticles into an extractor kit. In addition to comparable adsorption performance with commercial kits, MAGSi nanoparticles are also more cost-effective and simpler to prepare. Moreover, to date, most studies regarding reproducibility and scale-up of magnetic silica have focused on bead forms, while magnetic silica at a scale below 2 microns or in the nano-scale range has not yet been thoroughly explored. In addition, before proceeding with scale-up, it is crucial to repeat the synthesis to assess the various factors involved in the synthesis process. Ensuring the reproducibility of the synthesis method is a prerequisite for commercial and industrial applications, as it guarantees that each batch of nanoparticles will perform reliably. Therefore, this study aims to assess whether a specific sol-gel synthesis method for MAGSi nanoparticles can consistently produce high-quality nanoparticles with uniform physical properties. We repeated the synthesis protocol across five different batches and analyzed the size, morphology, crystal structure, and magnetic saturation to confirm the method's reproducibility before scaling up production. Subsequently, each batch was tested for its performance in RNA extraction from the SARS-CoV-2 virus, thereby establishing the method as a scalable and reliable approach for developing a cost-effective viral RNA extraction kit.

2 Methods

2.1 Materials

Magnetic nanoparticles (MAG) and magnetic-silica particles (MAGSi) were prepared using ferrous chloride tetrahydrate ($FeCl_2{\cdot}4H_2O$) 99.0 % and ferric chloride hexahydrate ($FeCl_3{\cdot}6H_2O$) 99.0 %, which were received from Sigma Aldrich. Ethanol analytical grade, sodium hydroxide (NaOH), HCl 37 %, and ammonia 25 % v/v were from Merck. As a silica source, we used tetraethyl orthosilicate (TEOS) 99.0 % (Sigma Aldrich). Deionized water (DI-water) was produced using Merck Millipore Simplicity-UV.

Extraction of RNA of SARS-CoV-2 has been conducted using samples from swab nasofaring/orofaring, which confirmed positive (Sample ID 6553 and 6554 provided by the Primate Research Center, IPB University), ethanol p.a 80% (Merck), the MagMAX™ Viral/Pathogen Nucleic Acid Isolation Kit (ThermoFisher Scientific), MAGSi samples, and nuclease-free water. The materials for RT-qPCR, including the Sensifast Probe No-Rox One Step Kit (Bioline-76001), were used, along with the 2019-nCoV RUO Kit (IDT), which contains a mixture of 3 primers, probes specific for N1 and N2 regions of SARS-CoV-2, and the RNase-P marker to verify the extracted RNA from the samples. Additionally, the 2019-nCoV N Positive Control and Hs RPP30 Positive Control were used.

2.2 Reproducibility of MAGSi Nanoparticles Synthesis

The preparation of magnetic nanoparticles (MAG) was carried out according to our previous study [4]. All MAGSi samples were prepared in a single batch to ensure high reproducibility and precise control over the co-precipitation process using magnetic iron oxide nanoparticles (MAG). The MAGSi samples were then prepared via a sonication method using an ultrasonic bath cleaner (Branson 1510, 40 kHz, 100 W) based on the optimal results achieved in our earlier research [4]. The detailed procedure is as follows: 0.25g of MNP was dispersed in 180 mL of ethanol and sonicated for 10 minutes. Then, 4 mL TEOS was gently added gradually to the MAG mixture, followed by 10 mL NH_4OH 25% and 10 mL DI-water. The addition of TEOS, NH_4OH, and DI-water was performed using a peristaltic pump. The solution was allowed to react for 3 hours in an ultrasonic bath at 40 °C, then aged overnight at ambient temperature. The solid was then separated using a neodymium magnet, then washed twice with DI water, followed by two ethanol washes, and then continued with DI water until the pH of the supernatant was 7. The MAGSi were redispersed in deionized water and stored in a sealed container (with a total volume of the dispersed sample of 50 ml). This procedure was repeated 5 times, with each batch named MAGSi-R1, MAGSi-R2, MAGSi-R3, MAGSi-R4, and MAGSi-R5. To determine the yield, a 1 mL aliquot from each batch was taken twice, dried, and weighed.

2.3 Characterization

To assess the quality of sample repetition, some characterizations have been conducted. The X-ray Difractometer (XRD, PANalytical) was used to analyze the crystallite phase of MNP and MAGSi samples. The functional groups were confirmed using Fourier Transform Infra-Red (FTIR, Bruker Tensor 27). The magnetic saturation of MAGSi was measured using a Vibrating Sample Magnetometer (VSM, OXFORD 1.2 H). The sample's morphology was observed using Scanning Electron Microscopy (SEM, JEOL 2300). Dynamic Light Scattering (DLS) using the Malvern Zetasizer Nano ZS was used to measure the size distribution of particles.

2.4 Repetition Test in Extraction of SARS-CoV-2 RNA using MAGSi and RT-PCR

Magnetic nanoparticle silica was produced in each batch at a density of 50 mg/mL. The MAGSi mixture was formulated by adding 530 μL of extraction buffer to 20 μL of MAGSi. For sample digestion, 200–400 μL of nasopharyngeal/oropharyngeal samples were mixed with 10 μL of proteinase-K, followed by the addition of 550 μL MAGSi mixture. The test samples were shaken at 1050 rpm for 2 minutes, incubated at 65°C for 5 minutes, and the MAGSi beads were subsequently separated by employing an external magnetic field within ten minutes.

The magnetic-silica beads were then washed using a wash buffer (1 mL) and agitated at 1050 rpm for a minute. Separation was then performed magnetically for 2 minutes. The washing step was performed twice with 1.0 mL and 0.5 mL of 80% ethanol, respectively. Afterward, the MAGSi beads were drained through shaking at 1050 rpm within 2 minutes. The RNA was eluted by introducing elution buffer (50–100 μL) and agitating at 1050 rpm for 5 minutes. Next process was incubation of the beads at 65°C for 10 minutes before being shaken again. Finally, the sample was isolated magnetically for about 3 minutes, and the RNA presented in the supernatant was examined at wavelengths 260 and 280 nm using a Nanodrop OneC Spectrophotometer (Thermo Scientific), followed by further examination using RT-qPCR.

Before starting RT-qPCR, two types of master mixes were made (master mix I and master mix II) with the following component: 1.8 μL primer mix (N1 for master mix I and N2 for master mix II), 0.4 μL of Ribosafe RNase inhibitor, 10 μL of Sensifast one-step kit mix, 0.2 μL of reverse transcriptase, and 8 μL of RNA obtained from the extraction. The RT-qPCR tests were conducted using a QuantStudio 5 thermocycler (Applied Biosystems) using the following conditions: reverse transcription was carried out at 45°C for 10 minutes, followed by initial denaturation at 95°C for 2 minutes. Amplification was then conducted through 40 cycles of the following thermal profile: denaturation at 95°C for 5 seconds and annealing at 55°C for 20 seconds, then extension and data collection.

3 Results and Discussion

The MAGsi characterization results will indicate the reproducibility of the synthesis process. To ensure the functionality of the MAGSi synthesis, a functional RNA extraction test is performed, followed by RT-qPCR testing. This functional test is not linearly related to the reproducibility of the synthesis process, as it is influenced by several conditions, from the RNA extraction preparation process to the RT-qPCR testing.

3.1 Evaluation of the Reproducibility of MAGSi Nanoparticles Synthesis

The mass percentage yield of the MAGSi series is presented in Figure 1. The yield percentage is used as a quality control parameter in this repetitive process and was measured for each batch. The yield presentation of the repetition process shows good results (above 80% yield). However, the MAGSi-R3 shows the lowest yield. Considering that the yield is mainly influenced by the effectiveness of the magnetic separation (the length of time the sample was attracted by an external magnet), it is crucial to carefully control the separation time. Additionally, the yield depends on factors such as particle size, colloidal stability, and the weighing process. To ensure consistency, the sample in the bottle should be shaken to achieve homogeneity before sampling and weighing. When using a vortex mixer for homogenization, a consistent speed and duration should be applied to all samples. Accurate pipetting of samples is also important for reliable results.

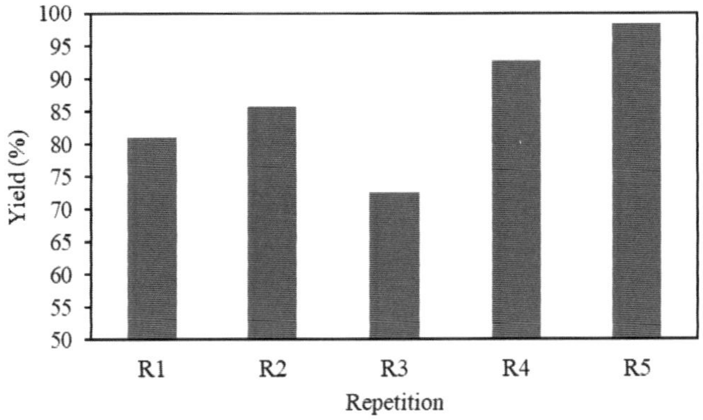

Fig. 1 The mass percentage yield of the MAGSi series (R = repetition)

To confirm the successful formation of silica coating on the magnetic material, we can identify functional groups in the sample using FTIR. The FTIR results for the repeated MAGSi samples are shown in Figure 2a. All curves display bands around 570 cm^{-1}, corresponding to the Fe-O stretching vibration related to the magnetite. All MAGSi samples also show the bands at 1070 cm^{-1} and 798 cm^{-1}, corresponding to the Si-O-Si asymmetric and symmetric stretching vibration, respectively, and a band around 960 cm^{-1} from the Si-O stretching vibration. Additionally, absorption at 468 cm^{-1} derived from the siloxane group, Si–O–Si, is observed. The bands around 1625 cm^{-1} and 3425 cm^{-1} are assigned to the O-H stretching and bending from the Si-OH or absorbed water molecules at the surface of the samples. These findings confirm that silica was formed in all repetition samples.

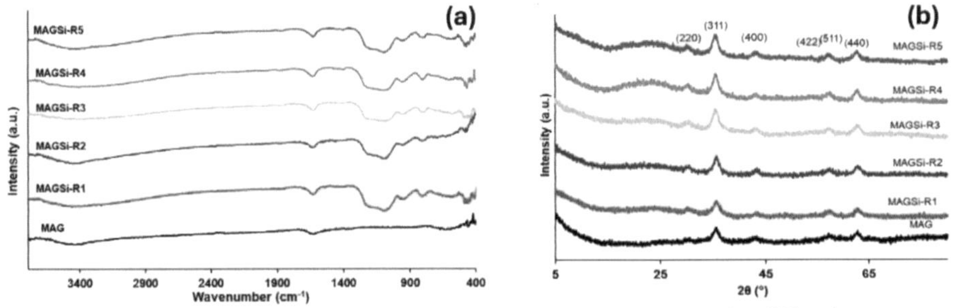

Fig. 2 FTIR spectra of MAG and MAGSi repetition series **(a)** and X-Ray powder diffraction pattern of MAG and MAGSi repetition series **(b)**.

XRD patterns of the samples are presented in Figure 2b. The XRD diffraction patterns of MAG showed the characteristic peaks of the cubic structure (COD 96-900-2360) [4]. The peaks of the diffractogram at 2θ 30.4°, 35.6°, 43.2°, 53.7°, 57.4°, and 62.8° correspond to the (220), (311), (400), (422), (511), and (440) crystal planes, respectively. The MAGSi samples present a wide diffraction peak at 2θ around 22 °–25 ° that belongs to amorphous silica [1,2]. It can also be observed that the diffractogram of all MAGSi samples exhibits similar diffraction patterns, indicating the successful and reproducible coating process of the silica to the magnetic nanoparticle surface without altering the structure of the Fe_3O_4 nanoparticles.

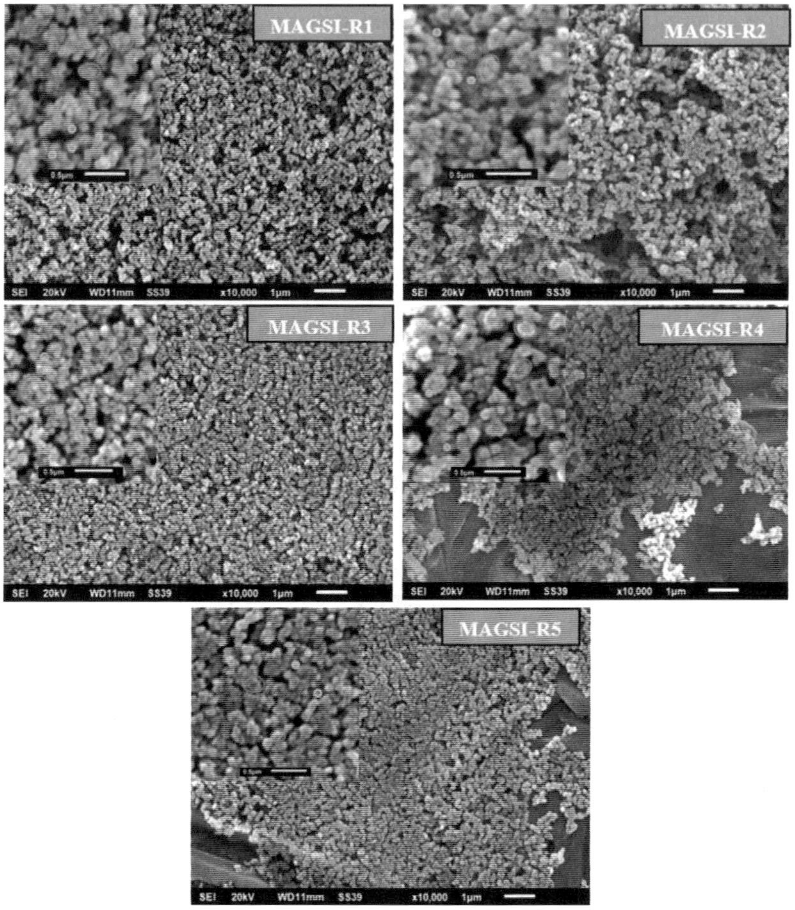

Fig. 3 SEM image of MAGSi repetition series.

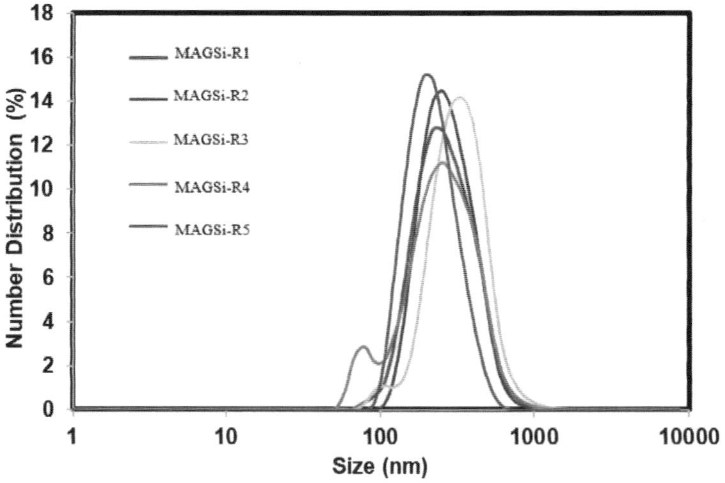

Fig. 4 Size distribution of MAGSi repetition series.

The morphologies of synthesized materials were examined using SEM, as presented in Figure 3. The SEM micrographs of all the MAGSi samples were obtained in irregular aggregates that contained spherical-shaped particles (shown by blue circles). The individual MAGSi particles have spherical shapes with particle sizes of around 50–70 nm. The particle distribution analysis of the samples was also conducted using dynamic light scattering (DLS), presented in Figure 4, giving a size of 280 – 530 nm. The MAGSi particle size on the DLS data shows a larger size than the SEM due to the agglomeration of the particles [5]. As most of the MAGSi series have the same size distribution based on Figure 4, it can be noticed that MAGSi-R3 has the largest size distribution, which may reduce its surface area for RNA adsorption. The homogeneity of the SiO_2 coating thickness on MAGSi can also affect the interparticle attraction forces which causes slightly varying hydrodynamic sizes in particle or particle aggregate sizes.

The process of repeating MAGSi synthesis provides a sufficiently good size distribution and similar morphological characteristics. This also indicates that the synthesis method using the ultrasonic bath technique provides uniformity in the shape and size of the particles in the sample. This finding is a crucial point, as it indicates a potential weakness in colloidal stability that could impact the nanoparticles' performance, despite the uniformity of the primary particles.

The magnetization measurement of the samples using a vibrating sample magnetometer is shown in Figure 5. The magnetic nanoparticles (MAG) possessed a high coercivity of 59 emu/g. While the saturation magnetization (Ms) value of all MAGSi series was lower than the MNP, due to the non-magnetic phase silica layer covering the MNPs, which led to a deficiency of total magnetic moment [9]. The Ms values of the MAGSi sample series are not significantly different, where the Ms values for MAGSi-R1, MAGSi-R2, MAGSi-R3, MAGSi-R4, and MAGSi-R5 are 10.07; 10.05; 13.00; 10.09; and 12.40 emu/g, respectively. With those Ms values, the MAGSi samples can still be easily attracted by neodymium external magnets, thus still suitable for molecules/RNA separation applications.

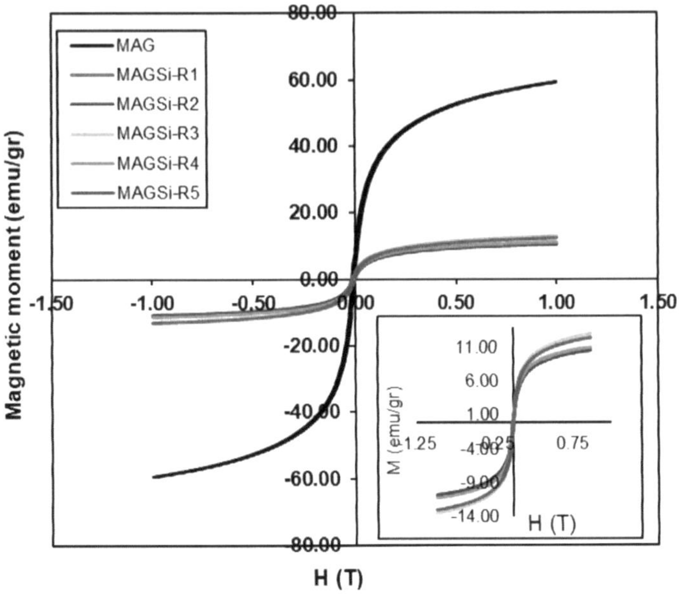

Fig. 5 Magnetization curves of MAGSi repetition series.

3.2 Evaluation of SARS-CoV-2 RNA Extraction by MAGSi Repetition Samples Through RT-PCR

MAGSi repetition samples have been evaluated for their performance for SARS-CoV-2 RNA extraction and confirmed using RT-PCR from a nasopharyngeal/oropharyngeal swab sample that has been confirmed positive for SARS-CoV-2. As mentioned in the method, to ensure comparability of the test, all the MAGSi series were set to an initial concentration of 50 mg/mL before being used for extraction. Table 1 shows the extracted RNA concentration of MAGSi repetition samples with the duplet test. The data show that all MAGSi samples have RNA extraction ability, and most can produce extracted RNA in a range of 10 – 20 ng/μL.

The RT-PCR data in Table 2 also confirm that SARS-CoV-2 RNA was successfully extracted by MAGSi nanoparticles, and the genes were also successfully amplified, with most of the CT values near the CT values of the positive control, which also indicates a high viral load due to lower than our RT-PCR amplification cycles setup (40 cycles) [10]. RNAse P target gene was used as a control for the SARS-CoV-2 RNA extraction process because RNAse P is a human gene present in all nucleated cells, and its detection confirms that the sample contains sufficient human cellular material and that the extraction and amplification steps worked properly [5,11,12]. Some of the repetition results showing NA values (not detected) or "false negative" for MAGSi-R3 and R5 may be due to factors originating from the MAGSi particles themselves or from the RT-PCR reaction of the extracted RNA. However, our data suggest that these results are more due to the RT-PCR reaction since in Table 1 the SARS-CoV-2 RNA can be detected in comparable concentration amounts, also based on the MAGSi characteristics data (FTIR, XRD, DLS, SEM, and VSM), MAGSi-R3 and R5 characteristics were not extremely different from the other series. These false negative results are not uncommon in RT-PCR due to several factors such as RNA degradation, low viral load, inadequate sample collection, or technical factors (pipetting errors, reagent inconsistencies, or minor differences in sample handling) [11–14]. However, based on the DLS data (Figure 5), MAGSi-R3 has the largest particle size distribution that may reduce the RNA adsorption ability due to lower surface area, and this can be reflected in Table 1, where RNA concentration from MAGSi-R3 was relatively lower compared to the other MAGSi particle series. Therefore, the use of duplicate tests was essential for revealing this crucial performance inconsistency, which might have been overlooked with a single trial.

Table 1 The concentration of SARS-CoV-2 RNA was extracted using MAGSi repetition samples.

Sample	RNA concentration (ng/μL)	
	test 1	test 2
MAGSi-R1	22.4	12.2
MAGSi-R2	32.2	26.1
MAGSi-R3	15.1	13.9
MAGSi-R4	19.7	14.7
MAGSi-R5	16.8	14.7

Table 2 RT-PCR results of extracted SARS-CoV-2 RNA using MAGSi repetition samples.

MAGSi Sample	Repetition of Test (RT)	Gene Targets*		
		N1 SARS-CoV-2	N2 SARS-CoV-2	Human RNase P
MAGSi-R1	**RT 1**	20.74	21.11	23.47
	RT 2	22.74	22.39	27.66
MAGSi-R2	**RT 1**	21.75	21.38	29.8
	RT 2	21.49	20.97	28.9
MAGSi-R3	**RT 1**	NA	NA	NA
	RT 2	20.55	19.57	23.11
MAGSi-R4	**RT 1**	21.8	22.66	NA
	RT 2	20.76	20.46	22.77
MAGSi-R5	**RT 1**	NA	NA	NA
	RT 2	27.83	28.36	NA
	Negative Control	NA	NA	NA
	Positive Control	21.62	21.48	NA

From the characterization data of reproducibility samples and their RNA extraction performance test, it can be summarized that the synthesis of MAGSi using the sol-gel method assisted by ultrasonic batch has quite good reproducibility. However, when it comes to the extraction test, the weakness of the MAGSi sample may lie in the unhomogenized particles due to aggregation. This can be mitigated by shaking the sample for a certain amount of time before use for extraction tests. This issue could also be addressed by coating with other molecules, such as surfactants or polymers, to further improve the colloidal stability of the final product [8,15].

4 Conclusions

Reproducibility in the synthesis of magnetic-coated silica (MAGSi) has been thoroughly investigated, as it is crucial for ensuring consistency in material properties before scaling up the process for the SARS-CoV-2 RNA extraction applications. The goal was to maintain consistent results across different batches, ensuring that key properties, such as size, morphology, magnetic properties, and functionality, remained stable. The sol-gel method assisted by ultrasonic bath processing was employed to produce MAGSi samples. Characterization of functional groups, crystalline phases, morphology, size distribution, and magnetic moments across different batches demonstrated consistent results. The performance of each repeat batch was also tested for SARS-CoV-2 RNA extraction, followed by RT-PCR, with each batch yielding positive results, confirming successful RNA extraction. Thus, as a reproducible synthesis protocol is foundational for successful large-scale production and commercialization, achieving reproducibility at the laboratory scale builds confidence in the feasibility of scaling up the process for the development of a SARS-CoV-2 extraction kit based on nano-magnetic-coated silica materials with further optimization, especially regarding colloidal stability and extraction robustness.

Acknowledgements

The authors would like to thank the Directorate for Nuclear Facility Management, National Research and Innovation Agency of Indonesia (BRIN), and Primate Research Center, IPB University, for providing the research facility. This research was funded by Research and Innovation Funding for Advanced Indonesia (RIIM) No. Contract: B-811/II.7.5/FR/6/2022, B-2103/III.2/HK.04.037/2022, and Decree of the Deputy for Research and Innovation Facilitation Number 27/II.7/HK/2023.

References

1 N. Capriotti, L.C. Amorós Morales, E. de Sousa, L. Juncal, M.L. Pidre, L. Traverso, M.F. López, M.L. Ferelli, G. Lavorato, C. Lillo, O. Vazquez Robaina, N. Mele, C. Vericat, P. Schilardi, A.F. Cabrera, S. Stewart, M.H. Fonticelli, P. Mendoza Zéliz, S. Ons, V. Romanowski, C. Rodríguez Torres, Silica-coated magnetic particles for efficient RNA extraction for SARS-CoV-2 detection, Heliyon **10**, e25377 (2024). https://doi.org/10.1016/j.heliyon.2024.e25377

2 G. Asab, E.A. Zereffa, T. Abdo Seghne, Synthesis of Silica-Coated Fe_3O_4 Nanoparticles by Microemulsion Method: Characterization and Evaluation of Antimicrobial Activity. Int. J. Biomater. **2020**, 4783612 (2020).
https://doi.org/10.1155/2020/4783612

3 K. Rodponthukwaji, C. Saengruengrit, P. Tummamunkong, A. Leelahavanichakul, P. Ritprajak, N. Insin, Facile synthesis of magnetic silica-mannan nanocomposites for enhancement in internalization and immune response of dendritic cells, Mater. Today Chem. **20,** 100417 (2021).
https://doi.org/10.1016/j.mtchem.2020.100417

4 A.D. Prasetya, M. Muflikhah, W.Z. Lubis, A. Insani, G.T. Sulungbudi, M. Mujamilah, U. Saepulloh, Development of Magnetic-Silica Particles and In-house Buffers Kit for SARS-CoV-2 and CDV RNA Extraction, Indones. J. Chem. **24**, 81-93 (2024).
https://doi.org/10.22146/ijc.83804

5 A.D. Prasetya, G.T. Sulungbudi, Muflikhah, Mujamilah, A. Insani, W.Z. Lubis, U. Saepulloh, Yuliana, E.D. Ayuningsih, Efficient Synthesis of Magnetic-Silica Particles for Nucleic Acid Extraction in Diagnostics Based on Washing-Free Magnetic Fine Nanoparticles, Korean J. Chem. Eng. **42**, 1757–1768 (2025).
https://doi.org/10.1007/s11814-025-00480-7

6 H. Fatima, K.S. Kim, Magnetic nanoparticles for bioseparation, Korean J. Chem. Eng. **34**, 589–599 (2017).
https://doi.org/10.1007/s11814-016-0349-2

7 M. Muflikhah, A.M. Ramadhan, M.C. Prihatiningsih, M. Mujamilah, A. Karo Karo, Synthesis and Characterization of Ferrofluid-Chitosan-Au Nanoparticles as Brachytherapy Agent Candidate, Indones. J. Chem. **21**, 891 (2024).
https://doi.org/10.22146/ijc.62191

8 A.P. Tiwari, S.J. Ghosh, S.H. Pawar, Biomedical applications based on magnetic nanoparticles:DNA interactions, Anal. Methods **7**, 10109–10120 (2015).
https://doi.org/10.1039/C5AY02334C

9 T.Q. Bui, H.T.M. Ngo, H.T. Tran, Surface-protective assistance of ultrasound in synthesis of superparamagnetic magnetite nanoparticles and in preparation of mono-core magnetite-silica nanocomposites, J. Sci.: Adv. Mater. Dev. **3**, 323–330 (2018).

https://doi.org/10.1016/j.jsamd.2018.07.002

10 R. Jaafar, S. Aherfi, N. Wurtz, C. Grimaldier, T. Van Hoang, P. Colson, D. Raoult, B. La Scola, Correlation Between 3790 Quantitative Polymerase Chain Reaction–Positives Samples and Positive Cell Cultures, Including 1941 Severe Acute Respiratory Syndrome Coronavirus 2 Isolates, Clin. Infect. Dis. **72**, e921–e921 (2021).
https://doi.org/10.1093/cid/ciaa1491

11 C. Vogels, D. E. Brackney, C. C Kalinich, I. M Ott, N. D Grubaugh, A. L Wyllie, SalivaDirect: RNA extraction-free SARS-CoV-2 diagnostics V.1, protocol.io, (2020). https://doi.org/10.17504/protocols.io.bhfej3je

12 F. Chuang, A-259 Detecting Human RNase P Gene Prevents False-Negative of SARS-CoV-2, Clin. Chem. **69**, hvad097.228 (2023).
https://doi.org/10.1093/clinchem/hvad097.228

13 Y. Zhang, C. Wang, M. Han, J. Ye, Y. Gao, Z. Liu, T. He, T. Li, M. Xu, L. Zhou, G. Zou, M. Lu, Z. Zhang, Discrimination of False Negative Results in RT-PCR Detection of SARS-CoV-2 RNAs in Clinical Specimens by Using an Internal Reference, Virol. Sin. **35**, 758–767 (2020).
https://doi.org/10.1007/s12250-020-00273-8

14 Y. Zhou, F. Pei, L. Wang, H. Zhao, H. Li, M. Ji, W. Yang, Q. Wang, Q. Zhao, Y. Wang, Sensitivity evaluation of 2019 novel coronavirus (SARS-CoV-2) RT-PCR detection kits and strategy to reduce false negative, PLoS ONE **15**(11), e0241469 (2020).
https://doi.org/10.1101/2020.04.28.20083956

15 J.G. Ovejero, A. Gallo-Cordova, A.G. Roca, M.P. Morales, S. Veintemillas-Verdaguer, Reproducibility and Scalability of Magnetic Nanoheater Synthesis, Nanomaterials **11**, 2059 (2021).
https://doi.org/10.3390/nano11082059

Examining active compounds in ginger as a potential treatment for respiratory syndrome: a molecular docking and dynamics study

Gani Purwiandono[1*], *Nisa* Nabila[1], *Muhamad* Busro[1], *Salmahaminati*[1], *Wiyogo* Prio Wicaksono[1], *Nurcahyo Iman* Prakoso[1], and *Maisari* Utami[1]

[1]Department of Chemistry, Universitas Islam Indonesia, Yogyakarta 55584, Indonesia

Abstract. Having a major worldwide influence, respiratory syndrome has caused extensive health problems. Although there is no sure cure for respiratory illness, unconventional therapies including ginger have attracted interest. This research analyzes the prospective interactions between numerous bioactive chemicals in ginger namely zingiberene, zingerone, geraniol, 6-gingerol, and beta-sesquiphellandrene and the major protease linked with respiratory syndrome (PDB ID: 6LU7). Among these, geraniol and 6-gingerol demonstrated the highest binding affinities, with binding energies of −7.12 kcal/mol and −7.22 kcal/mol, respectively. The RMSD analysis indicated that the complex formed between 6-gingerol exhibited highest stability complex. Despite these encouraging in silico results, further laboratory-based validation is necessary to substantiate their therapeutic potential.

1 Introduction

In the past few years, lung illnesses have become one of the biggest health problems in the world. In December 2019, the outbreak started in Wuhan, China [1]. It quickly grew into a global disaster that shook up healthcare systems, crowded hospitals, and caused a lot of social and economic problems [2]. One of the most important things that needs to be done right now is the creation of effective antiviral drugs, especially for people who are already affected [3]. Medications that have been used for other purposes, like remdesivir, favipiravir, and hydroxychloroquine, have been used to treat symptoms and lower viral load [4]. However, these treatments have shown inconsistent efficacy and, in some cases, caused serious side effects [5].

Because of these problems, scientists are still looking for other ways to treat illnesses, preferably ones that come from natural sources [5]. Herbal and plant-based treatments have been important in traditional healing systems for hundreds of years, and the COVID-19 pandemic has made them even more important. Ginger (Zingiber officinale) is one of these plants. It is used in traditional medicine all over the world because it has many health benefits [6]. A lot of research has shown that ginger has pharmacological effects, such as anti-inflammatory, antioxidant, antimicrobial, antiviral, and immunomodulatory qualities [7].

*Corresponding author: gani_purwiandono@uii.ac.id

Because of these qualities, ginger is a good choice to be studied more as a natural virus killer, especially for treating lung illnesses [8].

There are two main goals for this study. The first is to use molecular docking simulations to look at how well key compounds from ginger bind to and interact with the main protease of respiratory syndrome (PDB ID: 6LU7). The second goal is to look at their pharmacokinetic properties using Lipinski's Rule of Five. The study focuses on ginger's promise as a medicine, which is backed up by its long history of use in traditional medicine, wide range of biological effects, and easy access. The aim of this study was to develop a plant-based antiviral treatments for respiratory syndromes by finding chemicals that might stop viruses from replicating. The results also lay the groundwork for more tests to be done in the lab and for the possible creation of antiviral medicines made from ginger.

2 Computational Methods

This study employed molecular docking techniques to explore how active compounds found in ginger interact with the main protease of respiratory syndrome (PDB ID: 6LU7). A range of computational tools was used throughout the analysis, including AutoDock Tools (https://autodock.scripps.edu/), PyMOL (https://www.pymol.org/), and Discovery Studio (https://www.3ds.com/products/biovia/discovery-studio). There are several ginger-derived compounds under investigation included zingiberene, zingerone, geraniol, 6-gingerol, and beta-sesquiphellandrene. Protein structure data were retrieved from the Protein Data Bank (https://www.rcsb.org/), while the chemical structures of the ligands were sourced from PubChem (https://pubchem.ncbi.nlm.nih.gov/).

To assess the potential of these compounds as drug candidates, Lipinski's Rule of Five was applied (https://scfbio-iitd.res.in/software/drugdesign/lipinski.jsp). This rule suggests that, for a compound to be considered orally active, it should meet the following criteria: a molecular weight under 500 g/mol, no more than five hydrogen bond donors, no more than ten hydrogen bond acceptors, and a log P value not exceeding 5 [9].

The docking simulations were carried out using AutoDock Vina as a tool designed to predict how small molecules bind to a target protein by modeling the interaction process [10]. The primary protein of this analysis was the 6LU7 protease as a key enzyme in the virus's replication cycle. The simulations aimed to determine the binding affinity between this protease and each of the selected ginger compounds.

Molecular dynamics simulation was conducted using YASARA software (https://www.yasara.org/). The simulation mimicked cell conditions for 50 ns, a common duration for assessing protein-ligand complex stability. The simulation was executed using the md_run macro, and results were visualized using MD_analyze, MD_analyzeres, and MD_analyzed bind energy macros. A root-mean-square deviation value of 2 or less indicated validity [11]. The process involved inputting samples into the YASARA Dynamic program developed by Biosciences GmbH, followed by a macro-driven molecular dynamics simulation with controlled temperature (310 K) and pH (7.4). The md_run macro was set for 10,000 ps, using the AMBER14 forcefield, with snapshots saved every 25 ps. Energy potential, hydrogen bond counts, RMSD, and radius of gyration were analyzed using the md_analyze macro. RMSF analysis was performed via the md_analyzeres macro, and Molecular Mechanics Poisson-Boltzmann Surface Area analysis was done using the md_analyzebindenergy macro.

3 Results and Discussion

Molecular docking simulations were used to analyzed on several bioactive compounds derived from ginger, including zingiberene, zingerone, geraniol, 6-gingerol, and beta-sesquiphellandrene. These simulations provided valuable insights into how each compound interacts with the main protease of the respiratory syndrome virus (PDB ID: 6LU7, https://www.rcsb.org/structure/6LU7). The binding affinities result's expressed in kcal/mol, reflect the strength and stability of these interactions-where more negative values indicate stronger and more favorable binding [12]. Prior to docking, each compound was evaluated using Lipinski's Rule of Five to determine its potential as an orally active drug [10], as shown in Table 1. All investigated compounds, complied with Lipinski's criteria, indicating favorable pharmacokinetic characteristics. Pharmacokinetic behavior plays a crucial role in determining whether a bioactive compound with promising binding affinity can be further developed into a viable therapeutic agent to estimate drug-likeness and oral bioavailability All selected compounds met the criteria, confirming their suitability for oral administration and justifying their inclusion in the docking analysis.

Table 1. Lipinski's rule of five parameter

Compound	Parameter				
	Mol weight	Log P	Proton donors	Proton acceptor	Refractory Molar
	< 500 g/mol	< 5	< 5	< 10	40 – 130
Zingiberene	204	4.89	0	0	68.83
Zingerone	194	1.92	1	3	53.66
Geraniol	154	2.67	1	1	49.50
6-Gingerol	294	3.23	2	4	82.75
Beta-sesquiphellandrene	204	4.89	0	0	68.83

A comparative analysis with remdesivir, a known antiviral agent, showed that zingiberene and beta-sesquiphellandrene exhibited stronger binding affinities than remdesivir, which has a binding energy of −6.50 kcal/mol. This implies that against respiratory syndrome major protease, these ginger-derived compounds could have competitive inhibitory action. Table 2 shows that although having lower binding energies than remdesivir, zingerone, geraniol, and 6-gingerol still have some potential as antiviral drugs.

The results indicates that a strong interaction between geraniol and 6-gingerol with critical residues such as LEU, TYR, and ASN. These residues are correspond to the protease's catalytic function, indicating the ability of geraniol and 6-gingerol. This activity indicates that these chemicals can efficiently obstruct the protease's activity and affecting the viral propagation. The 2D and 3D interaction visualizations revealed that all compounds were well accommodated to the active site of the protease. The interactions effectively occupying the binding pocket and potentially its catalytic function.

Table 2. A comparative analysis of binding energy on respiratory syndrome main protease

Compound	Binding energy (kcal/mol)
Zingiberene	−4.91
Zingerone	−4.93
Geraniol	−7.12
6-Gingerol	−7.22
Beta-sesquiphellandrene	−5.70
Remdesivir	−6.50

Geraniol demonstrated a binding energy of −7.12 kcal/mol, signifying a strong interaction with the main protease of the respiratory syndrome virus. Geraniol demonstrates interactions with LEU220, TRP218, and LEU271 in simulation studies, as shown in Figure 1. The studies revealed that amino acids are play a crucial role for the cleavage of viral polyproteins during the replication process. The affinity of geraniol for these residues suggests a potential reduction in protease activity and viral replication. Among all the compounds tested, 6-gingerol demonstrated the highest binding affinity, achieving a binding energy of −7.22 kcal/mol.

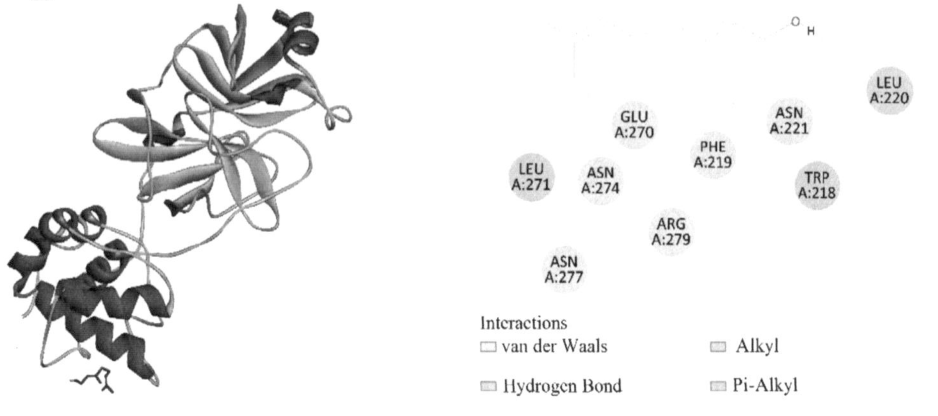

Fig. 1. Visualization of Geraniol - 6LU7 receptor compound in 3D and 2D

Among all the compounds, 6-gingerol exhibited the highest binding affinity, with a binding energy of −7.22 kcal/mol. Most likely geraniol, it interacted with key catalytic residues such as TYR239, LEU287, ASN277, and MET276, respectively as shown in Figure 2. These extended interactions contributed to the stable integration of 6-gingerol within the protease's active site, further supporting its potential as a protease inhibitor. The strong and stable interaction profile of 6-gingerol, combined with its low binding energy, suggests it has high potential as an inhibitor of the respiratory syndrome main protease, making it a promising candidate for further investigation.

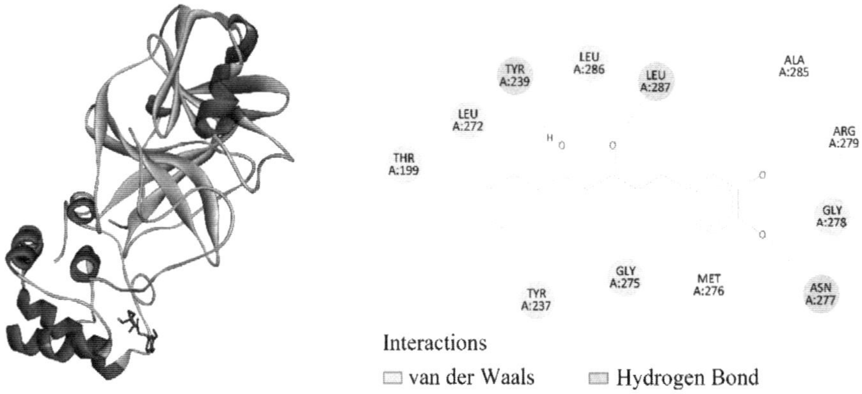

Fig. 2. Visualization of 6-Gingerol - 6LU7 receptor compound in 3D and 2D

In contrast, zingerone had a larger (less negative) binding energy of −4.93 kcal/mol, indicating a weaker interaction with the protease. There was hydrogen bonds created between zingerone and important residues GLN189 and HIS41; however, the stability of these interactions was not as high as that of the contacts formed by geraniol and 6-gingerol. It is possible that zingerone is not as efficient as the other chemicals because of its significantly weaker binding affinity. This is despite the fact that zingerone demonstrated some potential to inhibit the protease.

Zingiberene had a binding energy of −4.91 kcal/mol, similar to zingerone. While geraniol formed hydrogen bonds with ASP187 and TYR54, it did not interact with the critical catalytic residues like CYS145 or HIS41, which are essential for inhibiting the protease's activity. This lack of interaction with key residues diminishes geraniol's potential as an effective inhibitor of the protease.

Molecular Dynamics (MD) simulations, each with a 40 ns production run, were performed on the top-scoring ligand-target complexes to assess the stability and duration of their binding. Stable and long-lasting interactions would indicate the complexes' stability and, by extension, the small molecules' potential inhibitory effect [8]. Root-mean-square deviation (RMSD) analysis, a crucial tool for evaluating the stability of drug-target complex MD simulations, was conducted on all five identified complexes [13]. As depicted in Figure 3(a), the ligand's fit onto the protein in the 6-gingerol complex demonstrated consistent stability throughout the simulation, suggesting a relatively stable binding interaction. However, the Geraniol complex displayed fluctuating ligand-protein RMSD values (Figure 3(b)), indicating transient or less stable binding poses, potentially due to conformational changes in either the protein or the ligand during the simulation. This variability underscores the dynamic nature of protein-ligand interactions, where induced fit mechanisms may play a significant role [14].

Figure 3(c and d) indicated that the radius of gyration for 6-gingerol in complex with the drug was comparable to that with geraniol, reaching a maximum of 22.70 Å. The radius of gyration reflects changes in protein size during molecular dynamics simulations; a larger radius suggests a larger protein [15]. The figure illustrates these changes for the 6-gingerol complex with both geraniol and drug ligands. For the drug ligand, the radius of gyration fluctuated from the start until 3 ns, increasing from a minimum of 22.10 Å to 22.70 Å. In the 6-gingerol compound, a peak is observed at 18 ns before tending to stabilize. Conversely, the geraniol compound showed a fluctuating peak, indicating less stability in the protein-ligand complex compared to 6-gingerol. These findings align with the RMSD analysis, showing the dynamic nature of protein-ligand interactions throughout the simulation. Such fluctuations are significant for determining a ligand's potential as an enzyme inhibitor. This suggests that the interaction between geraniol was less stable than that with 6-gingerol.

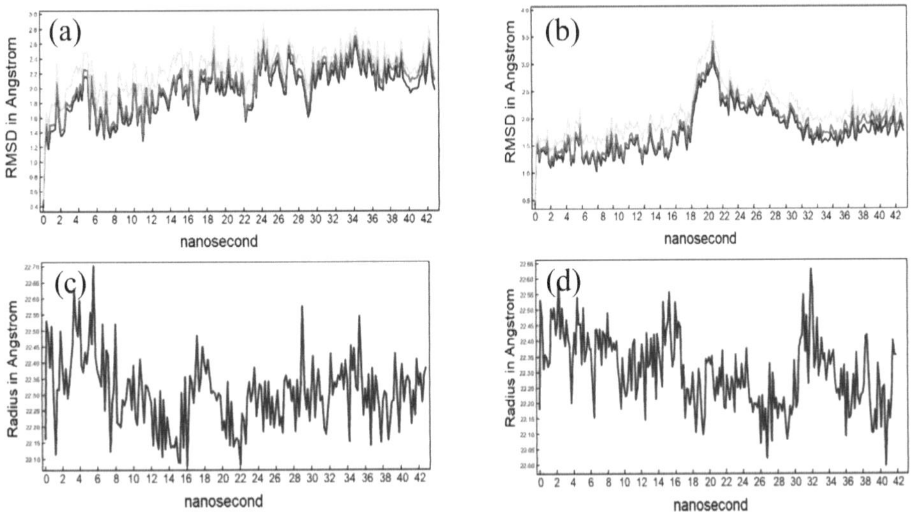

Fig. 3. illustrates the molecular dynamic simulations for the 6-gingerol and geraniol complexes. It includes analyses of protein compound complex RMSD for (a) 6-gingerol and (b) geraniol. The radius of gyration for the protein-compound complex for (c) 6-gingerol and (d) geraniol.

The simulations results indicated that the five ginger-derived compounds, geraniol and 6-gingerol showed the highest potential as natural inhibitors of the respiratory syndrome main protease. The results of this study indicated that these compounds have the potential to be investigated further as natural antiviral medicines for the treatment of respiratory syndrome therapy.

4 Conclusion

This molecular docking study examined five bioactive chemicals in ginger-zingiberene, zingerone, geraniol, 6-gingerol, and beta-sesquiphellandrene as natural respiratory syndrome major protease (6LU7) inhibitors. Geraniol and 6-gingerol showed the highest binding energies, −7.12 and −7.22 kcal/mol, respectively. These chemicals bound to TYR, HIS, and MET in the protease's active region, suggesting the limitation of viral multiplication. The molecular dynamics RMSD analysis indicated that the complex formed between 6-gingerol exhibited greater stability compared to the geraniol complex.

The potential anti-inflammatory implication of ginger in this study is indirectly associated with the inhibition of viral replication. The respiratory syndrome main protease plays a crucial role in viral polyprotein cleavage, which is essential for viral propagation and the subsequent activation of host inflammatory responses. Therefore, the computational observed protease inhibition provides a mechanistic basis for the potential anti-inflammatory effects of ginger. These findings highlight ginger-derived compounds as promising candidates for further experimental validation in antiviral and inflammation-related studies.

5 Acknowledgments

This study was supported by the Indonesia Endowment Fund for Education Agency (LPDP) with contract number: KEP-2/LPDP/LPDP.4/2021. The computational datasets are available from the corresponding author upon reasonable request.

References

1. Y. Cao, K. Cai, and L. Xiong, Coronavirus disease 2019: A new severe acute respiratory syndrome from Wuhan in China. Acta Virol. **64** (2020).
   https://doi.org/10.4149/av_2020_201
2. V.J. Clemente-Suárez, E. Navarro-Jiménez, L. Moreno-Luna, M.C. Saavedra-Serrano, M. Jimenez, J.A. Simón, and J.F. Tornero-Aguilera, The impact of the COVID-19 pandemic on social, health, and economy. Sustainability. **13**, 6314 (2021).
   https://doi.org/10.3390/su13116314
3. G. Uddin, O. Alam, and A. Serebrenik, A qualitative study of developers' discussions of their problems and joys during the early COVID-19 months. Empir. Softw. Eng. **27**, 117 (2022).
   https://doi.org/10.1007/s10664-022-10156-z
4. Y. Liang, X. Quan, R. Gu, Z. Meng, H. Gan, Z. Wu, Y. Sun, H. Pan, P. Han, and S. Liu, Repurposing existing drugs for the treatment of COVID-19/SARS-CoV-2: A review of pharmacological effects and mechanism of action. Heliyon. **10** (2024).
   https://doi.org/10.1016/j.heliyon.2024.e35988
5. D. Darmadi, D. Lindarto, J. Siregar, T. Widyawati, M. Rusda, M.M. Amin, F. Yusuf, P.C. Eyanoer, M. Lubis, and I. Rey, Study of the molecular dynamics stability in the inhibitory interaction of tenofovir disoproxil fumarate against CTLA-4 in chronic hepatitis B patients. Med. Arch. **77**, 227 (2023).
   https://doi.org/10.5455/medarh.2023.77.227-230
6. M.S. Azam, M.N. Islam, M. Wahiduzzaman, M. Alam, and A.A.K. Dhrubo, Antiviral foods in the battle against viral infections: Understanding the molecular mechanism. Food Sci. Nutr. **11**, 4444 (2023).
   https://doi.org/10.1002/fsn3.3454
7. F. Ayustaningwarno, G. Anjani, A.M. Ayu, and V. Fogliano, A critical review of Ginger's (Zingiber officinale) antioxidant, anti-inflammatory, and immunomodulatory activities. Front. Nutr. **11**, 1364836 (2024).
   https://doi.org/10.3389/fnut.2024.1364836
8. A. Andina, S. Sugiarto, and P. Pamungkas, Antiviral treatment of COVID-19. Int. J. Med. Sci. Heal. Res. **5**, 39 (2024).
   https://doi.org/10.70070/3kv7z420
9. C.A. Lipinski, Drug-like properties and the causes of poor solubility and poor permeability. J. Pharmacol. Toxicol. Methods. **44**, 235 (2000).
   https://doi.org/10.1016/s1056-8719(00)00107-6
10. S. Skariyachan, D. Gopal, A.G. Muddebihalkar, A. Uttarkar, and V. Niranjan, Structural insights on the interaction potential of natural leads against major protein targets of SARS-CoV-2: molecular modelling, docking and dynamic simulation studies. Comput. Biol. Med. **132**, 104325 (2021).
    https://doi.org/10.1016/j.compbiomed.2021.104325

11. G.P. Siregar, I. Parwati, T. Tjahjodjati, F. Safriadi, G.R. Situmorang, R. Yohana, and A.F. Khairani, Molecular Dynamic Stability Study of VEGF Inhibitor in Patients with Bladder Cancer. Acta Inform. Medica. **33**, 50 (2025). https://doi.org/10.5455/aim.2024.33.50-53
12. Б.Х. Абдул, А. Сарвари, М.Х. Хасанд, В.М. Абдул, П. Ниази, К.Б. Абдул, С. Седики, and У.М. Кака, The effects of antiviral therapeutic drugs for SARS-COV-2 on treatment outcomes and clinical parameters. Вестник КазНУ. Серия Экологическая. **79**, 136 (2024). https://doi.org/10.26577/eje.2024.v79.i2-013
13. A. Shamsi, M.S. Khan, D.K. Yadav, M. Shahwan, M. Furkan, and R.H. Khan, Structure-based drug-development study against fibroblast growth factor receptor 2: molecular docking and Molecular dynamics simulation approaches. Sci. Rep. **14**, 19439 (2024). https://doi.org/10.1038/s41598-024-69850-1
14. M.S. Djati, Molecular docking and dynamic simulation of sesamin from Piper crocatum Ruiz & Pav as an inhibitor of canonical NF-κB: Potential anti-inflammatory mechanism. J. Pharm. Pharmacogn. Res. **13**, 848 (2025). https://doi.org/10.56499/jppres24.2070_13.3.848
15. M.Y. Lobanov, N.S. Bogatyreva, and O. V Galzitskaya, Radius of gyration as an indicator of protein structure compactness. Mol. Biol. **42**, 623 (2008). https://doi.org/10.1134/s0026893308040195

Building pharmacovigilance-ready datasets from resource-limited settings: study of antiretroviral-associated ADRs in Melanesian patients

Siti Sitatul Mar'ah[1], *Zakky* Cholisoh[1*], and *Burhannudin* Ichsan[2]

[1] Faculty of Pharmacy, Universitas Muhammadiyah Surakarta, Jalan Ahmad Yani, Sukoharjo, 57169, Central Java, Indonesia
[2]Faculty of Medicine, Universitas Muhammadiyah Surakarta (UMS), Jalan Ahmad Yani, Sukoharjo, 57169, Central Java, Indonesia

Abstract. Adverse drug reactions (ADRs) remain a major challenge in antiretroviral therapy (ART), particularly in underrepresented populations where real-world safety data are limited. This study analyzed the type, severity, and management of ADRs associated with first-line ART among Melanesian people living with HIV/AIDS as a structured foundation for future data-driven pharmacovigilance. A cross-sectional observational study was conducted among adult patients attending the Voluntary Counseling and Testing clinic at Central Papua Regional Hospital, Indonesia. Sociodemographic, clinical, behavioral, and ART-related variables were systematically linked with ADR characteristics, including onset timing, severity grading, and management approaches. Of 143 patients, 142 (99.3%) experienced at least one ADR, generating 645 ADR events. Dizziness (35.2%), nausea (19.1%), and drowsiness (9.0%) were the most frequently reported reactions, with most events classified as mild to moderate. Hepatotoxicity represented the most clinically significant higher-grade ADR. Bivariate analysis showed that shorter ART duration (1–6 months) was significantly associated with increased odds of moderate–severe ADRs (OR 2.63; $p = 0.027$). The majority of ADRs were managed through over-the-counter self-medication (67.8%), while prescription-based interventions accounted for 30.0% of cases. This study highlights how systematically structured real-world safety data from underrepresented populations can support scalable, data-oriented pharmacovigilance frameworks and inform future AI- and machine learning–based drug safety surveillance.

1 Introduction

The widespread implementation of antiretroviral therapy (ART) has fundamentally transformed Human Immunodeficiency Virus (HIV) infection into a manageable chronic condition. Nevertheless, adverse drug reactions (ADRs) remain a major challenge to long-term treatment success, as they can reduce adherence, necessitate regimen modification, and

* Corresponding author: zakky.cholisoh@ums.ac.id

negatively affect quality of life among people living with HIV/AIDS [1,2]. Clinical studies across different regions consistently report ART-associated ADRs involving gastrointestinal, neurological, dermatological, and hematological systems, with considerable variability depending on drug regimens and patient characteristics [1–3].

Beyond pharmacological factors, sociocultural and behavioral determinants play an important role in shaping ART utilization and safety monitoring. In Indonesia, community knowledge and stigma toward HIV/AIDS have been shown to influence attitudes toward treatment and engagement with healthcare services [4]. Behavioral beliefs and attitudes toward Voluntary Counseling and Testing (VCT) also affect service uptake and continuity of care, which in turn determine opportunities for ADR detection and management [5]. Evaluations of ART implementation in Indonesian clinical settings further highlight heterogeneity in therapy outcomes and patient experiences, underscoring the importance of context-specific safety data [6].

Effective pharmacovigilance is therefore essential to ensure ART safety and optimize patient outcomes. However, conventional pharmacovigilance approaches, largely dependent on spontaneous reporting and small-scale observational studies, suffer from underreporting, inconsistent data quality, and limited scalability, particularly in resource-constrained settings [3,7]. These limitations are amplified in underrepresented populations, such as Melanesian communities, where systematic ADR documentation is scarce and often absent from global safety datasets.

Recent advances in health informatics, big data analytics, and artificial intelligence (AI) have highlighted the potential for transforming pharmacovigilance practices when sufficiently large, standardized, and interoperable datasets are available. Prior studies demonstrate that structured real-world data and algorithm-assisted approaches can enhance ADR standardization, signal detection, and safety evaluation when applied within appropriately scaled data environments [8–10]. However, the effectiveness of such approaches remains contingent on the availability of high-quality, well-characterized input data, particularly from populations that are currently underrepresented in pharmacovigilance research.

In this context, generating structured and systematically curated ADR data from underrepresented populations represents an essential preparatory step toward inclusive big data pharmacovigilance. The present study aims to characterize ADRs associated with first-line ART among Melanesian people living with HIV/AIDS, focusing on reaction types, causality, severity, timing, and management strategies. This study has the potential to provide a foundational real-world safety dataset that may support future integration into AI and machine learning enabled pharmacovigilance frameworks, thereby contributing to more equitable and generalizable drug safety knowledge

2 Methods

2.1 Study Design and Setting

A cross-sectional pharmacovigilance-oriented observational study was conducted among Melanesian people living with HIV/AIDS receiving first-line antiretroviral therapy (ART). The study was implemented at the Voluntary Counseling and Testing (VCT) clinic of Central Papua Regional Hospital (RSUD Paniai), Indonesia, during 2023. Ethical clearance was obtained from the Health Research Ethics Committee of Dr. Moewardi Hospital, Surakarta (Approval No. 609/IV/HREC/2023). This study was designed to generate structured real-world safety data suitable for aggregation into future big data pharmacovigilance systems, with a specific focus on underrepresented ethnic populations.

2.2 Study Design and Setting

The target population consisted of adult Melanesian patients (≥18 years) with confirmed HIV infection who were receiving first-line ART and attending routine follow-up at the VCT clinic. Patients were eligible if they demonstrated good adherence to ART as documented in medical records and pharmacy refill data. Patients were excluded if they had documented unstable psychiatric conditions or cognitive impairment that could interfere with reliable reporting of adverse drug reactions (ADRs). Written informed consent was obtained from all participants prior to data collection. Based on these criteria, a total of 143 patients were enrolled in the study.

2.3 Sample Size Determination

The minimum sample size was calculated using the proportion-based formula proposed by Lemeshow et al. [11], assuming a conservative estimated prevalence of ADRs among ART users, a 95% confidence level, and a margin of error of 5%. This calculation yielded a minimum required sample size of 140 participants, which was met and slightly exceeded in the present study.

2.4 Data Collection and Variables

Data were collected using a structured pharmacovigilance data collection form. Adverse drug reactions (ADRs) were defined based on ART drug monographs and treatment guidelines. All patient-perceived symptoms temporally associated with ART initiation, including mild (Grade 1) reactions were recorded as ADRs when no alternative clinical cause was documented. The recorded variables included: Sociodemographic characteristics: age, sex, marital status, education level, and employment status; Behavioral and clinical risk factors: smoking status, alcohol use, HIV transmission risk category; Antiretroviral therapy profile: ART regimen, duration of therapy, and concomitant medication use; Adverse drug reactions: type of ADR, time of onset after ART initiation, severity, causality, and management strategy. Data sources included patient interviews, medical records, and pharmacy dispensing records to ensure data triangulation and completeness.

2.5 Assessment of Adverse Drug Reactions

ADR identification was conducted through structured patient interviews combined with medical record review. Initial ADR screening was performed during routine clinic visits at the VCT service. Follow-up assessments were conducted periodically through structured telephone interviews to capture delayed or evolving ADRs. The causality of each suspected ADR was assessed using the Naranjo Adverse Drug Reaction Probability Scale, which classifies ADRs as definite, probable, possible, or doubtful based on standardized criteria [12]. The severity of ADRs was graded using the Hartwig and Siegel Severity Assessment Scale, categorizing reactions into mild, moderate, and severe levels, as described by Srinivasan and Ramya [13]. ADRs were further classified according to: Type of reaction (e.g., neurological, gastrointestinal, hematological); Time of onset, categorized as 1–3, 4–6, 7–9, and 10–12 months after ART initiation; Management approach, including self-medication and prescription-based interventions. Only patients experiencing at least one ADR were included in ADR-specific analyses.

2.6 Data Analysis

Data were analyzed descriptively using frequency distributions and percentages. Sociodemographic characteristics, ADR types, causality categories, severity grades, timing of occurrence, and management strategies were summarized in tabular form. Rather than inferential statistical testing, the analysis emphasized structured pattern recognition of ADR profiles to support future integration into large-scale pharmacovigilance databases. This approach prioritizes data standardization and reproducibility, which are critical for big data–driven drug safety monitoring in ethnically diverse and data-scarce populations. Bivariate analyses using chi-square tests and odds ratios were conducted to explore associations between ADR severity and clinical variables, with p value of <0.050 in the 95% confidence interval.

3 Results and Discussion

3.1 Population Characteristics and Data Structure Relevance

Among 143 Melanesian people living with HIV/AIDS receiving first-line antiretroviral therapy (ART), 142 patients (99.3%) experienced at least one adverse drug reaction (ADR) during the observation period. Table 1 summarizes the sociodemographic, clinical, and treatment-related characteristics of patients who experienced ADRs. Most patients were aged 12–35 years (90.8%), male (52.5%), married (69.0%), and receiving fixed-dose combination (FDC) of Tenofovir 300mg, Lamivudine 150mg, Efavirenz 600mg regimens (65.5%). The majority had been on ART for less than six months (78.2%), highlighting the early treatment phase as a critical window for ADR emergence. Smoking (26.8%) and alcohol consumption (20.0%) were present in a notable subset of patients, representing behavioral covariates that are relevant for multivariable and machine learning–based risk modeling. From a big data perspective, the integration of demographic, behavioral, and medication variables into a single analytical framework provides structured inputs suitable for downstream predictive analytics and pharmacovigilance signal detection [14,15].

Table 1. Dataset Composition and ADR Distribution Among HIV Patients Receiving First-Line ART (N = 142 ADR Cases).

Characteristic Category	Category Level	Number of ADR Cases (n)	% of All ADR Cases (n = 142)
Sex	Male	75	52.8%
	Female	67	47.2%
Age Group (years)	12–35	129	90.8%
	36–65	13	9.2%
Marital Status	Married	98	69.0%
	Unmarried	44	31.0%

Educational Level	No formal education / Not working	12	8.5%
	Primary school	11	7.7%
	Junior secondary	32	22.5%
	Senior secondary	61	43.0%
	Diploma / Bachelor	26	18.3%
Employment Status	Employed	102	71.8%
	Unemployed	40	28.2%
Behavioral Risk Factors	Smoker	38	26.8%
	Non-smoker	104	73.2%
	Alcohol user	29	20.4%
	Non-alcohol user	114	80.3%
Population Descriptor	Papuan (Melanesian)	142	99.3%

3.2 ADR Severity Profile and Clinical Patterns

A total of 645 ADR events were recorded and categorized by severity (Table 2). Based on Naranjo causality evaluation, ADRs were classified as probable (90%), possible (3%) and definite (7%). In terms of severity, ADRs were classified as Grade 1 (mild) or Grade 2 (moderate), with dizziness (35.2%), nausea (19.1%), and drowsiness (9.0%) being the most frequently reported manifestations. Hepatotoxicity accounted for the majority of higher-severity ADRs (8.7%), representing the most clinically significant safety concern in this cohort. The predominance of mild-to-moderate ADRs aligns with prior studies in ART-treated populations, where central nervous system and gastrointestinal symptoms commonly occur during early treatment initiation [3]. However, the relatively high cumulative frequency of ADR events per patient underscores the importance of longitudinal monitoring rather than reliance on single-point spontaneous reporting. Importantly, this graded severity structure enables transformation of clinical observations into machine-readable safety labels, which are essential for training supervised learning models in pharmacovigilance applications [10].

Table 2. Severity Levels of Adverse Drug Reactions (ADRs) in HIV/AIDS Patients at Central Papua Hospital.

No.	ADR Type	Grade 1 (Mild), n (%)	Grade 2 (Moderate), n (%)	Grade 3 (Severe), n (%)	Grade 4 (Life-threatening), n (%)
1	Dizziness	227 (35.2)	17 (2.6)	–	–
2	Nausea	123 (19.1)	42 (6.5)	–	–

3	Drowsiness	58 (9.0)	10 (1.6)	–	–
4	Hepatotoxicity	20 (3.1)	26 (4.0)	–	–
5	Insomnia	25 (3.9)	11 (1.7)	–	–
6	Anemia	–	15 (2.3)	–	–
7	Weight loss	–	8 (1.2)	–	–
8	Hallucinations	2 (0.3)	3 (0.5)	–	–

Values are presented as number of ADR events, n (% of total ADR events, n = 645). One patient may experience more than one ADR.

3.3 Temporal Distribution of ADR Events

A total of 645 ADR events were recorded across all patients, exhibiting distinct temporal patterns relative to ART initiation (Table 3). Dizziness was the most frequently reported ADR overall, with 244 events (37.8%), predominantly appearing within the first three months of ART. Nausea was the second most frequent ADR, followed by drowsiness and hepatotoxicity. The temporal profiles of ADR events align with previous observational studies showing higher ADR occurrence in the early treatment period, particularly within the first six months of initiating ART [15].

Table 3. Distribution of Adverse Drug Reaction (ADR) Events in HIV/AIDS Patients at Central Papua Hospital by Time to Onset During First-Line Antiretroviral Therapy (N = 645 ADR Events)

No	ADR Type	Total Events n (%)	1–3 Months n (%)	4–6 Months n (%)	7–9 Months n (%)	10–12 Months n (%)
1	Dizziness	244 (37.8)	102 (15.8)	93 (14.4)	45 (7.0)	4 (0.6)
2	Nausea	164 (25.4)	102 (15.8)	51 (7.9)	8 (1.2)	4 (0.6)
3	Drowsiness	79 (12.2)	37 (5.7)	23 (3.6)	18 (2.8)	–
4	Hepatotoxicity	56 (8.7)	4 (0.6)	20 (3.1)	32 (5.0)	–
5	Vomiting	36 (5.6)	18 (2.8)	17 (2.6)	–	1 (0.2)
6	Insomnia	36 (5.6)	17 (2.6)	10 (1.6)	7 (1.1)	2 (0.3)
7	Anemia	15 (2.3)	–	8 (1.2)	7 (1.1)	–
8	Weight loss	8 (1.2)	–	8 (1.2)	–	–
9	Pruritic rash	3 (0.5)	–	3 (0.5)	–	–
10	Hallucinations	2 (0.3)	2 (0.3)	–	–	–

11	Bone/muscle pain	2 (0.3)	–	2 (0.3)	–	–
12	General weakness	2 (0.3)	–	–	–	–

Values are presented as number of ADR events, n (% of total ADR events, n = 645). One patient may experience more than one ADR.

3.4 Severity of ADRs and Associated Factors

Bivariate analysis (Table 4) demonstrated that gender, age, ART regimen type, HIV clinical stage, co-infection, and comorbidity were not significantly associated with ADR severity (all $p > 0.05$). Duration of ART use was the only factor significantly associated with ADR severity. Patients receiving ART for 1–6 months had higher odds of experiencing moderate-to-severe ADRs compared with those treated for 7–12 months (OR 2.63; 95% CI 1.10–6.46; $p = 0.027$). This finding suggests that clinically relevant ADRs are more likely to occur during the early phase of ART initiation. The absence of significant associations with demographic and clinical variables indicates relatively consistent ADR severity profiles across patient subgroups. These results underscore the need for intensified pharmacovigilance and clinical monitoring during the initial months of ART.

Table 4. Bivariate Analysis of Factors Associated with the Severity of Adverse Drug Reactions (ADRs) among HIV Patients Receiving Antiretroviral Therapy

Variables	**Severity of ADRs**		**Bivariate Analysis**	
	Moderate–Severe n (%)	Mild n (%)	P-*value*	OR (95% CI)
Gender				
Male	60 (80.0)	15 (20.0)	0.576	1.255
Female	51 (76.1)	16 (23.9)		(0.565 – 2.785)
Age				
> 35 years	11 (84.6)	2 (15.4)	0.555	1.595
12-35 years	100 (77.5)	29 (22.5)		(0.334 – 7.608)
ART Regimen				
Fixed-dose combination FDC	44 (80.0)	11 (20.0)	0.816	1.143
Non-FDC	21 (77.8)	6 (22.2)		(0.372 – 3.511)
HIV Clinical Stage				
Stage III	39 (79.6)	10 (20.4)	0.766	1.138
Stage II	72 (77.4)	21 (22.6)		(0.487 – 2.656)
Co-infection				
Yes	35 (79.5)	9 (20.5)	0.790	1.126
No	76 (77.6)	22 (22.4)		(0.470 – 2.694)
Comorbidity				
Yes	10 (76.9)	3 (23.1)	0.909	0.924
No	101 (78.3)	28 (21.7)		(0.238 – 3.588)

Duration of ART Use				
1-6 months	92 (82.1)	20 (17.9)	0.027*	2.633
7-12 months	19 (63.3)	11 (36.7)		(1.098 – 6.460)

Data are presented as number (percentage). Odds ratios (ORs) were calculated using bivariate logistic regression. A p-value < 0.05 indicates statistical significance.

3.5 Management Strategies and Health System Implications

As shown in Table 5, most ADRs were managed through over-the-counter (OTC) self-medication (67.75%), while only 30% received prescription-based interventions and 2.2% received no treatment. Table 4 further details the specific medications used, revealing that paracetamol and antacids were the most frequently utilized self-medications, whereas hepatotoxicity and anemia were primarily managed through physician-prescribed therapies. This heavy reliance on self-medication reflects systemic challenges, including limited access to supportive care, resource constraints, and patient-driven coping strategies. From a pharmacovigilance standpoint, such patterns are often underreported in traditional surveillance systems, leading to incomplete safety signals [8-10]. Capturing both prescribed and non-prescribed ADR management provides a more realistic representation of real-world drug safety behavior and strengthens the external validity of pharmacovigilance datasets.

Table 5. Medications Used to Manage Adverse Drug Reactions (ADRs) in HIV/AIDS Patients at Central Papua Hospital.

A. Self-Medication (Over-the-Counter Use)			
No.	**ADR Type**	**Medication**	**n (%)**
1	Dizziness	Paracetamol	65 (10.1)
2	Nausea	Antacid	32 (5.0)
3	Nausea	Polysilane®	15 (2.3)
4	Dizziness	Ibuprofen	4 (0.6)
Total			116 (18.0)

B. Prescription Medications			
No.	**ADR Type**	**Medication**	**n (%)**
1	Hepatotoxicity	Curcuma	52 (8.1)
2	Anemia	Multivitamin (Caviplex®)	17 (2.6)
3	Pruritic rash	Cetirizine	3 (0.5)
4	Bone/muscle pain or weakness	Mefenamic acid	2 (0.3)

Total	74 (11.5)

Values are presented as number of ADR events, n (% of total ADR events, n = 645). One patient may experience more than one ADR.

3.6 Big Data Pharmacovigilance Perspective

Although this study involved a relatively small number of patients, it generated a structured and multidimensional dataset linking sociodemographic characteristics, ART regimens, ADR types, severity grades, timing of onset, and management strategies. The exceptionally high ADR prevalence (99.3%) reflects an intentionally inclusive pharmacovigilance approach that captured all patient-reported symptoms temporally associated with ART, including mild and transient reactions. While this approach may limit direct comparability with studies that apply stricter clinical thresholds, it provides a comprehensive representation of real-world ART tolerability during routine care. Importantly, the present work does not constitute a big data or AI-based analysis; rather, it demonstrates how systematically organized clinical safety data from underrepresented populations can serve as a foundational input for future large-scale pharmacovigilance systems when aggregated across centers and time. The inclusion of Melanesian patients addresses a critical gap in global drug safety datasets, which remain heavily skewed toward non-Indigenous populations and may contribute to bias in data-driven health models.

3.7 Implications for AI-Enabled Clinical Decision Support

The standardized classification of ADRs applied in this study, encompassing reaction type, severity, timing, and management, aligns with the core data elements required for clinical decision support systems (CDSS). Although the present analysis was primarily descriptive, bivariate analysis identified a significant association between shorter ART duration and higher ADR severity, indicating that the early treatment period represents a critical window for safety monitoring. When combined with electronic health records and laboratory data in future studies, similarly structured datasets may support risk stratification and targeted clinical follow-up. From this perspective, the value of the current study lies not in advanced analytics, but in demonstrating the feasibility of generating AI-ready pharmacovigilance data within routine clinical practice in resource-limited settings.

4 Conclusion

This study provides a detailed characterization of adverse drug reactions associated with first-line antiretroviral therapy among Melanesian people living with HIV/AIDS, a population that remains underrepresented in pharmacovigilance research. Most ADRs were mild to moderate in severity and occurred early after ART initiation, with neurological and gastrointestinal manifestations predominating. The high frequency of reported ADRs reflects comprehensive symptom capture rather than an excess of clinically severe events, and highlights the importance of clear operational definitions in ADR surveillance. Bivariate analysis indicated that ADR severity was significantly associated with shorter duration of ART use, underscoring the need for closer monitoring during the initial months of therapy.

Several limitations should be acknowledged. The one-month follow-up period may have preferentially captured early-onset ADRs, with late-onset toxicities likely underrepresented. In addition, although pharmacogenetic differences may contribute to interindividual variability in ADRs, no genetic data were collected, and such mechanisms should be considered hypothesis-generating rather than causal. Overall, this study demonstrates how systematically collected real-world safety data can strengthen pharmacovigilance in ethnically diverse and data-scarce populations and provide a foundation for future integrative and data-driven drug safety research.

The authors gratefully acknowledge the Director and healthcare staff of Central Papua Regional Hospital (RSUD Paniai), Indonesia, for their cooperation and support, and thank all patients who participated in this study. This research was funded by the Institute for Research and Innovation, Universitas Muhammadiyah Surakarta, through the Postgraduate Research Grant Scheme (Grant No. 303.19/A3-III/LRI/X/2023). The funder had no role in the study design, data collection, analysis, interpretation, or manuscript preparation. The datasets generated during the study are not publicly available due to ethical and confidentiality considerations but may be obtained from the corresponding author upon reasonable request and appropriate ethical approval. Mar'ah Siti Sittatul contributed to data collection, curation, and drafting of the manuscript; Ichsan Burhannudin contributed to methodology development, supervision, and critical review; Cholisoh Zakky contributed to study design, data analysis, interpretation, manuscript revision, and final approval. All authors approved the final manuscript and take responsibility for its content.

References

1. J.C. Mendes, P.F. Bonolo, M.D. Ceccato, E.A. Gomes, F.A. Acurcio, Adverse reactions associated with first-line regimens in patients initiating antiretroviral therapy. Eur. J. Clin. Pharmacol. **74**, 1077–1088 (2018). https://doi.org/10.1007/s00228-018-2472-y
2. S. G. Endalamew, S. K. Assefa, M. S. Getachew, F. G. Arage, D. K. Asgedom, B. Terefe, D. F. Teshome, Incidence of adverse drug reactions among HIV patients on antiretroviral drugs in Ethiopia : A systematic review and meta-analysis. AIDS Res. Treat. **40129733** (2025) https://pubmed.ncbi.nlm.nih.gov/40129733/
3. G.I. Eluwa, T. Badru, K. A. Agu, K. J. Akpoigbe, O. Chabikuli, C. Hamelmann, Adverse drug reactions to antiretroviral therapy: Incidence, type and risk factors in Nigeria. BMC Clin. Pharmacol. **12**, 14 (2012). https://doi.org/10.1186/1472-6904-12-14
4. I.T. Utami, F.R.S. Prakoeswa, N. Lestari, B. Ichsan, Relationship between knowledge level and community stigma toward HIV/AIDS in Indonesia: A literature review. J. Kedokt. Syiah Kuala **23**, 99–107 (2023). https://jurnal.usk.ac.id/JKS/article/view/24678
5. C. Nugroho, T.A.I. Kusumaningrum, Behavioral beliefs and attitudes toward VCT among MSM in Sukoharjo. J. Kesehat. **12**, 102–106 (2019). https://journals.ums.ac.id/index.php/JK/article/view/9766
6. A.Y. Nugraheni, R. Amelia, I.F. Rizki, Evaluasi Terapi Antiretroviral Pasien HIV/AIDS. J. Farmasetis **8**, 45–54 (2019). https://doi.org/10.32583/farmasetis.v8i2.567
7. A. Khan, M. Alanazi, H. Ben Khaled, S. Alanazi, F. H. Malik, Z. U. Rehman, A. Ali, M. I. Siddique, Pharmacovigilance in the era of digital health: Leveraging big data and artificial intelligence for enhanced drug safety monitoring. Asian J. Pharm. **19**, 141–152 (2025). https://doi.org/10.22377/ajp.v19i01.6077
8. M. A. Shamim, M. A. Shamim, P. Arora, P. Dwivedi, Artificial intelligence and big data for pharmacovigilance and patient safety. J. Med. Surg. Public Health, **3**, 100139 (2024). https://doi.org/10.1016/j.glmedi.2024.100139

9. K. Kreimeyer, M. Foster, A. Pandey, N. Arya, G. Halford, S.F. Jones, R. Forshee, M. Walderhaug, T. Botsis, Natural language processing systems for capturing and standardizing unstructured clinical information: A systematic review. J. Biomed. Inform. **73**, 14–29 (2017). https://doi.org/10.1016/j.jbi.2017.07.012
10. A. Nagar, J. Gobburu, A. Chakravarty, Artificial intelligence in pharmacovigilance: advancing drug safety monitoring and regulatory integration. Ther. Adv. Drug Saf. **16**, 20420986251361435 (2025). https://doi.org/10.1177/20420986251361435
11. S. Lemeshow, D.W. Hosmer Jr., J. Klar, S.K. Lwanga, Adequacy of sample size in health studies. John Wiley & Sons, Chichester (1990). https://tbrieder.org/publications/books_english/lemeshow_samplesize.pdf
12. C.A. Naranjo, U. Busto, E.M. Sellers, P. Sandor, I. Ruiz, E.A. Roberts, E. Janecek, C. Domecq, D.J. Greenblatt, A method for estimating the probability of adverse drug reactions. Clin. Pharmacol. Ther. **30**, 239–245 (1981). https://doi.org/10.1038/clpt.1981.154
13. R. Srinivasan, G. Ramya, Adverse drug reaction—causality assessment. Int. J. Res. Pharm. Sci. **1**, 606–612 (2011). https://ijrps.com/home/article/view/1220
14. World Health Organization, WHO pharmacovigilance indicators: A practical manual for the assessment of pharmacovigilance systems. WHO Press, Geneva (2020). https://www.who.int/publications/i/item/WHO-MVP-EMP-2015.03
15. I.R. Edwards, J.K. Aronson, Adverse drug reactions: Definitions, diagnosis, and management. Lancet **356**, 1255–1259 (2000). https://doi.org/10.1016/S0140-6736(00)02799-9

Green nanomedicine approach for diabetic wound healing: A Review Article

Aldyba Syaqilla Hafifah[1], *Prima* Aulia Putra[1], *Muhammad Rifqi Fadillah* Muslim[1], *Arba Pramundita* Ramadani[1], and *Lutfi* Chabib[1]*

[1]Master Program of Pharmacy, Faculty of Mathematics and Natural Sciences, Universitas Islam Indonesia, Yogyakarta, Indonesia

Abstract. Diabetes mellitus is a chronic disease that often leads to complications, such as diabetic foot ulcers that are difficult to heal. Conventional therapies remain limited due to microbial resistance, high costs, and adverse side effects. Therefore, more effective alternatives are needed. Plant extracts have antioxidant, anti-inflammatory, and antimicrobial activities. However, their bioavailability and stability are low. Nanoparticle technology can overcome these limitations by increasing the stability, penetration, and effectiveness of bioactive compounds. This study aims to examine the potential of plant extract-based nanoparticle formulations in accelerating diabetic wound healing. The research method employed a comprehensive literature review, utilising the Google Scholar and PubMed databases. Six articles met the inclusion criteria. The results showed that various plant extracts, such as Mentha piperita, Punica granatum, Bacopa monnieri, Althaea officinalis, and Ocimum sanctum, can act as reducing agents and sources of bioactive compounds in nanoparticle synthesis. These were shown to increase antimicrobial activity, accelerate wound contraction, and support tissue regeneration in diabetic mouse models. Thus, plant extract-based nanomedicines have the potential to be a promising therapeutic strategy for diabetic foot ulcers. However, further research is needed on toxicity, phytochemical mechanisms, and clinical validation.

1 Introduction

Diabetes mellitus is a chronic disease characterised by insufficient insulin production, leading to hyperglycaemia [1]. This condition is a global health issue that affects 830 million individuals worldwide [2]. Indonesia ranks fifth globally, with 19.47 million cases in 2021, projected to rise to 28.6 million by 2045 [3]. One of the major complications of diabetes is diabetic foot ulcers. These ulcers significantly increase morbidity, amputation risk, and impair patients' quality of life, requiring complex treatment [4]. Diabetic wound care has progressed from standard wound management and antibiotics to therapies based on growth factors and tissue engineering. Many topical treatments with antibacterial effects face challenges such as resistance from polymicrobial infections. Additionally, high costs, side

* Corresponding author: lutfi.chabib@uii.ac.id

effects, and poor efficacy in chronic cases remain significant barriers [5]. These issues underline the need for safer, more affordable, and effective alternatives. One such option is the use of plant extracts, which provide bioactive compounds with antioxidant, anti-inflammatory, and antimicrobial activities that support wound healing.

While plant extracts offer considerable potential as alternative therapies, their conventional use is limited by low bioavailability, solubility, and stability. To address these limitations, more efficient delivery technologies such as nanoparticle-based systems have been explored. Nanoparticle delivery systems protect active compounds from degradation, prolong release, increase penetration into target tissues, and may enhance therapeutic effects. Building on these advancements, several recent studies have reported success in combining plant extracts with nanoparticle systems to accelerate diabetic wound healing. Based on these observations, this article aims to comprehensively discuss the potential of plant extracts in nanoparticle formulations for the treatment of diabetic wounds.

Accordingly, this work presents a narrative literature review of experimental studies on plant extract–based nanoparticle formulations for diabetic wound healing, with an emphasis on methodological quality, study design, and translational implications.

2 Method

This study is a narrative literature review that synthesises experimental evidence on plant extract–based nanoparticle formulations for diabetic wound healing. Literature searches were conducted in PubMed and Google Scholar using combinations of the keywords “diabetic foot” or “diabetic ulcer” or “diabetic wound” and “plant extract” or “phytochemical” and “nanoparticle” or “solid lipid nanoparticle” or “nanomedicine. Eligible studies were limited to publications from 2020 to 2025. Screening was performed by applying the eligibility criteria to titles/abstracts and subsequently to full-text articles. Studies were included if they were full-text, open-access, original/experimental research articles that used plant extracts or herbal ingredients as the active component in topical nanoparticle formulations for diabetic wound healing. Studies were excluded if they were paid or not fully accessible, were review articles (including systematic reviews or meta-analyses), were incomplete publications (e.g., conference abstracts), or did not use natural/herbal ingredients as the primary active component.

To support interpretation of evidence quality, key methodological descriptors were recorded for each included study, including the experimental model, outcome measures, and statistical reporting elements aligned with ARRIVE 2.0 domains (e.g., randomisation, blinding, and sample size rationale when reported). A total of 460 records were identified, and six experimental studies met the eligibility criteria and were included, as depicted in Figure 1.

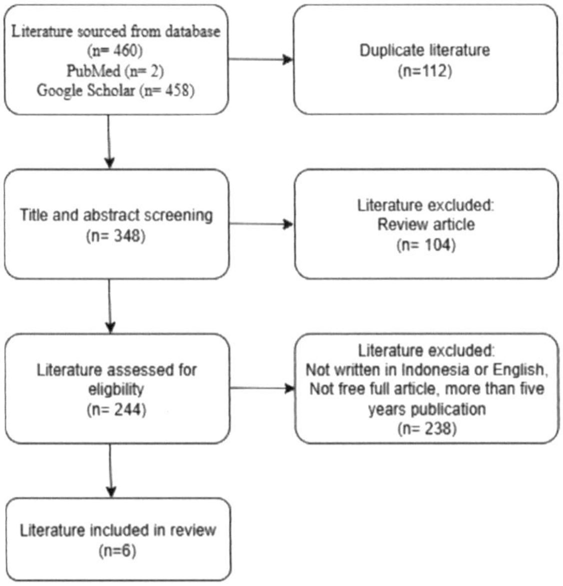

Fig. 1. Literature search and study selection process for this narrative review (PubMed and Google Scholar, 2020–2025): records identified (n = 460), screening and eligibility assessment, and studies included (n = 6).

3 Result and discussion

3.1 Green synthesis of nanoparticles

The synthesis of nanoparticles using plant extracts, also known as green synthesis, enables the use of plant-derived substances as both reducing and stabilising agents in nanoparticle formation, while also providing bioactive molecules that aid in diabetic wound healing. This method is widely used because it is environmentally friendly, cost-effective, and reduces chemical usage compared to conventional methods. Plant extracts facilitate nanoparticle synthesis by donating electrons and stabilising the newly formed particles, while their intrinsic pharmacological properties contribute to wound healing by promoting antioxidant, anti-inflammatory, and antibacterial responses. Based on Table 1, a study by Almasian et al. (2021) found that peppermint extract acted as both a nanoparticle synthesis mediator and a bioactive healing agent in a polyurethane/Pluronic F127-based nanofibre preparation containing gelatin nanoparticles. Similarly, *Bacopa monnieri* leaf extract enabled the reduction of copper ions to copper oxide nanoparticles (CuO-NPs) [8]. *Althaea Officinalis* flower extract reduced and stabilised zinc oxide nanoparticles (ZnO-NPs), later combined with clove oil in solid lipid nanoparticles (SLN) inserted into collagen sponges and *Ocimum sanctum* leaf extract enabled the formation of titanium dioxide nanoparticles (TiO2-NPs) [9–11]. In another case, pomegranate peel extract (*Punica granatum*) not only served as a reducing agent for the green synthesis of silver nanoparticles (AgNPs), but also provided healing benefits, which were compared with those of chemically synthesised nanoparticles [7].

Plant extracts used in nanoparticle preparations for diabetic wound healing exhibit phytochemical diversity and offer multiple mechanisms to enhance therapeutic effects. These extracts exert their effects through complementary pharmacological activities, including antioxidant, anti-inflammatory, and antibacterial mechanisms, all of which are crucial for healing diabetic wounds and promoting tissue regeneration. For example, polyphenols from pomegranate function as antioxidants and anti-inflammatories, while peppermint flavonoids and phenols exert antibacterial and antioxidant effects. *Bacopa monnieri*'s saponins and alkaloids aid in antidiabetic and anti-inflammatory responses. *Althaea Officinalis* flowers contribute flavonoids and phenols that promote antioxidant activity and tissue repair, and *Ocimum sanctum* leaves deliver flavonoids, tannins, and alkaloids with antioxidant and antimicrobial actions. Additionally, eugenol in clove oil enhances analgesic, anti-inflammatory, and antibacterial properties. The synergistic effect of these secondary metabolites, when combined within nanoparticle formulations, targets oxidative stress, inflammation, and infection, while providing a scaffold for cell growth and tissue repair.

Table 1. Summary of included experimental studies on plant extract–based nanoparticle formulations for diabetic wound therapy (2020–2025), including the plant source, formulation type, study design (in vitro/in vivo), primary outcomes, and reference.

No	Plant extract	Type of Preparation	Method	Results	Reference
1	Peppermint Extract	Polyurethane/Pluronic F127 Nanofibres with gelatin nanoparticles containing extracts.	In vitro studies (antibacterial activity against *Staphylococcus aureus* and *Escherichia coli*) and in vivo studies (using diabetic rat as subjects).	The formulation exhibit high antibacterial activity, accelerate wound healing, and significantly improve wound closure in diabetic rats.	[6]
2	*Punica granatum* peel extract,	Green-synthesised silver nanoparticles (AgNPs)	In vitro studies (antibacterial activity against *Staphylococcus aureus*) and in vivo studies (using diabetic rat as subjects).	The formulation promote the healing of infected wounds and reduce infection in diabetic mice.	[7]
3	*Bacopa monnieri* leaf extract	Copper oxide nanoparticles (CuO-NPs)	In vitro studies (antibacterial activity against 4 *Helicobacter pylori* strains) and in vivo studies (using diabetic rat subjects) to test antidiabetic,	This formulation demonstrate significant antimicrobial and antidiabeti	[8]

			analgesic, and anti-inflammatory activity.	c activity in rats.	
4	*Althaea Officinalis* Flowers Aqueous Extract	Chitosan-coated zinc oxide nanoparticles (ZnO-NPs)	In vitro studies (antibacterial against Gram-positive bacteria (*S. aureus* and *B. subtilis*) and Gram-negative bacteria (*E. coli* and *P. aeruginosa*) and in vivo studies (using diabetic rat subjects).	This formulatio n effective against bacteria and promotes wound healing in vivo.	[9]
5	*Ocimum sanctum* Lea f Extract	Titanium dioxide nanoparticles (TiO2)	Green synthesis, in vitro and in vivo characterisatio n, and evaluation of healing efficacy in specific animal subjects.	This formulatio n exhibit significant wound healing activity in diabetic mice.	[10]
6	*Althaea Officinalis* Extract with Clove Oil	Lipid nanoparticles in collagen sponges	In vitro studies (antibacterial and antifungal against *Pseudomonas aeruginosa*, *Escherichia coli*, and *Candida albicans*) and in vivo studies (using diabetic rat subjects).	This formulatio n demonstra ted strong antimicrob ial effects and successfull y accelerate d the healing of diabetic mouth ulcers in mouse subjects.	[11]

Although all included studies investigated plant extract–based nanoparticle formulations in the context of diabetic wound healing, the in vivo experimental designs differed across models and outcome frameworks, which affects cross-study comparability. In an infected cutaneous wound model, male Wistar rats were allocated to five experimental groups of six animals each, and wound progression was evaluated at days 2, 7, and 14 after topical treatment [7]. In another study, diabetic oral ulcers were induced in male Wistar albino rats (n = 40), with five treatment arms including PBS control, untreated ulcers, and lipid nanoparticle–based interventions embedded in collagen sponges, assessed through clinical ulcer scoring and tissue regeneration parameters [11]. Differences in wound location (dorsal skin versus oral mucosa), endpoint selection (photographic wound area reduction versus combined clinical and biomarker outcomes), and follow-up schedules indicate that efficacy findings should be interpreted within the specific experimental context of each model rather than treated as directly interchangeable evidence.

Statistical robustness also varied in reporting detail. Key reporting elements highlighted in ARRIVE 2.0, including randomisation procedures, blinding, sample size rationale, and complete statistical reporting, are essential for evaluating internal validity and reproducibility in animal-based wound healing studies [12]. The oral ulcer study used one-way ANOVA with Tukey–Kramer post hoc testing and a Student's t-test, with results presented as mean ± SD [11]. Explicit justification of sample size or power considerations was not consistently reported, which limits assessment of precision across formulations. These methodological and reporting differences should be acknowledged when drawing conclusions regarding comparative effectiveness and translational reliability.

3.2 Studies in vitro activity of green synthesized

In vitro assays across the reviewed formulations primarily evaluated antibacterial and antifungal activity, with performance influenced by the carrier matrix and the active constituents. Mentha piperita–loaded nanofibres showed strong antibacterial activity, reaching 99.9% against common wound pathogens such as *Staphylococcus aureus* and *Escherichia coli* [6]. A ZnO-NP–chitosan gel prepared using *Althaea Officinalis* extract inhibited multiple bacterial species, including *Bacillus subtilis* and *Pseudomonas aeruginosa*, and showed larger inhibition zones against Gram-positive bacteria [9]. Antifungal outcomes varied across formulations and test conditions. AgNP formulations derived from pomegranate extract in gellan gum and chitosan matrices were reported to lack activity against *Candida albicans* after 14 days [7]. Solid lipid nanoparticles containing *Althaea Officinalis* extract and clove oil were reported as active against *C. albicans* in the reported assays, including performance relative to ketoconazole [11]. Variation in phytochemical profiles, release characteristics, and assay parameters may contribute to differences in observed antifungal activity.

3.3 Studies in vivo activity of green synthesized

In vivo evaluations in STZ-induced diabetic mouse models indicate that plant extract–based nanoparticle systems support chronic wound repair across multiple formulation platforms. *Mentha piperita* loaded nanofibres achieved a 90.55% healing rate by day 14, supported by histopathological evidence of increased collagen deposition, fibroblast proliferation, and neovascularisation [6]. TiO_2 nanoparticle–based chitosan gel produced near complete healing by day 21 and was associated with epidermal regeneration consistent with normal tissue maturation [10]. Silver nanoparticles (AgNPs) and solid lipid nanoparticles (SLNs) were also associated with wound contraction in the reported models relative to control conditions [7,11].

Green nanoparticle formulations may support diabetic wound management through combined antimicrobial action, modulation of inflammatory responses, and delivery platforms that improve local retention of bioactive compounds. Metal-based nanoparticles can contribute antimicrobial effects through ion release, while ZnO-based systems have been associated with cytokine modulation in the reviewed studies [7–9]. Carrier platforms such as nanofibres and solid lipid nanoparticles can provide sustained release and reduce application frequency, and safety profiles were commonly supported by cytotoxicity and haemolysis evaluations [6,11].

3.4 Future Perspectives

Plant extract based nanomedicines represent a promising approach for diabetic wound therapy by combining antimicrobial and anti-inflammatory activities with delivery systems that improve the stability and tissue penetration of bioactive compounds. Current limitations include microbial biofilm resistance, an incomplete understanding of phytochemical mechanisms, and the continued reliance on animal models that do not fully replicate the clinical features of diabetic foot ulcers. Yacon (*Smallanthus sonchifolius*) leaves have been proposed as a potential source of phenolic and flavonoid compounds, although evidence for nanoparticle-based formulations remains limited.

4 Conclusion

The six experimental studies reviewed here indicate that incorporating plant extracts into nanoparticle-based topical systems can support diabetic wound healing through combined antimicrobial and tissue-regenerative effects [6–11]. Across the included formulations, in vitro assays commonly targeted bacterial pathogens relevant to infected wounds, and in vivo models reported improved wound contraction and histological repair features, including collagen deposition, fibroblast activity, neovascularisation, and epithelial restoration [6–11]. These findings support the view that formulation architecture, including the carrier matrix and the nanoparticles' physicochemical profile, contributes to local delivery performance and biological response.

The strength of the evidence depends on study design and reporting quality. Differences in animal models, wound sites, outcome definitions, follow-up schedules, and statistical reporting limit direct comparison of effectiveness across formulations. Future studies will benefit from clearer descriptions of allocation and blinding procedures, an explicit rationale for the sample size, and harmonised outcome measures, alongside standardised evaluation of toxicity and release behaviour. These steps would enable more reproducible results and more interpretable comparisons across studies and formulation types.

The authors extend their appreciation to the Kementerian Pendidikan Tinggi, Sains, dan Teknologi of Indonesia for funding this research work through project number 055/DirDPPM/70/DPPM/PTM-KEMDIKTISAINTEK/VI/2025.

References

1. S. A. Antar, N. A. Ashour, M. Sharaky, M. Khattab, N. A. Ashour, R.T. Zaid, E.J. Roh, A. Elkamhawy, A. A. Al-Karmalawy. Diabetes mellitus: Classification, mediators, and complications; A gate to identify potential targets for the development of new effective treatments. Biomed Pharmacother. **168**, 115734 (2023). https://doi.org/10.1016/j.biopha.2023.115734

2. World Health Organization (WHO), Health Topics: Diabetes. (2025). https://www.who.int/health-topics/diabevantes#tab=tab_1
3. IDF, Diabetes (2021). Available from: https://idf.org/our-network/regions-and-members/western-pacific/members/indonesia/
4. E. M. Matheson, S. W. Bragg, R. S. Blackwelder, Diabetes-Related Foot Infections: Diagnosis and Treatment. Am Fam Physician. **104**, 386–394 (2021). https://www.aafp.org/afp/2013/0801/p177-s1.html.
5. G. Boschetti, D. Sgarabotto, M. Meloni, M. Bruseghin, C. Whisstock, M. Marin, S. Ninkovic, M. Pinfi, E. Brocco. Antimicrobial resistance patterns in diabetic foot infections, an epidemiological study in northeastern Italy. Antibiotics. **10**, 4–11 (2021). https://doi.org/10.3390/antibiotics10101241
6. A. Almasian, F. Najafi, M. Eftekhari, M. R. S. Ardekani, M. Sharifzadeh, M. Khanavi, Preparation of Polyurethane/Pluronic F127 Nanofibers Containing Peppermint Extract Loaded Gelatin Nanoparticles for Diabetic Wounds Healing: Characterization, in Vitro, and in Vivo Studies. Evidence-based Complement Altern Med. **2021**, 1–16 (2021). https://doi.org/10.1155/2021/6646702
7. R. A. F. Scappaticci, A. A. Berretta, E. C. Torres, A. F. M. Buszinski, G. L. Fernandes, T. F. Dos Reis, F. N. de. Souza-Neto, L. F. Gorup, E. R. de Camargo, D. B. Barbosa. Green and chemical silver nanoparticles and pomegranate formulations to heal infected wounds in diabetic rats. Antibiotics. **10**, 1343 (2021). https://doi.org/10.3390/antibiotics10111343
8. S. Faisal, H. Jan, Abdullah, I. Alam, M. Rizwan, Z. Hussain, K. Sultana, Z. Ali, M. N. Uddin. In Vivo Analgesic, Anti-Inflammatory, and Anti-Diabetic Screening of Bacopa monnieri-Synthesized Copper Oxide Nanoparticles. ACS Omega. **7**, 4071–82 (2022). https://doi.org/10.1021/acsomega.1c05410
9. S. F. Elhabal, N. Abdelaal, S. A. K. S. Al-Zuhairy, M. F. M. Elrefai, A. M. E. Hamdan, M. M. Khalifa, S. Hababeh, M.A. Khasawneh, G.M. Khamis, J. Nelson, P.M. Mohie, R.A. Gad, A. Rizk, S.L. Kabil, M.K. El-Ashery, B.R. Jasti, N.A. Elzohairy, T. Elnawawy, F.E. Hassan, M.A. El- Nabarawi. Green Synthesis of Zinc Oxide Nanoparticles from *Althaea Officinalis* Flower Extract Coated with Chitosan for Potential Healing Effects on Diabetic Wounds by Inhibiting TNF-α and IL-6/IL-1β Signaling Pathways. Int J Nanomedicine. **19**, 3045–3470 (2024). https://doi.org/10.2147/IJN.S455270
10. M. Z. Ahmad, A. S. Alasiri, J. Ahmad, A. A. Alqahtani, M. M. Abdullah, B. A. Abdel-Wahab, K. Pathak, R. Saikia, A. Das, H. Sarma, S. A. Alzahrani. Green Synthesis of Titanium Dioxide Nanoparticles Using *Ocimum sanctum* Leaf Extract: In Vitro Characterization and Its Healing Efficacy in Diabetic Wounds. Molecules. **27**, 1–17 (2022). https://doi.org/10.3390/molecules27227712
11. S. F. Elhabal, A. M. Faheem, S. Hababeh, J. Nelson, N. A. Elzohairy, Y. F. Ibrahim, T.M. Ewedah, I. S. Mousa, K. M. Allam, A. M. E. Hamdan. Pharmaceutics. **17**, 1–36 (2025). https://doi.org/10.3390/pharmaceutics17050611
12. N. Percie du Sert, V. Hurst, A. Ahluwalia, S. Alam, M.T. Avey, M. Baker, W.J. Browne, A. Clark, I.C. Cuthill, U. Dirnagl, M. Emerson, P. Garner, S.T. Holgate, D.W. Howells, N.A. Karp, S.E. Lazic, K. Lidster, C.J. MacCallum, M. Macleod, E.J. Pearl, O.H. Petersen, F. Rawle, P. Reynolds, K. Rooney, E.S. Sena, S.D. Silberberg, T. Steckler, H. Würbel, The ARRIVE guidelines 2.0: Updated guidelines for reporting animal research. *PLoS Biol.* **18**, e3000410 (2020). https://doi.org/10.1371/journal.pbio.3000410

BIO Web of Conferences **226**, 03007 (2026) https://doi.org/10.1051/bioconf/202622603007
ISSTEC 2025

Exploration of endophytic fungi *meyerozyma guilliermondii* and its bioactive potential isolated from *ficus carica* l.

Annisa Fitria[1*], *Pinus* Jumaryatno[1], *Annas* Rabbani[2], *Ferry* Rimbawan[1], *Hanida* Destriana[1], *Nur* Siti Rahayu[1], *and Astri* Nur Azizah[1]

[1] Department of Pharmacy, Faculty of Mathematics and Natural Sciences, Universitas Islam Indonesia, Indonesia
[2]Faculty of Biology, Universitas Gadjah Mada, Indonesia

Abstract Endophytic fungi are plant-associated microorganisms known for producing bioactive secondary metabolites with antimicrobial and anticancer potential. This study evaluated endophytic fungi isolated from *Ficus carica* L. for their antimicrobial and anticancer activities. Fungal isolates were obtained from root, stem, leaf, and fruit tissues, and mycelial extracts were assessed for antimicrobial activity against bacterial and fungal pathogens, as well as for cytotoxicity against mammalian fibroblast cell. Phytochemical profiling was conducted using thin-layer chromatography (TLC), and molecular identification of bioactive isolates was performed by sequencing the internal transcribed spacer (ITS) region. A root-derived endophytic fungal isolate demonstrated significant antimicrobial activity, causing 62.29% cell death in *Staphylococcus aureus* at a minimum inhibitory concentration (MIC) of 31.25 µg/mL and 57.26% cell death in *Candida albicans* at an MIC of 125 µg/mL. Phytochemical analysis revealed the presence of terpenoids and alkaloids, and ITS sequencing identified the isolate as *Meyerozyma guilliermondii*. The extract exhibited no significant anticancer activity and showed no cytotoxic effects on normal fibroblast cells. These results indicate that *Meyerozyma guilliermondii* from *Ficus carica* is a promising source of non-cytotoxic antimicrobial agents.

1 Introduction

Antibiotic resistance is a growing global health threat, yet the discovery of new antimicrobial agents has declined over recent decades [1–3], highlighting an urgent need for alternative sources. Medicinal plants are rich reservoirs of bioactive compounds, but large-scale exploitation is limited by biomass requirements and cultivation challenges. Endophytic fungi, living as symbiont within plant tissues, offer a promising alternative, producing diverse secondary metabolites influenced by host–endophyte interactions and environmental factors, with activities including antimicrobial, antiviral, antimalarial, anticancer, and antioxidant effects [4-5]. *Ficus carica* L., widely used in traditional medicine, harbors endophytic fungi

*Corresponding author: annisa.fitria@uii.ac.id

with notable antimicrobial potential [6]. However, systematic evaluations of bioactivity and metabolite profiles across its tissues, particularly in Indonesian cultivars, remain limited. This study aimed to investigate the antimicrobial activity and metabolite composition of endophytic fungi from the roots, stems, leaves, and fruits of *F. carica*, identifying potential sources of novel antimicrobial agents.

2 Materials and methods

2.1 Plant collection and determination

Ficus carica L., 5 months old, was collected in July 2017 from Tegalrejo, Sleman, Yogyakarta, Indonesia. The species was identified based on morphological characteristics through comparison with published taxonomic literature at the Plant Systematics Laboratory, Faculty of Biology, Universitas Gadjah Mada (UGM), Yogyakarta, Indonesia.

2.2 Isolation and morphological identification of endophytic fungi

Roots, stems, leaves, and fruits were washed, cut into 2–3 cm segments, and surface-sterilized with 5.25% sodium hypochlorite for 2 min, followed by 70% ethanol for 2 min, and rinsed with sterile distilled water. The sterilized tissues were aseptically cut into approximately 1 × 1 cm^2 segments and plated on Potato Dextrose Agar (PDA) supplemented with chloramphenicol (100 µg/mL). Plates were incubated in the dark at 25–30°C for 5–14 days. Emerging endophytic fungi were subcultured on fresh PDA until pure isolates were obtained. Morphological identification was performed based on macroscopic colony characteristics at the Mycology Laboratory, Universitas Gadjah Mada, Yogyakarta, Indonesia.

2.3 Metabolite production of endophytic fungi

Five endophytic fungal isolates were aseptically excised from actively growing mycelia and inoculated into 125 mL of Potato Dextrose Broth (PDB) in 250 mL Erlenmeyer flasks. The cultures were incubated at 28°C on an orbital shaker at 65 rpm for 21 days. Following incubation, the fungal mycelia were separated from the culture broth by filtration. The culture broth extracted with ethyl acetate (EtOAc) at a 1:1 (v/v) ratio, with the extraction performed three times through liquid–liquid partitioning in a separatory funnel. The combined extracts were concentrated using a rotary evaporator at 50°C.

2.4 Identification of endophytic fungi metabolites by TLC

Compounds in endophytic fungi were identified using thin-layer chromatography (TLC). Silica gel GF254 plates were used as the stationary phase and eluted with a 1:1 mixture of ethyl acetate and chloroform as the mobile phase. The resulting spots were visualized using spray reagents to determine the classes of compounds present. Phenolic compounds were detected using $FeCl_3$, flavonoids with $AlCl_3$, terpenoids with anisaldehyde–sulfuric acid, and alkaloids using Dragendorff's reagent.

2.5 Test microorganism and cell line

The antimicrobial activity was evaluated against *Staphylococcus aureus* ATCC 25923, *Escherichia coli* 25922, and *Candida albicans* ATCC 10231 obtained from the Microbiology

Laboratory, Universitas Islam Indonesia. In vitro cytotoxicity was assessed using monkey kidney fibroblas cell line (Vero) obtained from the Parasitology Laboratory, Universitas Gadjah Mada.

2.6 Antimicrobial evaluation

The minimum inhibitory concentration (MIC) was determined using the broth microdilution method according to CLSI guidelines. Extracts were homogenized with 2% Tween and serially two-fold diluted in Mueller–Hinton Broth to concentrations of 25–0.625 mg/mL in 96-well microplates. Standardized inocula (5×10^5 CFU/mL) were added and plates were incubated at 37 °C for 24 h for bacteria and at room temperature for *Candida albicans*. MIC was defined as the lowest concentration showing no visible growth. The minimum bactericidal concentration (MBC) was determined by subculturing MIC wells onto Mueller–Hinton Agar and incubating at 37 °C for 24 h.

Cell death was assessed using the MTT assay with 3-(4,5-dimethylthiazol-2-yl)-2,5-diphenyl tetrazolium bromide (5 mg/mL). Microbial cells were treated with the extract, and MTT reagent was added to each well, followed by 4-hour incubation. The reaction was stopped with SDS solution, and plates were incubated overnight at room temperature in the dark. Absorbance was measured at 595 nm to quantify cell viability, and the half-maximal inhibitory concentration (IC_{50}) was determined to evaluate cytotoxic effects.

2.7 Cytotoxicity evaluation

The cytotoxic effect of endophytic fungal extracts was evaluated using the MTT assay on Vero cells in M199 medium. Each type of cells were seeded at 1×10^4 cells/well in 96-well plates and incubated at 37°C in a 5% CO_2 incubator for 24 hours. After reaching confluence, cells were treated with serial concentrations of endophytic fungal extracts and incubated for an additional 24 hours. Following treatment, the medium was removed, and MTT reagent was added to each well, followed by incubation for 4 hours. The reaction was stopped by adding SDS solution, and plates were incubated overnight at room temperature in the dark. Absorbance was measured at 595 nm using a microplate reader. Cell viability and cytotoxicity of the extracts were determined from the absorbance values, and IC_{50} values were obtained from dose-response curves.

2.8 Molecular identification of endophytic fungi

DNA analysis was carried out on the endophytic isolate demonstrating the highest antimicrobial activity among all tested isolates. DNA was extracted from fungal mycelia obtained from broth cultures using Chelex. Species-level identification was performed by PCR amplification of the ITS region using ITS1 (5'-TCCGTAGGTGAACCTGCGG -3') and ITS 4 (5'-TCCTCCGCTTATTGATATGC-3') primers under the following conditions: initial denaturation at 94°C for 5 min, followed by 30 cycles of denaturation at 94°C for 1 min, annealing at 52°C for 1 min, and extension at 72°C for 10 min. PCR products were verified on 1% agarose gels, and the resulting sequences were compared with reference sequences in the NCBI database using BLAST.

3 Result and discussion

3.1 Isolation and characterization of endophytic fungi

Isolation of endophytic fungi from the roots, stems, fruit and leaves of *Ficus carica* L. cultivated in Indonesia resulted in five distinct isolates as shown in Figure 1. These isolate were found from root (R), stem (S), fruit (F), leaf (L1 and L2). Macroscopic characterization was performed to observed colony morphology, including pigmentation, surface texture, zonation patterns, growth margins, presence of radial striations, and colony diameter. The isolates exhibited marked morphological variation, ranging from differences in colony color and texture to distinct growth patterns and zonation, suggesting phenotypic diversity among the recovered endophytes as shown in Figure 1. Such variation may reflect differences in taxonomic affiliation and ecological adaptation to specific host tissues. A comparative summary of these morphological characteristics is presented in Table 1.

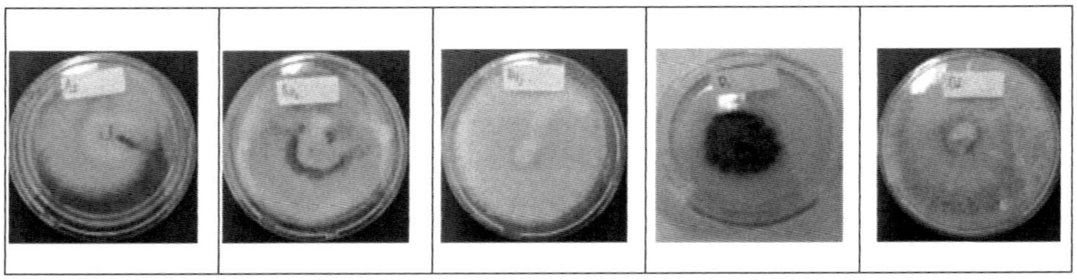
a b c d e

Figure 1. Morphological characteristics of endophytic fungal isolates. Isolates were obtained from different tissues of *Ficus carica* L.: (a) root, (b) stem, (c) fruit, (d) leaf and (e) leaf

Table 1. Macroscopic colony characteristics of endophytic fungal isolates.

Observation	Isolate				
Macroscopic	**R**	**S**	**F**	**L1**	**L2**
colony color	white	white	white	blacked dark green	white
zonation	ND	ND	ND	appeared	ND
colony texture	smooth	smooth	smooth	rough and granulated	smooth
growth zone	ND	ND	ND	appeared	ND
radial line	ND	ND	ND	ND	ND
colony diameter	7.5 cm	8 cm	8 cm	5.5 cm	9 cm

Macroscopic features are shown where detected; ND: no detectable

Differences in endophytic fungal isolates across plant tissues are shaped by environmental and host-related factors, including plant age and nutrient availability. Mature tissues support greater fungal colonization, which may contribute to increased metabolic diversity and the production of bioactive compounds with antimicrobial potential [7].

3.2 Metabolite profile of endophytic fungi

The metabolite profiles of endophytic fungal extracts were analyzed by thin-layer chromatography (TLC) as presented in Table 2. The results indicated that terpenoid and alkaloid found in each endophytic of each part of the plant. Terpenoids and alkaloids predominated among the metabolites detected in the endophytic fungi, as reported in earlier studies [8]. In *Ficus carica* L., phenolic acids and flavonoids dominate the secondary metabolite profile of the fruit and leaves, while triterpenoids, steroids, and volatile compounds occur at comparatively lower levels [9]. Endophytic fungi harbor biosynthetic gene clusters responsible for synthesizing diverse secondary metabolites through pathways independent of host plant metabolism, and can modulate key metabolic enzyme genes [10].

Table 2. Phytochemical profiling of endophytic fungal extracts by TLC

Compound class	Detection reagent	Endophytic fungi				
		R	S	F	L1	L2
Phenol	$FeCl_3$	0.15	ND	ND	ND	ND
Flavonoid	$AlCl_3$	0.9	ND	ND	ND	ND
Terpenoid	Anisaldehyde-sulfuric acid	0.2	0.25	0.4	0.3	0.3
Alkaloid	Dragendorff's reagent	0.3	0.3	0.5	ND	0.35

Rf values are shown where detected; ND: no detectable TLC spots observed under UV (254/366 nm) or after derivatization.

3.1 Antimicrobial activity of endophytic fungi

The extracts of endophytic fungi from root and stem exhibited the strongest activity against *Staphylococcus aureus*, reducing ≥50% of bacterial growth at their MIC, whereas activity against *Escherichia coli* was weak as presented in Table 3. Root, leaf, and fruit extracts showed notable antifungal activity against *Candida albicans*. These results are consistent with previous reports that while endophytic fungi such as *Penicillium oxalicum* and *Phoma herbarum* found in *Ficus sp.* display selective activity, being more effective against Gram-positive bacteria and fungi [11]. Previous studies have shown that *Ficus carica* exhibits broad-spectrum antimicrobial activity [12]. In contrast, its endophytic fungi display more selective bioactivity, showing weaker effects against general bacteria but strong and targeted activity against specific pathogens. The standard deviation value observed in the percentage of cell death reflects expected experimental variability among *in vitro* replicates, with values remaining consistently distributed around the mean. MIC values constitute the primary and most reliable indicator of antimicrobial activity, whereas percentage cell death is reported solely as supplementary data to support the MIC findings.

Table 3. Antimicrobial activity of endophytic fungal isolates against selected pathogens

Fungi	***S. aureus***		***E. coli***		***C. albicans***	
	MIC (µg/ml)	**Cell death (%)**	**MIC (µg/ml)**	**Cell death (%)**	**MIC (µg/ml)**	**Cell death (%)**
R	31.25	62.29±19.49	ND	ND	500	83.65±3.20
S	500	79.28±3.010	1600	51.88±25.27	ND	ND
L1	1000	41.10±18.53	1600	89.63±3.23	500	70.73±10.83
L2	ND	ND	ND	ND	2000	56.15±25.27
F	ND	ND	1600	42.81±18.02	500	74.00±11.80

MIC (µg/ml) and percentage of cell date values are shown where detected; ND: no detectable

3.2 Cytotoxic activity of endophytic fungi

The IC_{50} values (Table 4) indicate that none of the extracts exhibited significant cytotoxicity, and previous studies (data not shown) confirmed a lack of anticancer activity against MCF-7 cells. These findings suggest that the bioactive secondary metabolites of *Ficus carica* endophytes are selectively antimicrobial, particularly against Gram-positive bacteria and fungi rather than broadly cytotoxic. This finding demonstrates their potential as safe, targeted antimicrobial agents. The observed standard deviation value demonstrates expected experimental variability inherent to cytotoxicity assays, with values remaining uniformly distributed around the mean. The results demonstrated high IC_{50} values, indicating the absence of cytotoxic effects on Vero cells.

Table 4. Cytotoxic activity of endophytic fungal isolate on different cell lines.

Fungi	**IC_{50} (ppm)**
	Vero
R	570.76±11.34
S	2201.42±10.6
F	673.96±21.13
L1	742.32±16.17
L2	1277.54±19.01

3.3 Species detection of endophytic fungi

BLAST analysis of the ITS–rDNA sequences of the endophytic fungal isolates revealed a high similarity to *Meyerozyma guilliermondii* isolate R03 (GenBank accession no. MZ476182.1). The sequences showed 96% identity over an alignment length of 579 bp. These results indicate a close phylogenetic relationship between the isolated endophytic fungus and *M. guilliermondii*, supporting its tentative identification at the species level. Recent studies showed potential of endophytic fungi associated with plants. Whole-genome sequencing has revealed novel *Meyerozyma* species isolated from *Mangifera indica* in Malaysia, underscoring the ecological and metabolic diversity of this genus in fruit hosts. Within the genus *Ficus*, several endophytic fungi, including *Aspergillus* and *Penicillium*, have been reported to produce bioactive secondary metabolites with antimicrobial and anticancer potential. In addition, *Meyerozyma guilliermondii* isolated from pear fruit has demonstrated notable antifungal activity and enhanced antagonistic efficacy associated with host defense responses. Endophytic members of the genus *Penicillium* have also been detected in *Ficus carica* fruit, further supporting tissue-specific fungal colonization in this host. These findings indicate that the present study represents the first report of *Meyerozyma guilliermondii* isolated from *Ficus carica* L. cultivated in Indonesia, expanding current knowledge of endophytic fungal diversity within the Moraceae family [13-15].

4 Conclusion

This study represents the first report of *Meyerozyma guilliermondii* as an endophytic fungi isolated from *Ficus carica* L. cultivated in Indonesia. Five endophytic fungi isolated from the roots, stems, leaves, and fruit exhibited distinct morphological diversity and produced predominantly terpenoids and alkaloids. The extracts demonstrated selective antimicrobial activity, with root and stem isolates most effective against *Staphylococcus aureus* and root, leaf, and fruit extracts active against *Candida albicans*, while activity against *Escherichia coli* was weak. MIC values were the primary measure of efficacy, and no cytotoxic effects were observed in Vero cells. These findings highlight *M. guilliermondii* as a promising source of bioactive secondary metabolites with selective antimicrobial potential against Gram-positive bacteria and fungal pathogens.

References

1. Wen, J., Okyere, S. K., Wang, S., Wang, J., Xie, L., Ran, Y., Hu, Y., Endophytic fungi: An effective alternative source of plant-derived bioactive compounds for pharmacological studies. *J. Fungi* **8**, 205 (2022). https://doi.org/10.3390/jof8020205
2. Deshmukh, S. K., Dufossé, L., Chhipa, H., Saxena, S., Mahajan, G. B., Gupta, M. K., Fungal endophytes: A potential source of antibacterial compounds. *J. Fungi* **8**, 164 (2022). https://doi.org/10.3390/jof8020164
3. Adeleke, B. S., Babalola, O. O., Pharmacological potential of fungal endophytes associated with medicinal plants: A review. J. Fungi **7**, 147 (2021). https://doi.org/10.3390/jof7020147
4. Li, Z., Wen, W., Qin, M., He, Y., Xu, D., Li, L., Biosynthetic mechanisms of secondary metabolites promoted by the interaction between endophytes and plant hosts. *Front. Microbiol.* **13**, 928967 (2022). https://doi.org/10.3389/fmicb.2022.928967
5. K. K. Gupta, H. Chandra, A. Yadav, R. Prasad, S. J. S. Kalra, Fungal endophytes from medicinal plants acting as natural therapeutic reservoir. *Microbe* **3**, 100073 (2024). https://doi.org/10.1016/j.microb.2024.100073
6. H. Hasnat, S. Alam, S. A. Shompa, T. Saha, F. T. Richi, M. H. Hossain, A. Zaman, C. Zeng, C. Shao, S. Wang, P. Geng, A. A. Mamun, Phyto-pharmacological wonders of *genus Ficus*: ethnopharmacological insights and phytochemical treasures from natural products. *Saudi Pharm. J.* **32**, 102211 (2024). https://doi.org/10.1016/j.jsps.2024.102211

7. F.wu, D. Yang, L. Zhang, Y. Chen, X. Hu, L.Li, J.Liang. Diversity estimation and antimicrobial activity of culturable endophytic fungi from *Litsea cubeba* (Lour.) Pers. in China. *For.* **10**, 33 (2019). https://doi.org/10.3390/f10010033
8. S. Gupta, P. Chaturvedi, M. G. Kulkarni, J. Van Staden, A critical review on exploiting the pharmaceutical potential of plant endophytic fungi. *Biotechnol. Adv.* **39**, 107462 (2020). https://doi.org/10.1016/j.biotechadv.2019.107462
9. M. S. Yeasmin, S. A. Rahman, W. A. Hakeem, T. Abedin, M. Nur-E-Alam, Therapeutic potential and enhanced bioactivity of *Ficus carica* (fig) fruit: exploring bioactive compounds and nano-additive synergies. *S. Afr. J. Bot.* **187**, 12–31 (2025). https://doi.org/10.1016/j.sajb.2025.10.013
10. P. Toppo, L. L. Kagatay, A. Gurung, P. Singla, R. Chakraborty, S. Roy & P. Mathur, Endophytic fungi mediates production of bioactive secondary metabolites via modulation of genes involved in key metabolic pathways and their contribution in different biotechnological sector. *3 Biotech* **13**, 191 (2023). https://doi.org/10.1007/s13205-023-03605-z
11. M. M. M. AbdelRazek, A. M. Elissawy, N. M. Mostafa, A. Y. Moussa, M. A. Elshanawany & A. N. B. Singab, Bioactive secondary metabolites from fungal endophytes, *Penicillium oxalicum* and *Phoma herbarum*, associated with *Morus nigra* and *Ficus sycomorus*: an in silico study. *RSC Adv.* **49**, 36451–36460 (2024). https://doi.org/10.1039/d4ra06840h
12. T. A. Hajam & H. Saleem, Phytochemistry, biological activities, industrial and traditional uses of fig (*Ficus carica*): a review. *Chem. Biol. Interact.* **368**, 110237 (2022). https://doi.org/10.1016/j.cbi.2022.110237
13. M. I. Bin Rosali, S. M. Musa, S. K. Binti Kiram, J. T. Wen Yi, S. S. Wei & H. F. Ahmad, Whole-genome sequencing of a novel *Meyerozyma* sp. nov. fungi isolated from *Harumanis* mango (*Mangifera indica* L.) in Malaysia. *The Microbe* **7**, 100354 (2025). https://doi.org/10.1016/j.microb.2025.100354
14. L. Zhao, F. He, B. Li, X. Gu, X. Zhang, S. Dhanasekaran & H. Zhang, Transcriptomic analysis of the mechanisms involved in enhanced antagonistic efficacy of *Meyerozyma guilliermondii* by methyl jasmonate and disease resistance of postharvest apples. *LWT - Food Sci. Technol.* **160**, 113323 (2022). https://doi.org/10.1016/j.lwt.2022.113323
15. L. Abid, M. Smiri, E. Federici, B. Lievens, M. Manai, Y. Yan & N. Sadfi-Zouaoui, Diversity of rhizospheric and endophytic bacteria isolated from dried fruit of *Ficus carica*. *Saudi J. Biol. Sci.* **29**, 103398 (2022). https://doi.org/10.1016/j.sjbs.2022.103398

AUTHOR INDEX

AUTHOR INDEX

EDP Sciences – Web of Conferences
17, Avenue du Hoggar
Parc d'Activité de Courtabouf
BP 112
F-91944 Les Ulis Cedex A
France

ISBN 979-8-3313-3288-4

9 798331 332884

Eleventh International Conference on Advances in Computer-Human Interactions (ACHI 2018)

Held at DigitalWorld 2018

Rome, Italy
25 - 29 March 2018

IARIA